Engineering Materials

Andreas Öchsner · Christian Augustin (Eds.)

Multifunctional Metallic Hollow Sphere Structures

Manufacturing, Properties and Application

Springer

Prof. Dr-Ing. Andreas Öchsner
Department of Applied Mechanics
Technical University of Malaysia
81310 UTM Skudai, Johor
Malaysia
E-mail: andreas.oechsner@gmail.com

and

University Centre for Mass and Thermal
Transport in Engineering Materials
School of Engineering
The University of Newcastle
Callaghan
New South Wales 2308
Australia

Christian Augustin, M.A.
Universität Konstanz
LFE Groh
Universitätsstrasse 1
78467 Konstanz
Germany
E-mail: chr.augustin@gmail.com

or

hollomet GmbH
Grunaer Weg 26
01277 Dresden
Germany

ISBN 978-3-642-10152-6 e-ISBN 978-3-642-00491-9

DOI 10.1007/978-3-642-00491-9

Engineering Materials ISSN 1612-1317

Typeset & Coverdesign: Scientific Publishing Services Pvt. Ltd., Chennai, India.

Printed in acid-free paper

9 8 7 6 5 4 3 2 1

springer.com

Contents

1

Introduction

Andreas Öchsner[1] and Christian Augustin[2]

[1] Technical University of Malaysia, Faculty of Mechanical Engineering, 81310 UTM Skudai, Johor, Malaysia and
The University of Newcastle, Callaghan, New South Wales 2308, Australia
Andreas.Oechsner@gmail.com

[2] hollomet GmbH, Grunaer Weg 26, D-01277 Dresden, Germany and
University of Konstanz, Faculty of Humanities, Department of History, D-78457 Konstanz, Germany
Christian.Augustin@glatt.com

Nature frequently uses cellular and porous materials for creating load-carrying and weight-optimized structures. Thanks to their cellular design, natural materials such as wood, cork, bones, and honeycombs fulfill structural as well as functional demands. For a long time, the development of artificial cellular materials has been aimed at utilizing the outstanding properties of biological materials in technical applications. As an example, the geometry of honeycombs was identically converted into aluminum structures which have been used since the 1960s as cores of lightweight sandwich elements in the aviation and space industries. Nowadays, in particular, foams made of polymeric materials are widely used in all fields of technology. For example, Styrofoam® and hard polyurethane foams are widely used as packaging materials. Other typical application areas are the fields of heat and sound absorption. During the last few years, techniques for foaming metals and metal alloys and for manufacturing novel metallic cellular structures have been developed. Owing to their specific properties, these cellular materials have considerable potential for applications in the future. The combination of specific mechanical and physical properties distinguishes them from traditional dense metals, and applications with multifunctional requirements are of special interest in the context of such cellular metals. Their high stiffness, in conjunction with a very low specific weight, and their high gas permeability combined with a high thermal conductivity can be mentioned as examples. Cellular materials comprise a wide range of different arrangements and forms of cell structures. Metallic foams are being investigated intensively, and they can be produced with an open- or closed-cell structure, cf. Fig. 1.1. Their main characteristic is their very low density. The most common foams are made of aluminum alloys. Essential limiting factors for the utilization are unevenly distributed material parameters and relatively high production costs.

Several textbooks cover the topic of cellular materials in general and give an introduction to the whole range of physical properties and possible applications. The books by L.J. Gibson and M.F. Ashby (Cellular Solids: Structures & Properties) [3], M.F. Ashby et al. (Metal Foams: A Design Guide) [1], and H.-P. Degischer and B. Kriszt (Handbook of Cellular Metals: Production, Processing, Applications) [2] are recognized

A. Öchsner, C. Augustin (Eds.): Multifunctional Metallic Hollow Sphere Struc., EM, pp. 1–3.
springerlink.com

a) b)

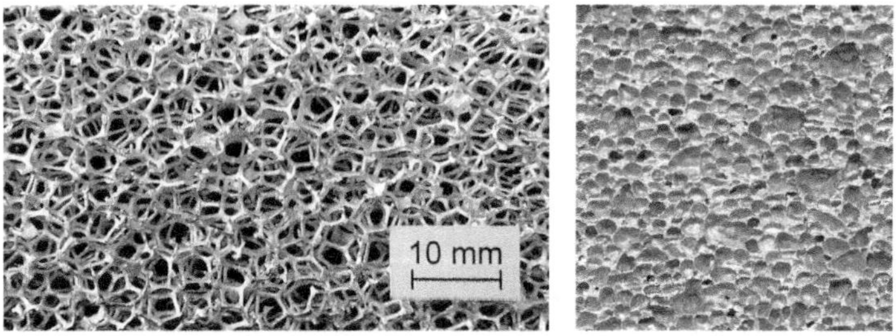

Fig. 1.1. Aluminium foams: a) Open-cell Duocel®; b) closed-cell Alporas®

as standard works on this topic and give the most comprehensive general overview of cellular and porous materials. Cellular materials made of ceramics are treated in the book by M. Scheffler and P. Colombo (Cellular Ceramics. Structure, Manufacturing, Properties and Applications) [5] and the thermal properties of cellular and porous materials are treated in detail in the recent monograph by A. Öchsner, G.E. Murch and M.J.S. de Lemos (Cellular and Porous Materials: Thermal Properties Simulation and Prediction) [4].

Metallic hollow sphere structures (MHSS) are a new group of cellular metals characterized by easily reproducible geometry and therefore consistent physical properties. A new powder metallurgy based manufacturing process enables the production of metallic hollow spheres of defined geometry. This technology brings a significant reduction in costs in comparison to earlier applied galvanic methods and all materials suitable for sintering can be applied. EPS (expanded polystyrol) spheres are coated with a metal powder-binder suspension by fluid bed coating. The green spheres produced can either be sintered separately to manufacture single hollow spheres (cf. Fig. 1.2 a)) or be precompacted and sintered in bulk (cf. Fig. 1.2 b)) thus creating sintering necks between adjacent spheres. Various further joining technologies such as soldering and adhering can be used to assemble the single hollow spheres to interdependent structures. For example, adhering is an economic way of joining and therefore can be attractive for a wide

a) b)

Fig. 1.2. a) Single hollow spheres of different diameter; b) different hollow sphere structures

range of potential applications. Another important advantage is the possible utilization of the physical behaviour and morphology of the joining technique as a further design parameter for the optimization of the structure's macroscopic properties for specific applications.

The idea of this book on *Metallic Hollow Sphere Structures* is to introduce this new composite material and to cover different important physical characteristics, i.e. the mechanical, thermal, acoustic and further properties, in detail from different points of view and for different types of hollow sphere structures and composites. Latest experimental results, numerical prediction methods and analytical modelling are covered by the contributions.

The editors wish to thank all the chapter authors for their participation and cooperation which made this text possible. Special thanks to Ms. Alison Russel for the English corrections of several chapters.

Finally, we would like to thank the team at Springer, especially Ms. Carmen Wolf and Dr. Christoph Baumann, for their excellent cooperation during the whole phase of the project.

References

1. Ashby, M.F., Evans, A., Fleck, N.A., Gibson, L.J., Hutchinson, J.W., Wadley, H.N.G.: Metal Foams: A Design Guide. Butterworth-Heinemann, Boston (2000)
2. Degischer, H.-J., Kriszt, B.: Handbook of Cellular Metals: Production, Processing, Applications. Wiley-VCH, Weinheim (2002)
3. Gibson, L.J., Ashby, M.F.: Cellular Solids: Structures & Properties. Cambridge University Press, Cambridge (1997)
4. Öchsner, A., Murch, G.E., de Lemos, J.S.: Cellular and Porous Materials: Thermal Properties Simulation and Prediction. Wiley-VCH, Weinheim (2008)
5. Scheffler, M., Colombo, P.: Cellular Ceramics. Structure, Manufacturing, Properties and Applications. Wiley-VCH, Weinheim (2005)

2

History and Production of Hollow Spheres

Christian Augustin[1] and Wolfgang Hungerbach[2]

[1] hollomet GmbH, Grunaer Weg 26, D-01277 Dresden, Germany and
University of Konstanz, Faculty of Humanities, Department of History,
D-78457 Konstanz, Germany
Christian.Augustin@glatt.com
[2] hollomet GmbH, Grunaer Weg 26, D-01277 Dresden, Germany
Wolfgang.Hungerbach@hollomet.com

Abstract. This chapter will deal with historic aspects and offer an insight into the origins of Hollow Sphere manufacturing processes up until the present day, highlighting various interesting Hollow Sphere technologies and ideas from old patents. The second part of the chapter will concentrate on production and process application of the present day Hollow Spheres.

Keywords: History of Hollow Spheres; Production of Hollow Spheres.

2.1 History

2.1.1 The Term "Hollow Sphere"

The term σφαίρα, Hollow Sphere, literally translated to "round shells" dates back to ancient Greece, used in philosophy, mathematics, geography and astronomy. In Plato's astronomic philosophy Timaeus, the shells are depicted with stars hanging from them and rotating around the world. Aristotle in his metaphysics, as well as Plutarch refer to the ancient greek terminology. The Roman scholars adapted the mathematical, astronomical term into Latin, but literally translating it to generally meaning *globe* or in scientific terms *sphere*.

The meaning and understanding was being ever expanded by younger scholars and authors, with the Latin term "sphaerae" mainly used in the new-high-german with regards to philosophy, mathematics and astronomy. Metaphorical connotations of either the Latin "sphaerae" or the German version "Sphäre" cropping up in literary works of Goethe, Lessing, Wienand and Schiller. During this time the German term "Hohlkugel" was introduced, replacing the Greek σφαίρα i.e. the Latin sphaerrae with "Hohlkugel" and becoming a technical term in many areas. The term "Hohlkugel" – for example - is described as a "hollow" filled cannonball in Grimms' dictionary [1].

2.1.2 Patents as a Fundus of Science History

The term "Hollow Sphere" has a long history, having established itself in recent years into a cellular material type with regard to engineering science. In the following we

A. Öchsner, C. Augustin (Eds.): Multifunctional Metallic Hollow Sphere Struct., EM, pp. 5–30.
springerlink.com

will present this approximate 100 year old technology-concept with various specific examples we take out of patents, highlighting the term "Hollow Sphere" has a quite accurate and precisely meaning. Of all the patents from a database searched by the terminus "Hollow Sphere" only a small amount are found that do not represent the "Hollow Sphere" concept in terms of geometrical properties. Of the approximate 150 relevant patents obtained from the Deutsche Patent- and Markenamtes (DPMA) [2] we have handpicked and researched a number of interest.

The patents have varying levels of standard, with some patents coming from hobby thinkerers, who patent their idea without any basic market requirement. Other patents however are quite obviously developed for concrete commercial reasons. Further patents can be found at the rear of this book. Christian Augustin will offer a more precise detailed history of the Hollow Spheres development in his dissertation 2009 [3].

Fundamentally there are three basic technologies regarding the production of Hollow Spheres since 1906:

- Cupping with moulds, mostly for the production of Hollow Spheres with large diameters.
- Spray-drying or a similar technology, for the production of smaller Hollow Spheres.
- Other innovative technologies such as chipping and build-up of separate parts.

We will now underscore these basic technologies by displaying some samples of patents describing the production of Hollow Spheres.

2.1.3 Cupping with Moulds

The first comprehensive Hollow Sphere Patent US 861 403: "Method or Process of making Hollow Metal Balls" [4] was filed on 17.08.1906 by the American Otto Spahr, who later on in the Patent (issued 30.07.1907) claimed to be able "... to make a pressed metal ball the internal and external surfaces which are absolutely concentric at all parts.", using the well known "cupping" or also known in German as "Zugdruckumformen" moulding technique. According to Spahr's patent a stamp and punch formed out of a round metal disk or a "mass of metal" the Hollow Spheres within several production stages. In the first step a hemisphere with an elongated edge is punched from metal sheet: "through the agency of punches and dies so constructed and operated that a disk or mass of metal to be operated upon is first caused to assume a hemispherical shape and then successive cylindrical shapes." This "hemispherical cup" is further treated: "I then cause the outer or open edges of the cylindrical part to be gradually thinned so that any longitudinal section thereof is of a wedge-like nature. I then by successive steps cause the thin edge to be turned inward until the mass of metal assumes a spheroidal shape; finally compressing the metal into a spherical shape with a limited projection of such metal extending beyond the outer surface of the sphere, which projection of metal is ultimately forced inward by pressure until the completed ball assumes the desired symmetry." Then the Hollow Sphere cup is closed, "...of the ball is now inserted and forced downwards until the upper end is flush with the upper end of the die 6 (cf. Fig. 2.1), thereby causing the material to be firmly set or united together so that the ball is completed and assumes the conformation shown in cross section in Fig 11 (cf. Fig. 2.1), which will be noted is absolutely symmetrical throughout." During the final production stage the "hemispherical cup" is compressed into a Hollow Sphere, with any metal leftovers being bent and moulded inwards until "the mass assumes the confirmation of a perfect sphere".

The following patent specification diagrams present Spahr's method in Fig. 2.1.

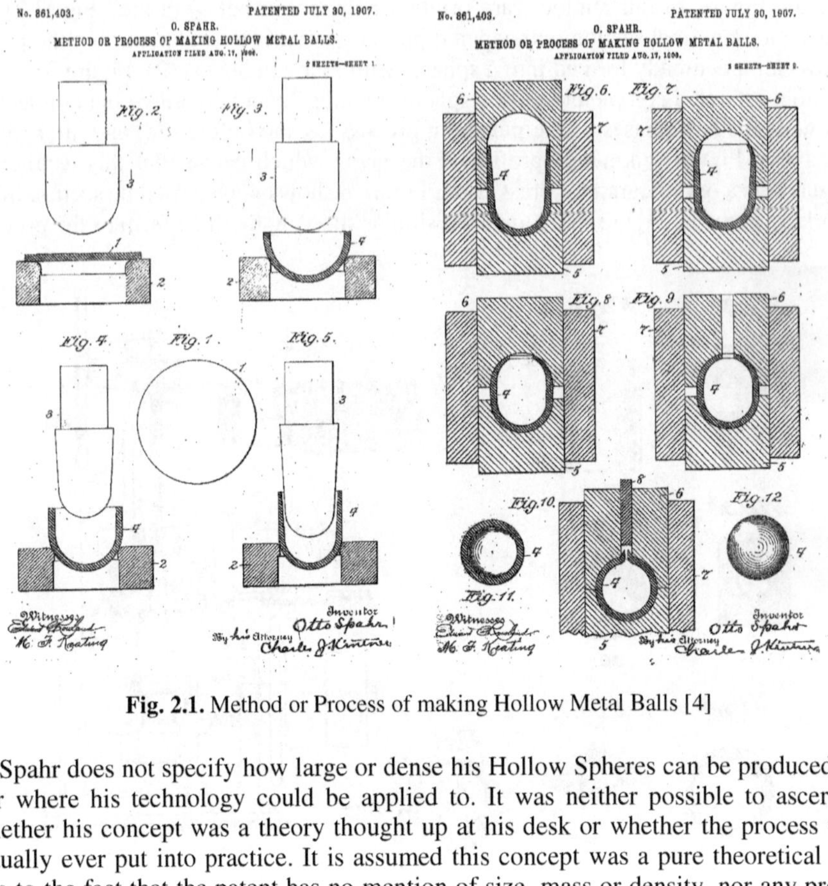

Fig. 2.1. Method or Process of making Hollow Metal Balls [4]

Spahr does not specify how large or dense his Hollow Spheres can be produced to, nor where his technology could be applied to. It was neither possible to ascertain whether his concept was a theory thought up at his desk or whether the process was actually ever put into practice. It is assumed this concept was a pure theoretical one due to the fact that the patent has no mention of size, mass or density, nor any prototypes described.

Spahr's fundamental technology was taken by many other inventors, such as John W. Schatz. In his United States Patent No. 955 698, 25.05.1910 [5] he focuses on a clean connection of the middle part, the closing part, of spheres which are produced according to the engineering process mentioned above. He introduces a "metallic plug" to be inserted into the hemispherical cup to form a Hollow Sphere. Others, in the meantime concentrated on improving the production process, e.g. the "Seamless Hollow Ball Company" (with its Patent US 1278914, priority from 12.04.1917 [6]), founded in Baltimore between 1917 and 1921. Unfortunately there is no mention of any commercial use in the patents, only generally mentioning ball valves and furniture applications.

The "Seamless Hollow Ball Company" however developed the technology and transferred it to Europe within years, demonstrated in their European patents: Austrian Patent No. 103454 of 16.09.1924 "Verfahren und Vorrichtung zur Herstellung von Hohlkugeln aus Metall" [7] and the Swiss Patent No. 112433 of 16.09.1924, 18.30 p.m. "Verfahren und Einrichtung zur Herstellung von metallenen Hohlkugeln" [8]. These patents describe the mechanism which enables to punch out of a pipe a Hollow Sphere in one single process step. The following diagram explains the technology. Figure 1a (all numbers are original numbers in below printed Fig. 2.2) presents a longitudinal section with tapered pipe; Figure 2 is a similar cut, but the picture shows a pipe already formed into a sphere, with both openings (12). Figure 3 is a dissection through a completely closed sphere. Figures 3a and 3b are vertical dissections showing the two phases of the punching process i.e. the before (3a) and after (3b) of the press. Figure 4 is a side-profile of the press, which carries out this technology. Figure 5 is a part diagram of press plate. Figure 6 shows a magnified dissection of two working grooves. Figure 7 is somewhat similar showing the final form of the groove.

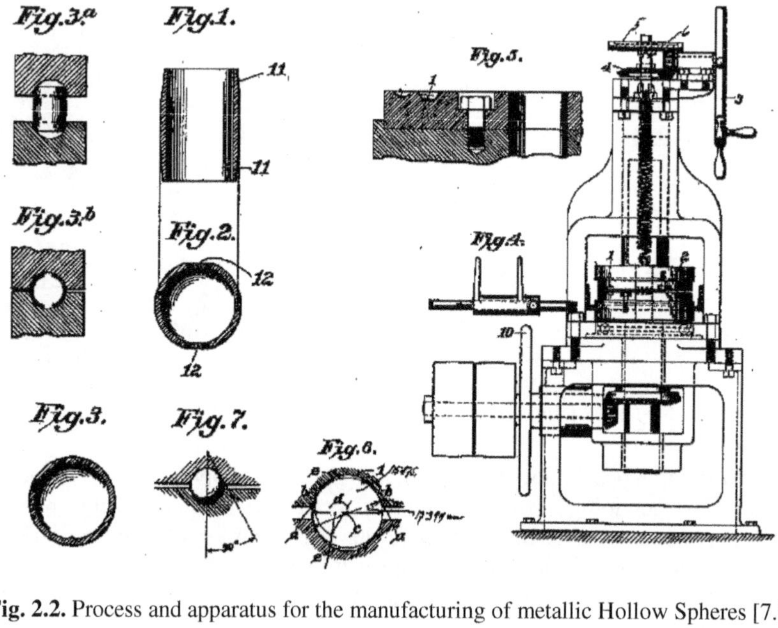

Fig. 2.2. Process and apparatus for the manufacturing of metallic Hollow Spheres [7.8]

Hollow Spheres are still nowadays being produced in this similar technology, with the company Isometall [9] being a perfect example.

2.1.4 Spray Drying

A further technology in the production of Hollow Spheres is spray-drying; the majority of patents only suiting a very few Hollow Sphere fabrication, which are mostly characterized by a large size spectrum of manufactured Hollow Spheres. Due to the hard to control quality of this specific Hollow Sphere technology some patents use the term "spheroid particles" rather than "Hollow Spheres". Spray drying in conjunction with Hollow Spheres became relevant towards the middle of the last century, where such technology was mostly used to produce ceramic Hollow Spheres. The technology is however not only limited to that of ceramics, as mentioned in some patents, but also works with metals (for example DE 100 39 320 of 07.08.2000, "Verfahren zur Herstellung von Hohlkugeln" [10]). One of the oldest patents is US 2 543 987 of Stark Ceramics "Method of Producing Porous, Lightweight, Burned Argillaceous Material" (registered 23.08.1947) [11] in which "thin free falling stream vertically through a furnace at a temperature sufficient to immediately produce expanded, porous, light weight lumps of burned argillaceous material" are described. Principally dealing with the classic spray-drying technology through which Hollow Spheres are produced by drying. The first part of the patent describes the spray-tower design. The machinery described produces Hollow Spheres as a specific material filler and aggregate for the construction industry: "for the use as and aggregate for concrete, or as a grog for the making of bricks or other burned clay bodies or forms or as filler for rubber of plastic materials, from a raw material such as clay or shale." The following diagram (cf. Fig. 2.3) from the patent shows a continuous spray process, where "S" is the ceramic binder suspension and 12 the reservoir. The ceramic binder suspension enters the spray-tower 10, where dry hot air is loaded (through the nozzle 11) through gravity and the opening 15. This is how S becomes S' and finally turns into the Hollow Sphere S'', exiting the spray tower on a conveyor belt.

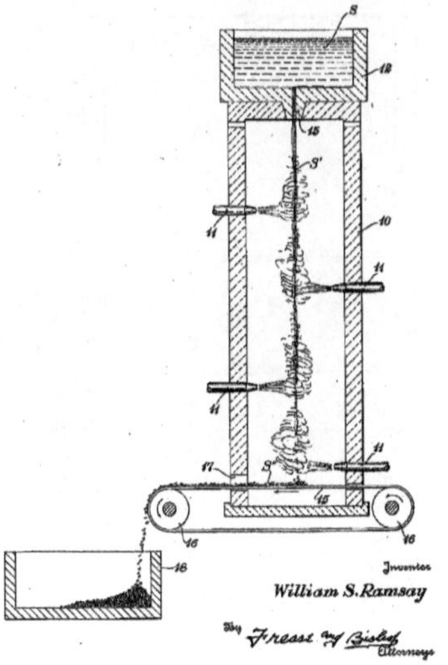

Fig. 2.3. Spray drying Apparatus for a "Method of Producing Porous, Lightweight, Burned Argillaceous Material" [11]

A similar technology and understanding of Hollow Spheres is also mentioned in the German Reichspatent 604 691, dated 27.04.1932 of the Champion Spark Plug Company, Toledo, Ohio, USA [12]. In this patent a technology for the production of small spheres made particularly from ceramic is mentioned; using the underlying technology of spray drying of ceramic materials to produce granulate or Hollow Sphere parts, which are finally put into a "Formteilautomat" (moulding part tool) and moulded into "Förmling" (composite shapes). Fundamentally the patent conveys the fabrication of ceramic parts, but at the same time highlights that "any fluids, solutions, emulsions or suspensions that have the ability to dry can be manufactured into spheres using this technology, with the possibility of either adding a hardener or further heat treatment to a material, which hardens through either heat or compaction." Foundations of this technology are "that with a constant spray and drying process a suspension or solution can be transformed into small spheres, either poured into forms or moulded through bonding pressure and compacted, creating a self-carrying "Förmling".

The patent technology of mechanically producing various different sized "Composite shapes", "Förmlinge", is due to the patent the safest possible routine with a minimal failure rate compared to previous mechanical moulding tools that failed to work, and forced the process to be manually carried out on an old fashioned hand-press, with disadvantages of the pour and mould technologies proving to be more problematic the larger the moulding was intended on. "A perfectly constructed drying process and spray injection can form considerable sized spheres. The sphere size, be it large of small can be manipulated, enabling an array of different sizes." The

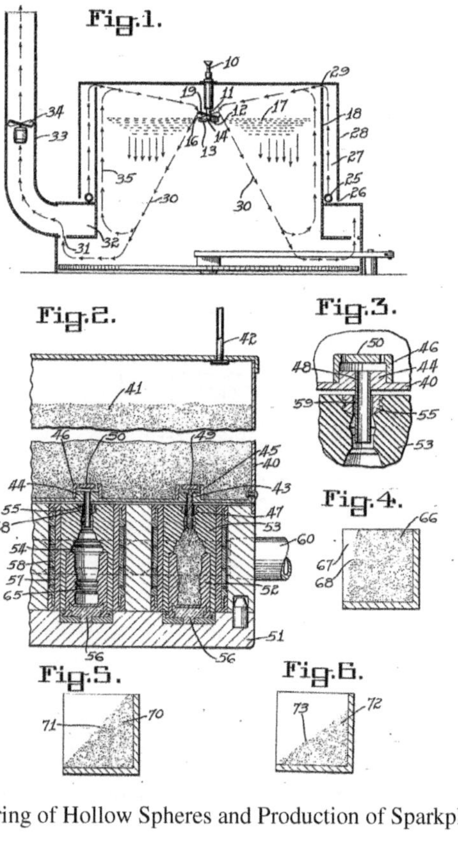

Fig. 2.4. Manufacturing of Hollow Spheres and Production of Sparkplugs out of them [12]

spheres are squashed in the machine, leading to the compacted "moulding" being considerably denser than that of a milled powder product. Advantage of this patent is the ability to manipulate the end products' material characteristics: "Moulding pressure practically does not require much force, which otherwise damages the spheres. Slight pressure will keep the particles intact and leave some equally distributed hollow areas, enabling a passing of gasses both in and out, especially during the first burning phase, which improves the control of the burning process".

Technological advantages of this patent are mentioned, e.g. in the wide variety: "The patent highlighting further technological examples, such as coal or graphite resistances mixed with powdered coal and a binding agent", made possible through the exact controllability of matter density, adjustation and uniform distribution of electric resistance. Proper core-size control and moulding pressure of differing resistances enables moulded parts to be produced in uniform, with the same size and shape. Despite the only technology mentioned being that of spark plug production, the patent does highlight the diverse applications and processes the technology has to offer.

Principally the patent has more to do with the explanation of how the patent wise described machine produces Hollow Spheres rather than the Hollow Sphere itself, which can be clearly seen in the diagram. In this diagram (cf. Fig. 2.4), Figure 1 illustrates the spray-drier production of the Hollow Sphere. Figure 2 illustrates the "moulding part tool" and the production of sparkplugs from Hollow Spheres.

2.1.5 Further Technologies

There are further technologies other than cupping and spray-drying of Hollow Sphere Production. On 30.07.1948 at 06:45 pm a patent (Priority in Austria 31.01.1948) was written up at the Eigenössischen Amt für geistiges Eigentum by Baumat AG, Zürich [13]. This patent (No 266315) dealt with "The production of ceramic Hollow Spheres for use in the construction industry", where for example a ceramic Hollow Sphere is described, as being produced from clay. These Hollow Spheres can be characterized displaying good pressure resistance, low weight, and proving to be of benefit in the construction industry and the production of light-weight cement.

The Hollow Spheres are produced in such a way that the clay, or any other ceramic is moulded into a tube, compressed and separated in a clamp, and forms consistent seamless Hollow Spheres. The following diagram 2.5 shows a clamp.

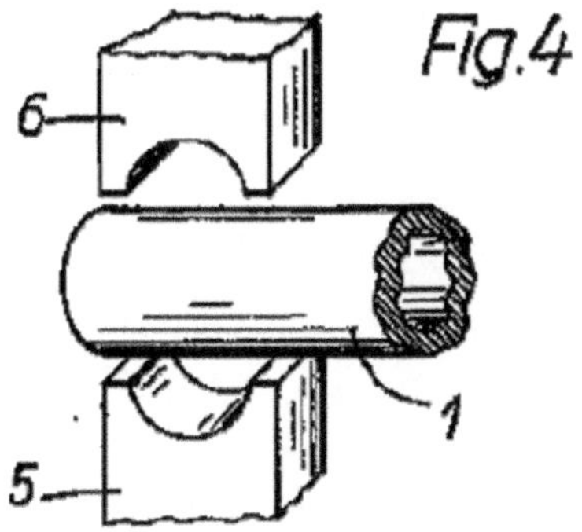

Fig. 2.5. Clamp to produce Hollow Spheres for the construction industry [13]

The patent also mentions that the Hollow Spheres may also be produced using chains or belts, rotated over grinders. The chains or belts have a ring rather than hemispherical notches. Fundamentally this alternative concept is the same, but enables a continuous production on a larger scale. Therefore the author mentions the additional ability to punch tubes parallel or even use wave-like plates, with the dip of the wave glued to the plate. Hollow Spheres manufactured in this procedure were qualitatively not particularly convincing regarding roundness and shell thickness- and consistency, but may suffice for certain applications.

Following compression, during the drying and partially burning process, the patent describes how the Hollow Spheres are subjected to a surface treatment, i.e. rounding the Hollow Spheres. This rounding process is however not described in the patent in detail, but rather highlights the ability to blend burnable aggregates such as coal, charcoal etc. into the Hollow Sphere which combust during the burning process and generates a porous shell. A further factor in the patent is the introduction of material, which in combination with water turns to gas, thus an additional porous surrounding concrete. Principally, the technology deals with a clamp-like concept, which enables a constant mass production of Hollow Spheres, either from just one clay tube or several at a time, or from the wave-like sheets previously mentioned.

Further genius techniques of how to produce the Hollow Sphere can be found in further patents, such as the Deutsche Reichspatent No. 694 903 prepared by Toyo Tokushu Immono Kabushiki Kaisha, dated 17.05.1938 and titled: “Process and

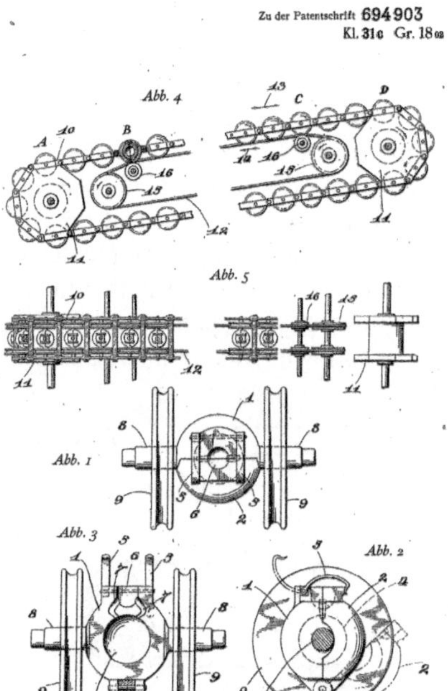

Fig. 2.6. Process and mechanisms to the production of metal Hollow Spheres [14]

mechanisms to the production of metal Hollow Spheres" [14]. The invention concerns a procedure and a manufacturing device for metallic Hollow Spheres in cooled centrifugal casting forms, which swivel around their axle. Liquid metal is poured into a spherical shape. Due to the special reservoir the melted metal stocks in the upper part of the form and closes it (cf. Fig. 2.6, Abb. 3, 7). Centrifugal energy is produced by rotation of the form, with the liquid metal forming the outer wall of the form. While the spherical shape turns around all axles, the liquid metal cools and solidifies into the form of a Hollow Sphere. Ideally (i.e. with even and constant rotation of the forms around their axles) the liquid metal is poured into a spherical shape, and after cooling forms a smooth and immaculate Hollow Sphere. The patent primarily describes a machine for such a continuous procedure, by means in which liquid metal can be brought into a rotating Sphere, and finally hardens in to a Hollow Sphere. This machinery (cf. Fig. 2.6) consists of a belt and arranged ball-like shapes. During the process the belt turns the shapes around their axle.

The patent diagram describes the following: "Procedures for manufacturing metallic hollow balls in cooled, over their axle rotating centrifugal casting forms, thereby characterized that in the gate openings of the forms the remaining liquid metal is brought to the cast in metal quantity to solidifying and then only the forms are set in motion." And: "Device for the execution of the procedure… thereby characterized that forms between two parallel, endlessly rotating belts start rotating only after the remaining metal in the gate has been solidified." The patent does state that the manufacture of Hollow Spheres using cooled forms is well known, but claims to improve the existing technology. "Up to now the forms were locked with a screw bolt, thumbscrews etc. The locking of the forms after filling in the liquid metal, by hand for instance, is not only dangerous, but also requires relatively long; while the metal already partly solidifies; posing questions as to a regular spherical shape. Besides, such a process is obviously not suitable for mass production." A mass production is possible applying this technology: "The process is more continuous when following the steps of the invention. Where filling the forms, and their rotation takes place in one uninterrupted step. Even opening, taking the poured balls out and relocking the forms for the admission of a new batch takes place sequentially according to invention, with mass production made possible in the simplest way." Obviously large commercial usability was assigned during patent application, as it was announced parallel in the USA (2184257 "Process for manufacturing chilled hollow ball", registration 12.05.1938, dispatching 19.12.1939), France (838357 "Procédé appareil destinés à la fabrication the billes creuses", registration 23.05.1938, dispatching 03.03.1939) and in Canada ("Chilled Hollow ball Manufacture/Fabrication de Billes Creuses Trempees", published 26.08.1941). Content wise the patents are equivalent; the inventor being identified as Shinsaku Nakagawa in the US American patent.

A further linear path is described by the US Patent No. 2,553,759 of 20.02.1946, issued to the Carborundum Company, Niagara Falls, NY, claiming a patent on "High temperature insulating refractories and pore-containing bodies or hollow spheroids useful in forming such refractories or for use as porous aggregates" [15]. The claims are formulated as follows: "The process of forming light weight, cellular insulating refractory bodies which comprises producing hollow, substantially spherical sticky globule of combusting material and sintering the finely-divided particles of refractory material in the shell together; forming a shaped body from a plurality of so called hollow pellets with a small amount of vitrifiable ceramic bonding material, and firing the body to mature it." The Hollow Spheres are manufactured in an apparatus, which carries out the spray drying process in an in-vitro process manner, on a conveyor belt system. The following Figure 2.7 at Fig. 1 and Fig. 2 from the patent specification, illustrates the patentwise manufactured Hollow Spheres; Fig. 3 is a sketch of the production process. Fig. 4 and Fig. 5 illustrate possibilities of subsequent Hollow Sphere treatment, with the patent demonstrating the possibility to either calcinate or deep inject the Hollow Spheres.

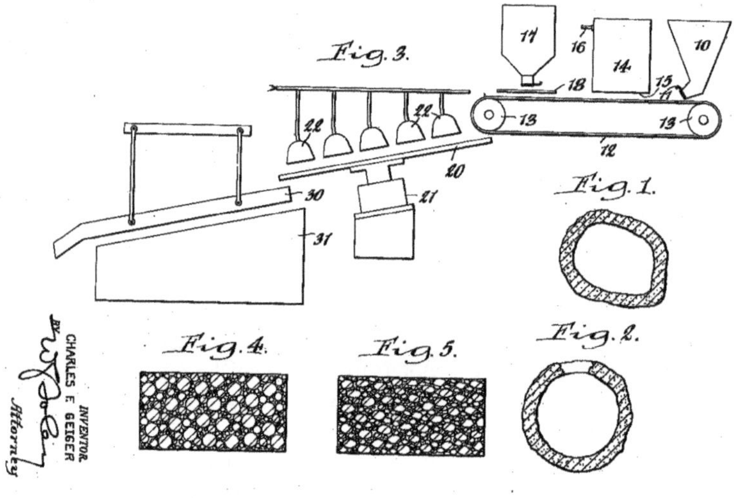

Fig. 2.7. Producing Hollow Spheres attaching powder on a sticky core [15]

In the tool described above the powder, which later on in the process forms the Hollow Spheres is released on to a conveyor belt (number 12 in Fig. 2.7). This belt is in motion and passes a container, from which small droplets are released (number 14 in Fig. 2.7). These droplets hit the powder on the conveyor belt. The viscous drops keep their shape after being released on to the powder, and are further moistened and powdered in transit and under the powder dispenser (illustrated and marked in 17). The conveyor belt finally arrives at a vibrating filter tray (number 30) which compresses the powder-shell and rounds the drops off, where the spheres are subsequently sorted using a sieve. This procedure causes a pulvrige skin to develop around the viscous material drop core, enabling a defined size and shell thickness in the development of the Spheres: "The size of the combustible adhesive cores is so chosen that the coated globules formed by building up a layer around the core, ranging in thickness from about 5 to 30 thousands of an inch, of finely-divided refractory material may be in the size range desired." These "globules" can either be injected and sintered together (the patent speaks of sintering, but in actual fact means calcinating), or individual Spheres sintered and later on injected or glued together (see in Figure 2.7: Fig. 4 and Fig. 5). The inventor however not putting too much emphasis on a precise distribution of shell thickness, porosity nor size distribution as the patent primarily offered a solution for the construction materials industry, which becomes clear by the descriptive materials: “Satisfactory hollow spheroids may be made in according with the present invention from practically any refractory material. Especially good results may be obtained with mullite, kyanite, aluminia, forsterite, clay or kaolin grog or clay itself”. Despite the technology having been invented and patented, it is unclear whether theory was ever put into practice, and Hollow Spheres were actually manufactured on an experimental or industrial level using this procedure. The patent nevertheless exhibits interesting process engineering, innovation and finesse with the combustion of the coated inner core a good example. Other similar patents do of course exist, in these however, the material is mostly burned in order to establish porosity, not to lose the core.

2.1.6 Specific Hollow Sphere Production Routines

Another manufacturing process for Hollow Spheres is in the patent specification GB 1083993 (with the German priority DE 1202910 of 07.05.1963) [16]. The Patent entitled “Device for a Production of a Hollow Sphere” describes the production of a Hollow Sphere to bring in radioactive substances and a moderator and/or fertile material. Thereby the patent explains how a solid body can be turned into a Hollow Sphere. The patent refers to the machine, which carries out this process by turning a Hollow Sphere out of a fully solid ball. This procedure is mainly suitable for larger Hollow Spheres and it is however very time and cost intensive. In Figure 2.8, Fig. 1 shows the machine, Fig. 2 the geometrical basic assumptions, which describe the movements of the milling head in the Hollow Sphere.

FIG. 1.

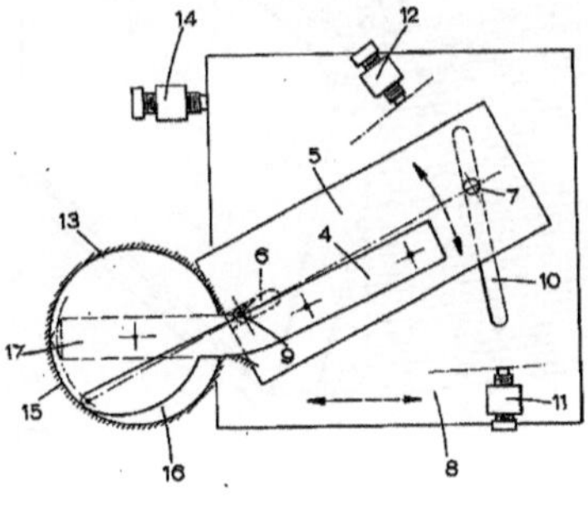

FIG. 2.

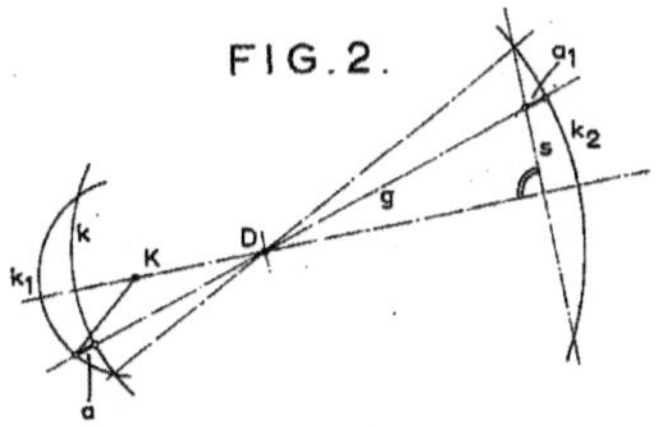

Fig. 2.8. Hollow Sphere being milled out of a solid ball [16]

Not milling but gluing two halves is the core idea of another patent. The main aim of producing seamless Hollow Spheres using pre-fabricated plastic hemispheres was layed out in DE 1 168,058 of 08.07.1960 [17]. The Patent is entitled "Welding body device for connecting two Hollow body halves, e.g. to Christmas tree spheres or such". In the patented tool two plastic hemispherical shells are inserted, which in turn are welded together through pressure and heat. Pressure and heat is generated by turning the sphere halves against one another: "This procedure is thus the basic concept of welding two hemispherical shells together by means of pressure and heat. The thermoplastic raw material softens with the heat, and automatically welds the two halves together". The following diagram (cf. Fig. 2.9) is taken from the patent and illustrates the relatively complex welding-machine.

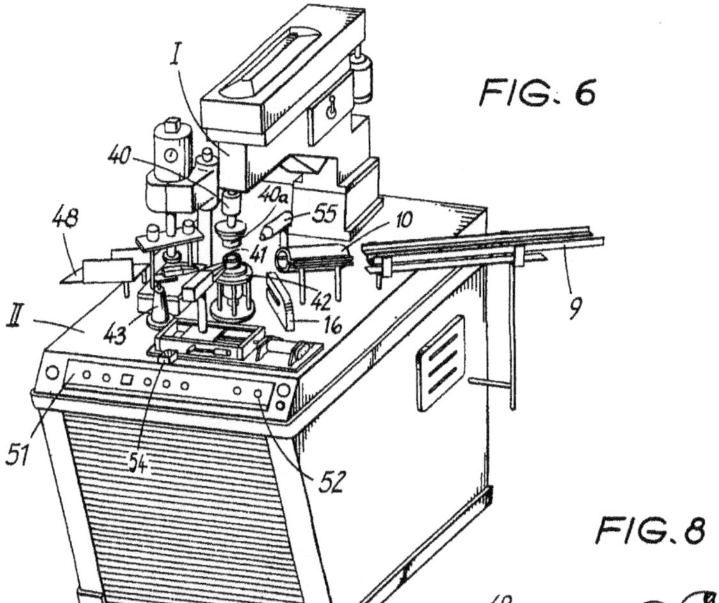

Fig. 2.9. Welding body device for connecting two Hollow body halves, e.g. to Christmas tree spheres or such [17]

The Inventor of the patent "Hollow Sphere and Structural of Element for Constructing same" USA patent No. 3,691,704, submitted to 19.05.1970 [18] also wants to construct Hollow Spheres out of individual parts. This patent has the challenge of producing Hollow Spheres out of separate preformed parts. The following illustration (cf. Fig. 2.10) shows the procedure and the partial elements.

Put into practice the invention is that of "a toy, a container, a gas-filled balloon, or building, for example". The Hollow Sphere consists of a number of relatively small

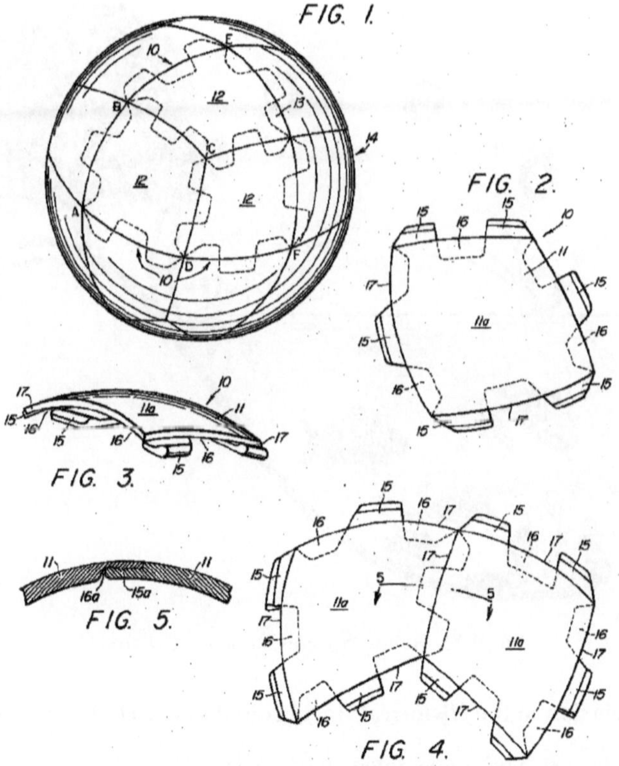

Fig. 2.10. A toy, a container, a gas-filled balloon, or building, for example [18]

sections of almost identical design. These parts can be made of different materials, for instance "inject molding of a plastic material or by molding a foamed plastic material such as "Styrofoam".

A further and somewhat other example is patent No. 231,541 of the Patent Office of the German Democratic Republic (DDR) of 13.12.1984, where a Hollow Sphere for a deodorant stick is described ("Hollow ball for the production of roll-on deodorants") [19]. The purpose of producing this plastic Hollow Sphere with a wall thickness of 1 mm "by means of blowing, rotating, spraying or centrifuge procedure" was primarily to reduce material use. The exact manufacturing technology of this patent however still remains somewhat indefinite. But it is an interesting fact that the patent refers to state-of-the-art technologies to produce such Hollow Spheres for Deodorants.

Patent DE 198 46 784 A1, of 10.10.1998 [20] mentions a porous disintegrating dough Hollow Sphere to attract fish. The Sphere is filled with fish food, thrown into the water and sinks to the bottom of the lake. After a certain period of time the Hollow Sphere begins to dissolve and releases the fish food. According to invention this "controlled, punctual" feeding of the fish enables a more efficient possible catch for hobby and sport anglers. The following picture 2.11 is taken from the patent.

Fig. 2.11. Hollow Spheres attracting fish, too [20]

2.1.7 Conclusion to the Historical Background of the Hollow Sphere Concept

These few patents have highlighted the broad variety of the term Hollow Sphere which connects all these projects not only by name but also by concept, even if the production and application concepts differ. Remarkable however is how the developed procedures are time and time taken up again and continuously improved; partially by means of technology transfers, further development and new and improved concepts of existing inventions. For instance, the concept to produce Hollow Spheres out of a pipe by means of deep-drawing and forming was fundamentally thought of for metals, later on however developed and adapted to produce ceramic Hollow Spheres. All in all it can be said that Hollow Sphere manufacturing techniques mainly orientate themselves around the function they have to carry out. It is also remarkable that only a few patents face the challenging task of describing a universal applicable principle of different sized metallic and ceramic Hollow Spheres on a production scale. This may be a reason why the concept of hollow spheres as a cellular material has not been establish as a broadly used technology compared to metallic foams as an example.

2.1.8 The Path to "Modern" Hollow Spheres

During the 1980s and 1990s, cellular materials within material sciences became a major theme of research. With the main concern of developing new materials for lightweight construction, and to fulfill specific material requisitions, Hollow Spheres became of special interest with regards to cellular materials in the research and development focus of the scientific community. Hollow Spheres are however a special case regarding cellular materials, where commercial use, and an economical large scale manufacturing process of the Spheres or their Structures is of greatest interest. Because only this enables Hollow Spheres to get out of the lab and into applications.

A universal production method was developed in 1982 by Manfred Jaeckel, where he set up a comparatively easier routine, which enabled the production of highly customised Hollow Spheres at a larger scale. For this development, the rapid technical evolution in the 60's, 70's and 80's within polymers, of metal (alloys) and high performance ceramics was an advantage. Thus, he was able to change the initial Hollow Sphere production process to that of today; and enabling the truly multifunctional and customized Hollow Sphere. A first step for this was patent DE 32 10 770 of 24.03.1982 [21] by Manfred Jaeckel, where he describes "metallic, essentially spherical light-weight particles, as well as their use and production process", which can be characterized as being "inside hollow and have closed or porous walls possess." The described process is a wet-chemical production, where a polymer core is wet-chemically metalized. The procedure is accomplished by means of a stirrer. The stirrer "mixes the granulates in the liquid bath, so that the whole surface of the core is exposed to the metallization procedure. This dead metallization can be carried out up to reaching the desired total layer strength of the metal, initially only a thin conducting layer can be applied..." with the metallization finally being continued to the desired layer thickness electrolytically. In such a way coated cores, which consist of a plastic core (e.g. foam material), are decomposed "pyrotolytic at temperatures at for instance 400°C in such a way." The Hollow Spheres are sintered in the subsequent manufacturing. Jaeckel identified several application scopes for his invention, such as:

- "the production of molded articles as fillers of plastic, hardening or sinterable materials ",
- "Production of molded articles cast in a form with liquid metal",
- "metallic light-weight parts (...), as wadding for molded article formations with closed or open-porous cavities, and in particular the creation of light weight molded articles from basic metallic materials are suitable",
- "super-light sintered molded articles..., suitable for applications such as filter technology and sandwich slabs in the aerospace industry".

Jaeckel also sees his invention in the catalysis carriers range. He quotes, that as a result of a controlled manufacturing process and the finished product "a large contact surface for chemical reactions arises."

According to the patent, the spheres' size is influenced by the implemented core and the metal powder used. More generally spoken Jaeckel sees a possible economic production: "... in a size of approximately 0.5 to 3 mm" with a "shell thickness of 5 to 15 μm", resulting in "a middle specific weight of the light body... 0.29-0.35 g/cm³". These statements of size refer to the first wet-chemically metalized core and its shell, which can be built up in a further process. In the patent Jaeckel describes both, the product and the manufacturing apparatus required. Jaeckel's ground breaking invention mainly combines two separate points, which can be found in older patents: a) pyrolysis of the core and b) two stage coating, stage one being through wet chemical methods (electroless and electrolytic) to build up a definite shell size.

Jaeckel was confident about the potential of his invention, and applied to the North German Affinerie AG in order to further develop his technology. Hence, it is no surprise that the disclosure writing of the North German Affinerie AG, Hamburg, No. 36 40 586 of 27.11.1986 [22] describes the next Hollow Sphere development step, being entitled "Procedures for the production of hollow balls or their connection with walls of increased firmness". In this patent he points out how on one "foamed polymer" layers of metallic powder can be constructed. The patent mentions wet chemically metalized cores which are coated with a second layer using a powder slurry, which can be applied e.g. by a fluid bed apparatus on the polymer carrier. The coating apparatus is not named but only a "Fluid bed apparatus" can be used for such a process. By means of this technology a self-supporting shell is constructed on a foamed core, and not – as pointed out before in DE 32 10 770 [21] - built up using a wet chemical process. To follow, the polymer core is pyrolised at a temperature of 400°C for instance, and afterwards the shell is being sintered to a solid state. The patent also describes that compact or porous shells can be produced, depending upon the particle, binder, type and temperature of the sintering. Jaeckel mentions "closely sintered shells made out of metal or ceramic material, like Al_2O_3". Furthermore the patent describes different kinds of treatments to manufacture Hollow Sphere Structures, e.g. "by shaking consolidated, if necessary, using minor pressure on the green sphere to interconnect the spheres" as well as the treatment of single spheres in order to glue them together or sinter them into blocks at a later stage".

Jaeckel recalls the steps of his invention in an private email to the authors: "1. Production of Cu-coated polystyrene balls according to DE 32 10 770 [21] in his cellar laboratory at home, and development of a carbonyl iron suspension. 2. Sending these two components to AEROMATIC, Switzerland and requesting them to coat the Cu-coated polysterene balls. 3. Once he had received the finished coated polystyrene balls, they were sintered in a production furnace with positive results at the Ringsdorff, Bonn, only made possible due to a former research colleague being employed at the company. 4. Based on the results I had the courage to finalize DE 36 40 586 [22] at the North German Affinerie". Research at the North German Affinerie was continued following this personal commitment. Jaeckel recalls: "That a laboratory size fluid bed coater was purchased at the North German Affinerie to improve the technology of patent DE 36 40 586,8 [22] and widen the range of applications."

The advancement of this patent is DE 37 24 156 [23], filed 22.07.87 entitled: "Procedure for manufacturing metallic or ceramic hollow spheres". The patent's underlying problem definition is to develop: "an economical and easy to handle procedure for the production of metallic or ceramic Hollow Spheres with a dense or micro-porous shell of high stiffness." In this patent the process described is a clear link to the further developed patents, using the Fluid Bed: "The carriers with a diameter of 1 to 8 mm are placed into a fluid bed and are sprayed with a hydrous suspension consisting of binder and metallic or ceramic particles. In such a way coated and afterwards within the fluid bed dried particles then have a dry powder-coating, a so called "greensphere", with a sufficient and stable stiffness to handle them. In the next step the core is pyrolized at a temperature in the range of 400 to 500°C; then the oven is heated up between 1000 to 1500°C in order to sinter the Hollow Sphere in motion or stationary", further on: "The dried powder layers possess sufficient firmness, so the coated, essentially spherical foam particles can be subjected to a pyrolytic decomposition process without the powder-shell losing its spherical shape."

Generally spoken, the patent is a more specific form of DE 36 40 586 [22], with the key-advantage being reduced manufacturing steps, as the wet-chemical process is no longer required. Jaeckel recalls: "While doing coating experiments with the department head of the pilot plant station at the North German Affinerie, he asks me why we don't coat the cores with a self supporting powder coat instead of wet-coating a Cu layer first and skip the electrolysis. I responded: "that should be possible, if I have some time to find a suitable polymer binder." That's how the idea of DE 37 24 156 [23] was developed at North German Affinerie and submitted to the German patent office in Munich about two years after the initial DE 36 40 586.8 [22]."

The production of metal sponge bodies is also described in a similar procedure. First, a sponge-structure is moistened with a metal-powder suspension, then dried and finally sintered. This section is very briefly mentioned in the patent and does not include any illustrations, which highlights the important developments in the performance of binder-systems. What connects the described manufacturing process of sponges and Hollow Spheres is at the very front the use of a suitable binder to connect the metal or ceramic powder to the (organic) carrier.

The major development of the technology was outlined by Jaeckel, namely the use of the fluid bed coater for coating a foamed core. It is very interesting to see how this technology was developed step by step, which is clearly documented in the patents. Tasks later on arose in manufacturing the Hollow Spheres according to the patent and development at the North German Affinerie was stopped, as during this time the fluid bed apparatus technology was not advanced enough, Glatt GmbH developed a special fluid bed apparatus [24] and improved the process engineering together with the Fraunhofer IFAM of Dresden, enabling a reliable large scale production.

Details regarding the industrial scale production of Hollow Spheres are mentioned the next paragraphs.

2.2 An Advance Technology Platform

Hollomet GmbH, a subsidiary of Glatt GmbH is pleased to introduce a wide spectrum of Hollow Spheres. Hollomet produces metallic and ceramic Hollow Spheres based on the hollomet-manufacturing Process and market them under the brand names "globomet" and "globocer". In the following paragraphs we will give an insight into the materials required for the production of Hollow Spheres. We also describe the industrial manufacturing process in general, followed by highlighting the possibilities of mechanical treatment.

The development of the Hollow Sphere production process has been made in cooperation with the Fraunhofer IFAM Dresden over several years. Yet hollomet focuses on the industrial production of hollow spheres and the product development. Together with customers we develop solutions along their needs based on metallic or ceramic Hollow Spheres. Several years of research and development were needed to be able to fulfill customer's needs.

The Fraunhofer IFAM Authors Andersen et al. describes in their paper published 2000 at an early stage [25] a new method to produce metallic Hollow Spheres from arbitrary metals and alloys. They give a general overview about the technology and the properties of metallic Hollow Spheres. Fiedler and Öchsner [26] give in their paper an excellent overview on different kind of Hollow Sphere structures and the impact that manufacturing techniques have on the material parameters. They mainly focus on partially bonded metallic Hollow Sphere structures on the one hand and syntactic metallic Hollow Sphere structures on the other hand. They address the mechanical properties of adhesively bonded metallic Hollow Sphere structures and conduct finite element analyses in order to determine Young's modulus and the initial yield stress dependence on the loading direction. Fiedler and Öchsner describe a numerical analysis of the different kind of Hollow Spheres using a three dimensional unit cell model with a primitive cubic arrangement. The PhD Theses of Thomas Fiedler describes in-depth and in comparison to other cellular metals the properties of metallic Hollow Sphere structures. He puts the main focus on linear elasticity, plasticity and thermal conduction [27]. In comparison to metallic foams metallic Hollow Sphere Structures have the huge advantage that due to the high reproducibility a simulation becomes possible.

2.2.1 Materials Used for Production

Polystyrene Core. A core is used in the production of the hollow sphere. For both metallic and ceramic Hollow Spheres a solid un-foamed polystyrene ball is usually used, which is, prior to the coating process placed into a foamer and expanded (EPS = expanded polystyrene). Grain sizes may vary during the expansion process. Sieving the polystyrene spheres is necessary for a narrow size distribution of the final product.

Coating. Various metal powders can be applied, their ability to be sintered being the only requirement. The grain size of the metal powder determines the minimum coat thickness possible. In order to achieve a solid integral Hollow Sphere after sintering, the coat should approximately be five times larger than the grain size of the metal powder being used. Initial preparation of the metal powder (e.g. water and gas atomization) considerably effects production costs, as well as technical specifications of the end-product. So far various materials such as iron, stainless steel, titanium and molybdenum have been well proven. Hollow Spheres of Alloys can also be produced in situ. Therefore a suspension of different metal powders are simultaneously sprayed on the core, the formation of the alloy takes place during sintering. The approach of combining different metal powders to spheres is not only a more cost effective means of production, but also enables an amalgamation of low and high-cost powders. Using this technique one may improve the technical specification and could reduce the costs of production in comparison to spraying an alloy-powder.

Binder. Prior to coating the metal powders are suspended in a binder-water mixture. Selecting the appropriate binder system depends on certain criteria such its pyrolysis, its gluing ability and its rheological properties. The binding forces glue the metal particles onto the core and interconnect the particles. The binder agent is generally kept at a minimum in the suspension.

2.2.2 Coating

The suspension recipe of metal powder, water and a certain amount of binding agent is weighed and poured into a suspension-container, where the ingredients are stirred, forming a consistent suspension, avoiding any sedimentation. A certain amount of polystyrene spheres, according to the recipe are placed into the spray chamber of the coating apparatus (as described in Patent DE 197 50 042 C2 [24]), where they are fluidized and rotated. The pump releases an exact amount of suspension through a nozzle onto the rotating and fluidized polystyrene spheres. The following Figure 2.12 shows the manufacturing process.

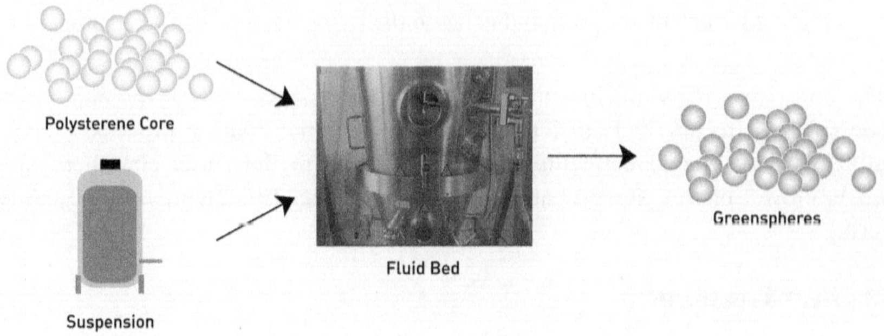

Fig. 2.12. Manufacturing Process of green spheres using a fluid bed coater

The following factors should be considered:

- The spherical shape is affected by the precise moisture content of the spray. An overly moist process will cause agglomeration, whereas a process that is too dry will cause the spheres' surface to become uneven and fixation on the core is poor.
- Consistency and density can be controlled by process parameters of the coating apparatus.
- Droplet size discharged from the spray nozzle also contributes to consistent application of the suspension. An over accelerated spray process of smaller droplets will cause an increased amount of dust particles in the spray chamber, whereas agglomeration will occur if the droplets are too large.

The outer diameter of the Hollow Sphere is affected by the material being sprayed and the geometry of the equipment. For cost reasons, large-volume production metal Hollow Spheres are generally produced by this method are in a range of 1 and 15 mm in diameter. Shell thickness can be varied, depending on the grain size and grain spectrum of the metal powder. Stronger, thicker shells with further increased solidity and robustness are always possible.

Outer shell characteristics such as porosity, density and grading are influenced by the coating and sintering process. The following microscopic shell dissections (cf. Fig. 2.13) highlight the Hollow Spheres' wide technological spectrum.

The hollomet technology furthermore enables a consistent and reproducible shell structure as well as an invariable constant Hollow Sphere size; advantageous compared to other cellular materials in both mass-produced articles made of Hollow Spheres and for simulating these materials.

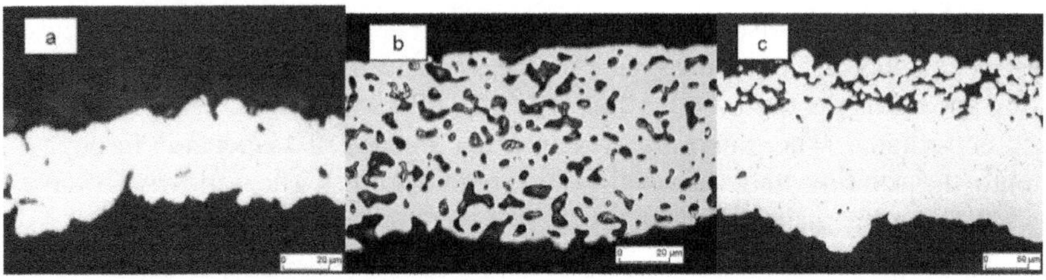

Fig. 2.13. Shell dissections: a) thin dense shell, b) porous shell, c) graded shell

The constant airflow during the manufacturing process enables the coated spheres (green-spheres) to dry and sufficiently harden after the coating process. Once sufficiently hardened they are pneumatically ejected from the spray chamber. Now the green Hollow Spheres are prepared for the final heat treatment, i.e. debinding and sintering.

2.2.3 Heat Treatment

A further important stage in the production of metal Hollow Spheres is the heat treatment, where the green-spheres are transformed into solid Hollow Sphere shells by sintering. The basic concept of heat treating Hollow Spheres is very similar to that of metal injection moulding (MIM). Heat treatment is carried out in two phases. During the first stage the polystyrene core and the binding agent are removed by pyrolysis. In the second phase the aggregation of the metal powders to a solid metal takes place, i.e. the end product is sintered and hardened.

2.2.4 Debinding

During the de-binding process all organic parts within the green-spheres, i.e. the EPS core and the binder are pyrolised. For metallic systems this step is usually carried out under inert gas like hydrogen or endo-gas or in vacuum enabling an optimized temperature control and a reduced risk of oxides building up within the shell. Furnace temperatures and heating rates depend on the amount and kind of binder material and EPS used in the green-spheres. Once the de-binding process has been completed the previous green-spheres turn a brownish colour, which is the result of small metal powder particles having been bound together, hence the spheres are now called brown spheres. These brown spheres are fragile structures out of loose interconnected metal particles.

2.2.5 Sintering

Sintering follows the debinding process, where single powder particles amalgamate to form the outer hollow sphere's shell. Sintering occurs when powder particles are heated up to about 80% of their melting temperature. This process step is carried out in vacuum or inert gas to avoid surface reactions of the metal powder. Such surface reactions as oxidation may make the sintering process impossible by inhibiting the surface diffusion, the basic mechanism responsible for sintering.

Shell characteristics: Affected by the sintering parameters, the Hollow Spheres' shell can be manipulated to reach specific physical and chemical characteristics. Temperature control, as well as the green-spheres' chemical composition and the combination of protective gases all play a part in the consistency of the outer shell. Porosity in the outer shell can also be manipulated through temperature control, enabling the production of either permeable or gas-tight shells.

The following picture 2.14 summarizes the processing of the Hollow Spheres.

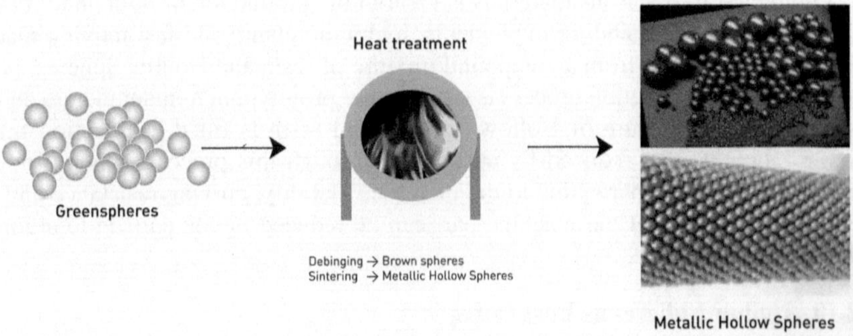

Fig. 2.14. Producing Hollow Spheres: green spheres to solid metallic Hollow Spheres

2.2.6 Moulding

The wide spectrum of applications for Hollow Spheres makes different kinds of molding techniques necessary.

2.2.7 Gluing

For example, single spheres coated in glue can be brought into forms with different shapes and sizes. It is also possible to fill the glue-coated Hollow Spheres into an arbitrary complex unit and activate the glue after filling. One might reach a higher resistance of the unit together with a significant weight-reduction. The glue does not only connect the spheres, but additionally acts as an extra reinforcement in the structure and simultaneously retaining its light weight. Obviously this technique is limited to room temperature applications such as crash absorption or the formation of lightweight sandwich structures, especially with CFC-composites as facesheets.

2.2.8 Sintering

A further production process is with the use of an industrial hydraulic moulding press. This press shapes the green spheres into a component like form. To ensure a perfect bond between each green-sphere gases are introduced during the compacting phase. The hot gasses work as a foamer and expand the polystyrene cores. So the metal-binder suspension on the core interconnects with its neighboring coated spheres. Once compressed into shape, the hydraulic moulding press ejects the moulded form, which then goes on to debinding and sintering. The advantage of this method is that the joints between the spheres exhibit the same characteristics as the spheres themselves, so this is best used for high temperature applications like mufflers in hot exhaust gas ducts. This moulding process also allows to produce complex shapes for a near net-shape process, which reduces the amount of machining after sintering. However since shrinkage during sintering is not accurate enough a final machining step is usually necessary.

2.2.9 Hollow Spheres as Compounds

The combination of various materials is ideal in the production of light and vibration-free components. Demanding implements, hydraulic-clamp and fast moving machine parts can be produced from a compound mixture of resin and Hollow Spheres, poured into a cast. One production process is to use rapid prototyping to manufacture an outer form in which a mixture of Hollow Spheres and resin is filled. Production time of complex tools can be reduced by up to 80% through this process. When producing fast moving machine parts, due to the increased stability, curving resistance and light weight, a 25% mass of the machine part can be reduced in comparison to aluminum counterparts.

2.2.10 Hollow Spheres as Lost Cores

A further possible application exists within the use of casting construction units. Here a Hollow Sphere structure is for example integrated as a "lost core" in a cast part. Thus simplifying the production process by the omission of the de-coring and reaches a weight reduction.

2.2.11 Brazing

Brazing can be used as a high temperature glue to either bond the spheres together or to bond sintered Hollow Sphere structures to facesheets. Temperatures of up to 1000°C are possible, depending on the braze used.

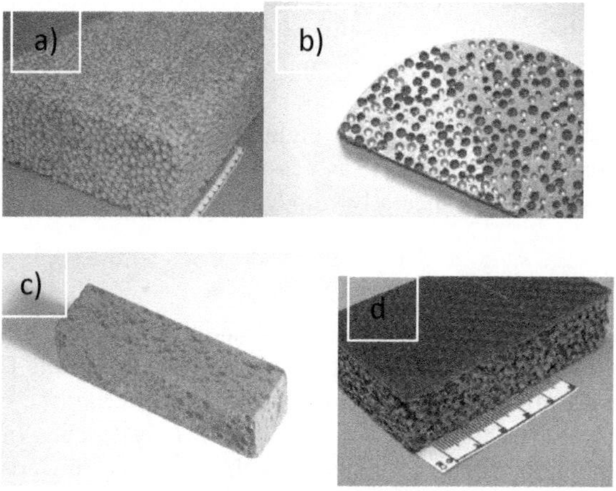

Fig. 2.15. a) sintered Hollow Sphere structure, b) Hollow Spheres cast in aluminum, c) Hollow Spheres cast in resin, d) Hollow Spheres as a sandwich structure

The following pictures (cf. Fig. 2.15) show different kind of above described types of Hollow Sphere Structures: a sintered metallic Hollow Sphere structure, metallic Hollow Spheres cast in aluminum, metallic Hollow Spheres cast in resin and metallic Hollow Spheres as a sandwich structure.

2.2.12 Machining

The metallic Hollow Sphere structures can be machined comparable to the well-known mechanical machining processes used for steel. In a SAB (Sächsische Aufbaubank, Germany) sponsored project carried out by Fraunhofer IFAM, Glatt Systemtechnik and the Institute of Surface and Measurement Technology, Technical University Dresden (Günter, Popp, Gerighausen) different Hollow Sphere materials and structures were analyzed with emphasis on how to machine them. Each experiment and result was documented in a database, available upon request from hollomet Company. Regarding this investigation metallic Hollow Sphere materials with differing mechanical treatment processes were examined. Determining measured variables for the Hollow Sphere Structures are mechanical properties such as: sphere diameter, structural density, structural connection (sintered, glued), the sintering material and the corrosion behavior. The processes are grouped into conventional mechanical treatment and cutting technology.

2.2.13 Conventional Mechanical Treatment

Standard Tool Drilling. Structures can be drilled at high speed (in this case at 6300 rpm) and pressure. Despite an insubstantial cut most components become deformed along the point of entry. The inner spheres are cut, crushed and compressed by the drill, resulting in the hole being considerably wider than that of the drill head diameter itself. A drill head with a diameter of 10 mm only fulfills the limit "rough to very rough" of the general tolerance according to DIN ISO 2768 T1. Due to their mechanical stability drilling the green-spheres is limited to impossible. Special cutting tools did not prove to be any more advantageous than that of a standard drill.

Turning. The turn part components range between a "mid to rough" tolerance in the general DIN ISO 2768 T1 classification, with most Hollow Sphere particles situated on the incision being deformed and/or compressed or quarried out.

CNC Milling. The top surface of the mould is generally crushed, compacted and partially chipped, with a ridge along the periphery of the structure and within the dissected spheres. Variation of the basic size is estimated at approximately 0.5 mm of the grooves. This is due to the Hollow Spheres being cut, compressed and covered during milling. The milled and sintered plane, free-formed surfaces and grooves fulfill a "medium" tolerance according to DIN ISO 2768.

2.2.14 Cutting Technology

Band Saw. The width of the flat joint is about 0.9 mm. A ridge is formed, particularly where the saw exits. The dimension deviation and evenness of the cut piece lie in the general tolerance of DIN ISO 2768. The band saw cutting technique is by far a more cost effective option compared to hot diamond wire cutting for example.

Diamond Wire Saw. To saw sintered structures the blade has a diameter of approximately 0.3 mm with a minimum strain and load on the tool. The flat joint is also approximately 0.3 mm. During the process water penetrates into the Hollow Spheres situated at the cut edge, leading to a buildup of rust, only avoidable through a final drying process, resulting in further costs. The characteristic evenness of the cut surface and its flush surface is influenced by the two-way direction of the cutting wire. A slight wave of approximately 0.2 mm in the direction of the blade is visible, each wave representing the wire's shift in direction.

Laser cutting. The cut width is approximately 0.2 mm and is not visibly thermally affected. Obviously, the thicker the material, the wider is the cut. Annealing color, caused by the temperature of approximately 300°C was detected on the base.

Water jet cutting. Hollow Sphere structures can be cut either with a pure water jet or an abrasive. The cut measuring an approximate width of 0.5 mm – 1 mm, depending on whether an abrasive is used or not. For example, a material thickness of 8 mm had an approximate cut-width of 0.2 mm, where as a material density of 22 mm approximately 1 mm.

Wire cut EDM. The width of an EDM cut is between 0.25 mm and 0.3 mm, the feeder rate being determined by the automated cut speed. During the cut the de-ionised water seeps into the Hollow Sphere structure, in turn leading to a rust build-up on the outer surface, only avoidable through a final drying process, resulting in further costs.

2.2.15 Costs

Complex, high-grade metal powder is required in the production of Hollow Spheres and their structures. Due to up to 50% of the manufacturing costs being that of the metal powder, the Hollow Sphere and its structures are higher-priced than normal steel plates.

A profitable application of Hollow Spheres and Hollow Sphere structures is feasible once this new material carries out various functions. Such as sound proofing applications, with the resistance to corrosion and heat, the ability to withstand vigorous currents and being maintenance friendly are of great advantage. Expensive installation is avoided due to the simple self-supporting construction, its resistance to stresses and strains guarantees a maintenance-free durability; all money saving factors to consider.

Light-weight metallic Hollow Spheres are notably preferential in fast moving machine parts for their superior flexibility and shock absorption compared to heavier aluminum-alloy counterparts. An impressive manufacturing economization of 80% was made in the production of intricate components with the hollow sphere's technology.

2.2.16 Development Status

After ten years of Hollow Sphere technology development and proof of concept for various applications, it has been ascertained that production of this cellular material is both assured and reproducible [28] up to industrial production. Hollomet GmbH, a Dresden affiliate of the Glatt Group, manufactures the Hollow Sphere and Hollow Sphere structures on an industrial scale. A list of applications can be found on the hollomet homepage [29] or in the Cellmet News [30]. At the end of this book, in chapter 11 Christian Augustin describes some ongoing research projects.

References

[1] Grimm, J.: Deutsches Wörterbuch. DTV München (1877 ND 1999)

[2] Deutsches Patent- und Markenamt (accessed 2008), `http://www.dpma.de`

[3] Augustin, C.: Prozess und Produktinnovationen aus philosophisch-historischer Sicht am Beispiel Nicolaus Copernicus und metallischer Hohlkugeln. University of Konstanz, Germany (2009)

[4] Spahr, O.: Method or Process of making Hollow Metal Balls. US Patent 861 403 (1906)

[5] Schatz, J.W.: Process of Making Hollow Metallic Balls. US Patent No. 955 698 (1910)

[6] Gruver, J.H.: Seamless Hollow Ball Company: Sheet Metal Hollow Ball and Method of Making the Same, US Patent 1 278 914 (1917) cf., Verfahren zur Herstellung von Hohlkugeln. AT 92 463 (1920)

[7] Seamless Hollow Ball Company: Verfahren und Vorrichtung zur Herstellung von Hohlkugeln aus Metall. AT 103 454 (1924)

[8] Seamless Hollow Ball Company: Verfahren und Einrichtung zur Herstellung von metallenen Hohlkugeln. CH 112 433 (1924)

[9] `http://www.isometall.de` (accessed 2008); the production process is described, `http://www.isometall.de/html/kugelherstellung.html`

[10] Müller, H., Neuhäuser, M., Bliedtner, J., Institut für Fügetechnik und Werkstoffprüfung: Verfahren zur Herstellung von Hohlkugeln. DE 100 39 320 (2000)

[11] Ramsay, W.S., Stark Ceramics: Method of Producing Porous, Lightweight, Burned Argillaceous Material. US 2 543 987 (1947)

[12] Champion Spark Plug Company: Verfahren zur Herstellung von Formlingen, insbesondere aus keramischem Stoff. German Reichspatent 604 691 (1932)

[13] Baumat, A.G.: Verfahren zur Herstellung von als Zuschlagstoffe für die Bauindustrie verwendbaren Hohlkugeln aus keramischen Stoffen. CH 266 315 (1948)

[14] Toyo Tokushu Immono Kabushiki Kaisha: Verfahren und Vorrichtung zum Herstellen metallener Hohlkugeln. German Reichspatent 694 903 (1938)

[15] Geiger, C.F., Carborundum Company: Method for making Refractory Bodies and Products Thereof. US 2 553 759 (1946)

[16] Nukem Nuklear Chemie und Metallurgie GmbH, Device for a Production of a Hollow Sphere. GB 1083 993 (1964)

[17] Schmidt, M.: Schweißvorrichtung zum Verbinden von zwei Hohlkugelhälften, z.B. zu Christbaumkugeln od. dgl. DE 1 168 058 (1960)

[18] Novak, P.: Hollow Sphere and Structural of Element for Constructing same. US 3 691 704 (1970)

[19] Kühn, R., Umlauf, K., VEB Ingenieurbüro für Rationalisierung Kosmetik: Hohlkugel für Roll-on-Erzeugnisse. GDR 231 541 = DD 231 541 (1984)

[20] Unterländer, J.: Fisch-Lockfutter-Hohlkugel. DE 198 46 784 (1998)

[21] Jaeckel, M.: Metallische, im wesentlichen kugelförmige Leichtkörperteilchen, sowie Verwendung und Verfahren zu ihrer Herstellung. DE 32 10 770 (1982)

[22] North German Affinerie: Verfahren zur Herstellung von Hohlkugeln oder deren Verbunden mit Wandungen erhöhter Festigkeit. DE 36 40 586 (1988)

[23] Jaeckel, M., Smigilski, H., Norddeutsche Affinerie AG: Verfahren zum Herstellen von metallischen oder keramischen Hohlkugeln. DE 37 24 156 (1987)

[24] Bretschneider, F., Bruno, P., Brückner, J., Glatt Systemtechnik, KWS Kleinwanzlebener Saatzucht AG: Einrichtung und Verfahren zur Herstellung eines schüttfähigen Produktes mit einer Hüllschicht. DE 197 50 042 C2 (1997)

[25] Anderson, O., Waag, U., Schneider, L., Stephani, G., Kieback, B.: Novel Metallic Hollow Sphere Structures. Adv. Eng. Mater. 2(4), 192 (2000)
[26] Fiedler, T., Öchsner, A.: On the anisotropy of adhesively bonded metallic Hollow Sphere structures. Scripta Mater. 58, 695 (2008)
[27] Fiedler, T.: Numerical and Experimental Investigation of Hollow Sphere Structures in Sandwich Panels. Mater. Sci. Found 44 (2008)
[28] Anderson, O., Hungerbach, W., Stephani, G., Studnitzky, T.: Inno.zellmet - a concerted approach towards the application of non-foam type cellular metals. Mater. Sci. Forum 539-43(2), p. 1890 (2007)
[29] http://www.hollomet.com (accessed 2008)
[30] http://www.metalfoam.net (accessed 2008)

3

Geometrical Properties of Hollow Sphere Structures

Andreas Öchsner[1] and Thomas Fiedler[2]

[1] Technical University of Malaysia, 81310 UTM Skudai, Johor, Malaysia and The University of Newcastle, Callaghan, New South Wales 2308, Australia
Andreas.Oechsner@gmail.com

[2] The University of Newcastle, Callaghan, New South Wales 2308, Australia
Thomas.Fiedler@newcastle.edu.au

Abstract. This chapter presents in the first part the basic methods on how two single spheres can be connected. The cases of a point contact, neck region and matrix embedded are distinguished. A general equation for the calculation of the volume of the neck region between two spheres is analytically derived. In the second part of the chapter, simple model arrangements of spheres, i.e. primitive cubic, body-centered cubic, face-centered cubic and hexagonal are presented. Based on a general introduction of characteristic quantities such as relative density and different volume fractions, equations and reference numbers for characterisation of the model arrangements are collected. A few comments on random arrangements conclude the chapter.

Keywords: Sphere packing, Periodic arrangement, Sphere joining, Bond volume, Density of the arrangement, Volume fraction.

3.1 Introduction

Cellular solids are materials with high porosity which are divided into distinct cells. The boundaries of these cells are typically made of solid polymer, ceramic or metal, while the internal regions are air cavities. Cellular materials therefore exhibit densities which are typically below 10% of their corresponding base material. Two special cases of cellular materials are metallic foams and sponges. Metallic foams like Alporas® originate from a liquid and are characterised by closed cells. Metallic sponges such as M-Pore® are characterised by an interconnected porosity of open cells. Metal foams and sponges have cells with a certain dispersion in their cell size and shape which influences the physical properties. In contrast to cellular metals, porous metals contain a multitude of microscopic pores and the densities of these materials are in the range of their base material(s). It should be mentioned here that the definitions of metallic foam, sponge and porous metals are not mutually exclusive.
The basic unit of a hollow sphere structure (HSS) is a single sphere which is from a mechanical point of view an ideal load-carrying element. These *single* spheres can be connected by different methods which significantly influence the macroscopic properties of a hollow sphere *structure*.

Figure 3.1 a) shows schematically a connection of two single spheres where the join is symbolically illustrated as a spring in order to highlight that the connection itself codetermines the macroscopic properties of the structure. In addition to the connection

A. Öchsner, C. Augustin (Eds.): Multifunctional Metallic Hollow Sphere Struc., EM, pp. 31–46.
springerlink.com

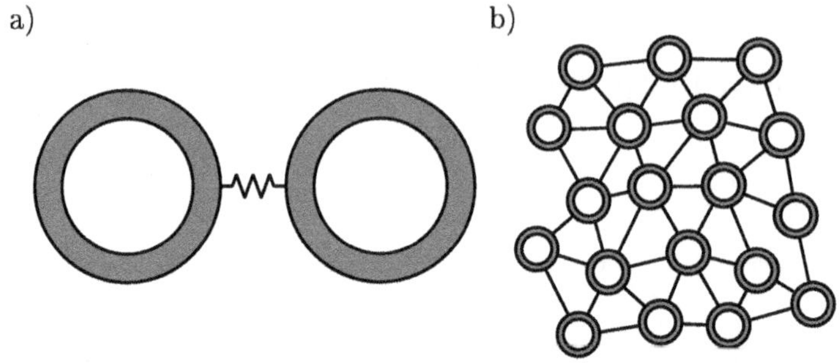

Fig. 3.1. a) Connection of single spheres; b) Multi-sphere network

of single spheres, the spatial arrangement of the spheres, i.e. the formation of a spatial network, determines the properties on the macroscopic level, cf. Fig. 3.1 b). The aim of this chapter is to provide a basic understanding on the geometrical peculiarities of the connection of single spheres and their arrangement in space. Emphasis is given to idealised geometries and structures which are commonly used to analytically and numerically model the macroscopic properties of hollow sphere structures.

3.2 Connection of Single Spheres

3.2.1 Basic Configuration

Single hollow spheres can be joined in three main procedures. Sintering of metallic precursor spheres by applying heat and pressure is a first method. Examples of sintered metal hollow sphere structures are given in references [1, 2] for titanium and 316L and in [19, 30] for steel and titanium. The contact areas of the spheres (cf. Fig. 3.2 a)) are to a certain extent flattened or even curved which complicates the generation of computational models. A simplified model of the contact area is shown in Fig. 3.2 b) where two spheres are considered to be in point contact and retain their original shape. The thickness of the sphere shell is given by h while the outer radius of the sphere is indicated by R.

A second method for joining hollow spheres is to bond the spheres using a liquid phase which forms a neck region between the spheres. Such neck regions can be formed by adhesives or soldering agents which during a certain period of the joining process are in the liquid state. A typical example is a hollow metal sphere structure (pure nickel and nickel alloys) bonded with epoxy or by brazing described in [5]. Figure 3.3 shows an idealised neck region where the minimum distance between both spheres is reduced to zero in a single point (so-called neck region with point contact). In addition, it can be seen that the neck region is tangent to the hollow spheres in points which are characterised by the bond angle θ.

However, the minimum distance between two spheres does not need to reduce completely to zero. In [7], a minimum distance between 0.12 and 0.13 mm is reported for adhesively bonded spheres of diameter 1.0 mm. A schematic representation of such a

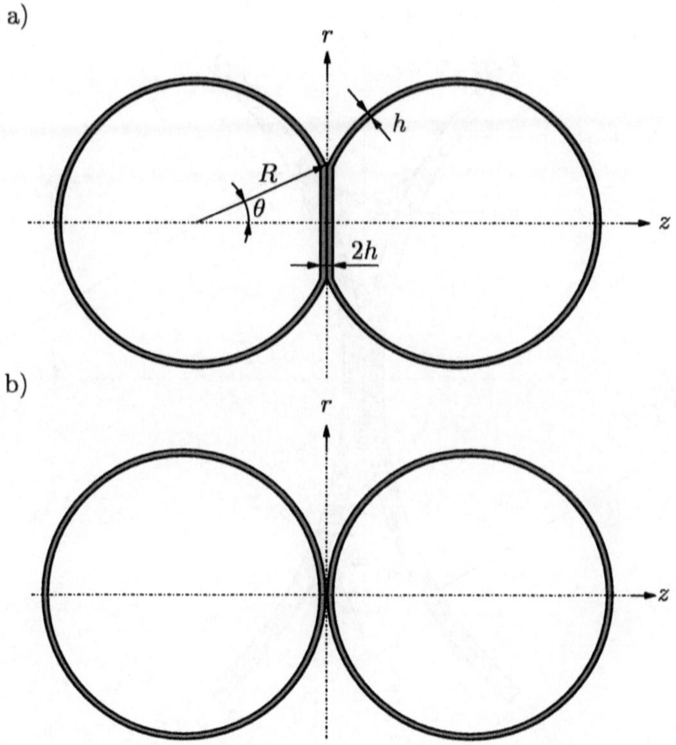

Fig. 3.2. Simplified models for sintered hollow sphere structures: a) flattened contact region; b) point contact

neck region is shown in Fig. 3.4 where the minimum distance between the spheres is indicated by $2d$.

As a result of the third manufacturing method, hollow spheres are included in solid matrices as in the case of classical fibre reinforced composites (e.g. class or carbon fibres in epoxy resin). This can be achieved by infiltrating the interstitial spaces in a three-dimensional network of hollow spheres with a liquid resin such as epoxy or a low melting temperature metal. Typical examples are matrices of magnesium or aluminium filled with hollow ceramic spheres [21], 45 and 270 μm ceramic microspheres in nominally 1350, 5083 and 6061 aluminium alloy matrices [25], hollow aluminium spheres in an aluminium matrix [26, 27] or corundum based (0.5-1 mm) macro-hollow-spheres and aluminium-silicate Fillite (5-300 μm) micro-hollow-spheres in an epoxy resin [4].

Such structures are commonly known as syntactic foams. The adjective 'syntactic' is used in this context to indicate a 'constructed' foam, as opposed to the normal foaming process which results in a random arrangement of cavities in a single material [23]. Figure 3.5 shows a schematic representation of a syntactic foam.

Table 3.1 summarises the different joining methods of single spheres and collects several representative references.

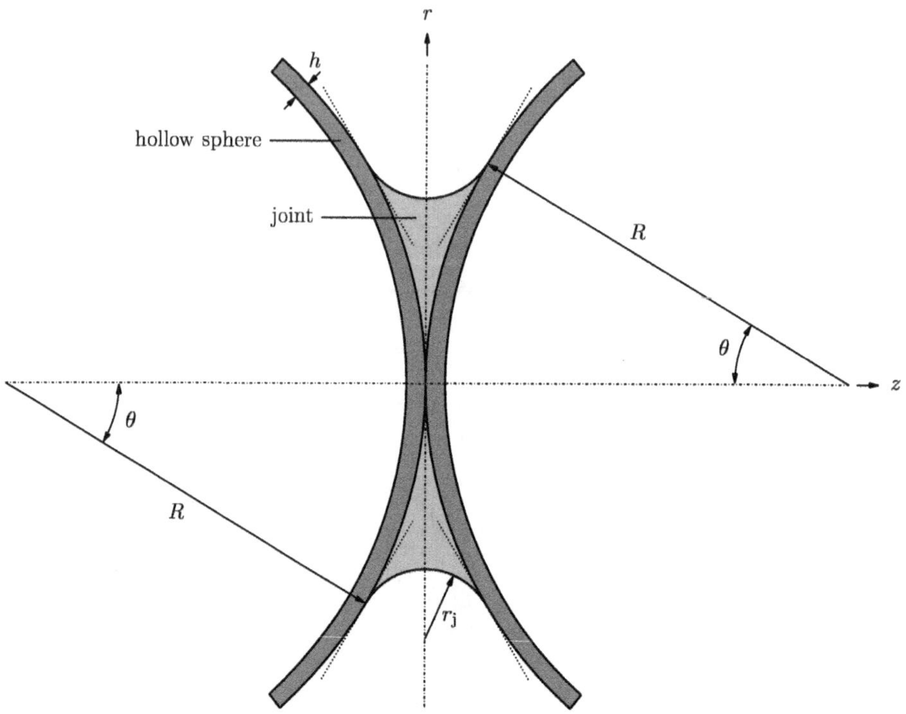

Fig. 3.3. Hollow spheres with a neck region in point contact; joined by a liquid-based method

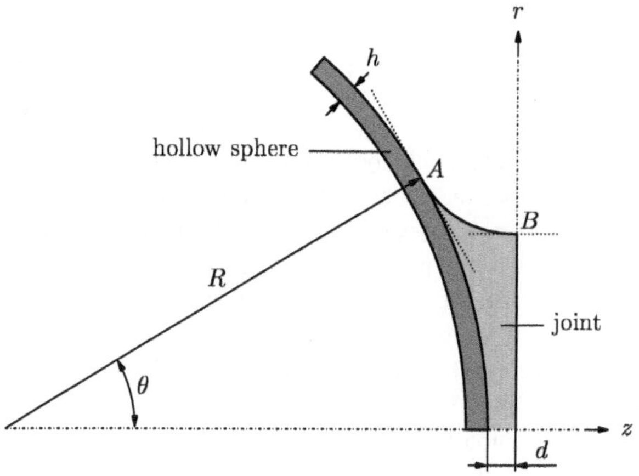

Fig. 3.4. Hollow spheres with a neck region at a minimum distance of $2d$; joined by a liquid-based method

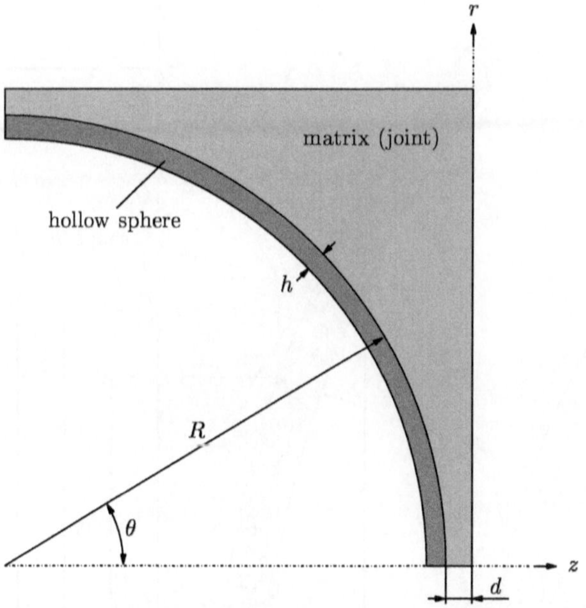

Fig. 3.5. Syntactic hollow sphere structure; matrix made of cast resin or low melting temperature metal

Table 3.1. Different modifications of the connecting area of single spheres

Joining Process	Reference	Figure
Sintering	[1, 2, 19, 30]	3.2
Bond neck	[5, 19]	3.3, 3.4
Infiltrating	[4, 5, 16, 17, 18, 21, 25, 26, 27]	3.5

3.2.2 Calculation of the Bond Volume

In the following section, the volume of the bond material which forms a neck will be derived. The general configuration of a neck region is shown in Fig. 3.6 where the z-axis is the axis of revolution and the plane perpendicular to the z-axis and located in point $(z = 0, r = 0)$ is a reflective symmetry plane. The minimum distance between the two spheres in the neck region is equal to $2d$. The approach in the case of a point contact $(d = 0)$ is given in [28] where a numerical evaluation is applied for the bond volume determination. This section provides analytically derived relationships for the general case $d \geq 0$.

Evaluating the right triangle OFA, the adjacent (segment OF) can be expressed as $R \cdot \cos\theta$ and thus, the side length a of triangle ADC as

$$a = d + R \cdot (1 - \cos\theta). \tag{3.1}$$

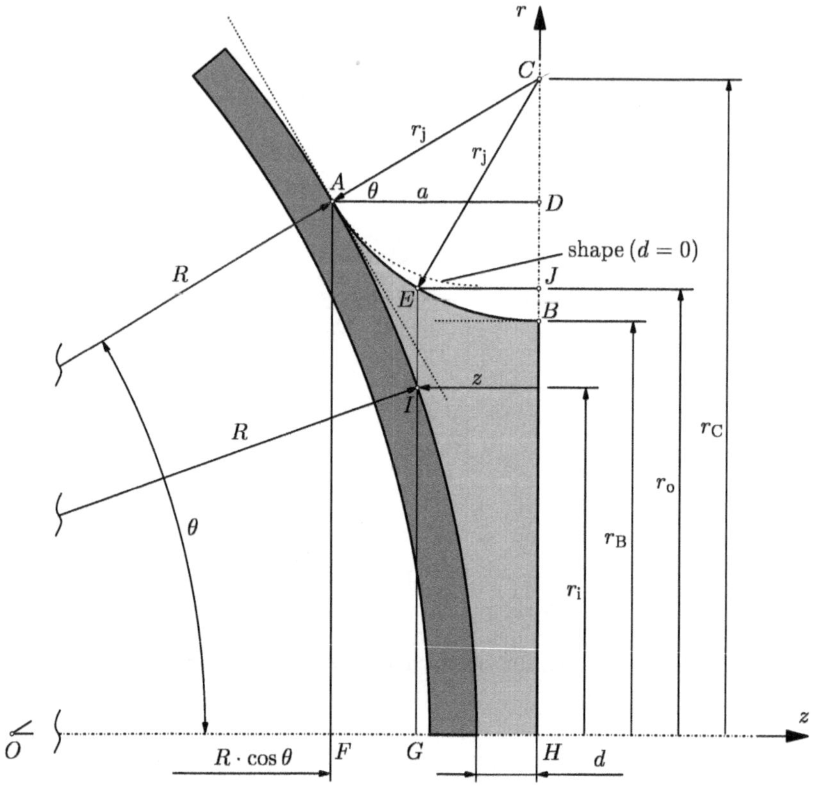

Fig. 3.6. Details of the half bond region with geometric parameters for bond volume calculation

Looking at the right triangle *ADC* and considering Eq. (3.1), the length of the hypotenuse r_j can be expressed as:

$$r_j = R \cdot \left(\frac{1}{\cos\theta} - 1 \right) + \frac{d}{\cos\theta}. \tag{3.2}$$

From triangle *OHC*, one can conclude that $r_C/(R+d) = \tan\theta$ and the side length r_C can be calculated according to

$$r_C = R \cdot \tan\theta + d \cdot \tan\theta\,. \tag{3.3}$$

Applying Pythagoras' theorem to the triangle *OGI* gives $R^2 = (R+d-z)^2 + r_i^2$ which can be rearranged to obtain

$$r_i = \sqrt{R^2 - (R+d-z)^2}\,. \tag{3.4}$$

In the same way, one can conclude from the right triangle *EJC* that $r_j^2 = (r_C - r_o)^2 + z^2$ which gives under consideration of Eq. (3.3) and (3.2)

$$r_{\mathrm{o}} = R \cdot \tan\theta + d \cdot \tan\theta - \sqrt{\left[R\left(\frac{1}{\cos\theta} - 1\right) + \frac{d}{\cos\theta}\right]^2 - z^2}. \qquad (3.5)$$

The bond volume in an element dz is a ring-like object with inner radius r_{i} (segment GI) and outer radius r_{o} (segment GE) at at distance z from the reflective symmetry plane. To calculate the total bond volume between two spheres of distance $2d$, it must be considered that the inner radius is zero in the range $0 \leq z \leq d$. Thus, the integration must be split into two parts:

$$V_{\mathrm{Bo}}(R,\theta,d) = 2\pi \cdot \int_0^d r_{\mathrm{o}}^2 \mathrm{d}z + 2\pi \cdot \int_d^a (r_{\mathrm{o}}^2 - r_{\mathrm{i}}^2)\mathrm{d}z. \qquad (3.6)$$

Introducing the expressions for the inner and outer radius, i.e. Eqs. (3.4) and (3.5), in Eq. (3.6) gives after some mathematical transformations the following closed-form solution for the entire bond volume (note that Fig. 3.6 shows only the half neck region) as:

$$\begin{aligned} V_{\mathrm{Bo}}(R,\theta,d) = & \frac{2\pi}{\cos^2\theta} \cdot \big[2a \cdot (R+d)^2 - 2aR \cdot (R+d) \cdot \cos\theta \\ & - (a^2 \cdot (R+d) + d^2 \cdot (R + \tfrac{1}{3}d)) \cdot \cos^2\theta \\ & - \tan\theta \cdot (R+d) \cdot (R \cdot (1 - cos\theta) + d)^2 \cdot \arcsin(\cos\theta) \\ & - \tan\theta \cdot (R+d) \cdot (R \cdot (1 - cos\theta) + d)^2 \cdot \sin\theta \cdot \cos\theta\big]. \end{aligned} \qquad (3.7)$$

Figure 3.7 presents a graphical representation of the bond volume, Eq. (3.7). It can be seen that the bond volume disproportionally increases for larger distances d.

As a numerical example, one may take $R = 1.35$, $d = 0.12$ and $\theta = 24.25°$ to obtain a volume of $V_{\mathrm{Bo}} = 18.11 \cdot 10^{-2}$. It should be noted here that the radius r_{j} should in all

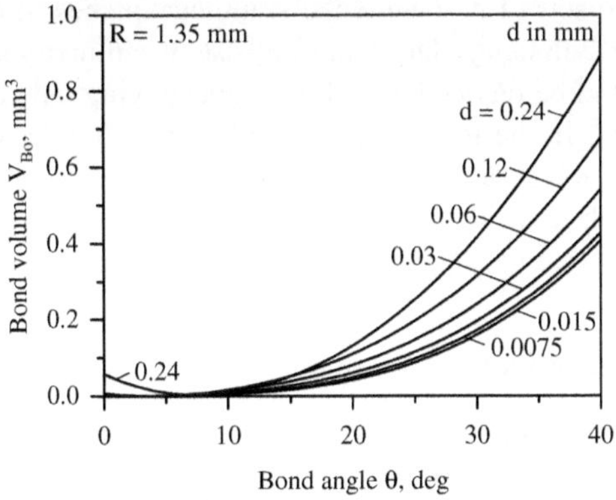

Fig. 3.7. Bond volume as a function of the bond angle and different distances between neighbouring spheres $2d$

cases not be greater than the radius r_C. The volume for the special case that the two spheres are in point contact (cf. Fig. 3.3) in the origin ($z = 0, r = 0$) can be obtained for $d = 0$ from Eq. (3.7) as:

$$V_{Bo}(R,\theta,d=0) = 2\pi R^3 \cdot \left(\frac{1}{\cos\theta} - 1 \right)^2 \cdot (1 - \tan\theta \cdot \arcsin(\cos\theta)) . \tag{3.8}$$

Taking $R = 1.35$ and $\theta = 24.25°$, Eq. (3.8) gives a volume of $V_{Bo} = 6.99 \cdot 10^{-2}$. Sanders and Gibson give in [28] the following equations to approximate the bond volume as

$$V_{Bo}|_{d=0} = 2R^3 \cdot \left(1.68 \cdot \theta^3 - 0.2072 \cdot \theta^2 + 0.01055 \cdot \theta - 9.77\right) \cdot 10^{-6} , \tag{3.9}$$

and

$$V_{Bo}|_{d=0} = 2R^3 \cdot 2.74 \cdot 10^{-7} \cdot \theta^{3.41} , \tag{3.10}$$

with a correlation coefficient of 1.0 for Eq. (3.9) and 0.9995 for Eq. (3.10) in relation to the numerically evaluated integration of the bond volume.

In the case of syntactic hollow sphere structures, the matrix or bond volume can easily be calculated and is considered in section 3.3.2.

3.3 Spatial Arrangements of Spheres and Morphologies

In real hollow sphere structures the spheres are randomly arranged and most numerical and analytical approaches require a simplification of the real arrangement. A common modelling approach is to periodically arrange the spheres in space groups as in crystallography [6]. A simplified two-dimensional model of such periodic crystal system is shown in Fig. 3.8. It can be seen that two principal sets of symmetry systems, i.e. the $0°/90°$ and the $\pm 45°$-system, exist. The thin dotted squares represent the possible unit cells for the two sets of perpendicular symmetry planes. If the loading is applied parallel to one of the symmetry planes, this consistent symmetry inherent in the loading enables the unit cell to be reduced even further, occupying only one-fourth of the area of the entire unit cell. In the following, we will restrict the derivation of equations in most cases to the $0°/90°$-system as shown in Fig. 3.8 a).

The difference in the consideration of the $0°/90°$ or the $\pm 45°$-system on mechanical properties has been investigated e.g. in [9, 15].

3.3.1 Basic Configurations

The cubic system (or isometric) is a space group where the unit cell (UC) is in the shape of a cube. The three Bravais lattices which form the cubic crystal system are, cf. Fig. 3.9: simple or primitive cubic (PC), body-centered cubic (BCC) and face-centered cubic (FCC). The primitive cubic system (cf. Fig. 3.9 a)) consists of one lattice point (or one sphere centre in the case of HSS) on each corner of the cube. Each sphere

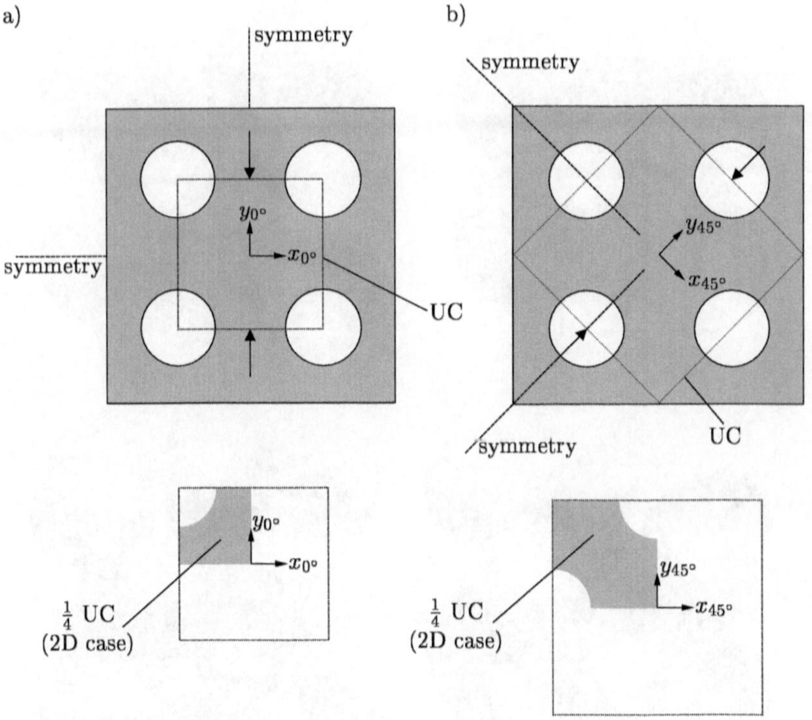

Fig. 3.8. Orientation of the symmetry planes and UCs for primitive cubic 2D topologies: a) $0°/90°$-symmetry planes; b) $\pm 45°$ symmetry planes

at the lattice points is then shared equally between eight adjacent cubes, and the unit cell therefore contains in total one sphere: $\frac{1}{8} \cdot 8$. The body-centered cubic system (cf. Fig. 3.9 b)) has one sphere centre in the center of the unit cell in addition to the eight corner points. It has a contribution of 2 sphere centres per unit cell $(\frac{1}{8} \cdot 8 + 1)$. Finally, the face-centered cubic system (cf. Fig. 3.9 c)) has sphere centres on the faces of the cube of which each unit cube gets exactly one half contribution, in addition to the corner sphere centres, giving a total of 4 spheres per unit cell (($\frac{1}{8}$ for each corner) $\cdot\, 8$ corners $+(\frac{1}{2}$ for each face) $\cdot\, 6$ faces).

In hexagonal close packing (HCP), layers of spheres are packed so that spheres in alternating layers overlay one another (cf. Fig. 3.9 d)). As in face-centered cubic (or cubic close packing), each sphere is surrounded by 12 other spheres. Looking at a single plane, the centres of the spheres are arranged in a hexagonal lattice (staggered rows, like a honeycomb), and each sphere is surrounded by 6 other spheres. The internal angles of the regular hexagon are all $120°$ and each of the six sides is a line of symmetry with a length of $l = \frac{2}{\sqrt{3}} \cdot (R + d)$. The area of such a regular hexagon of side length l is given by $1.5\sqrt{3} \cdot l^2$.

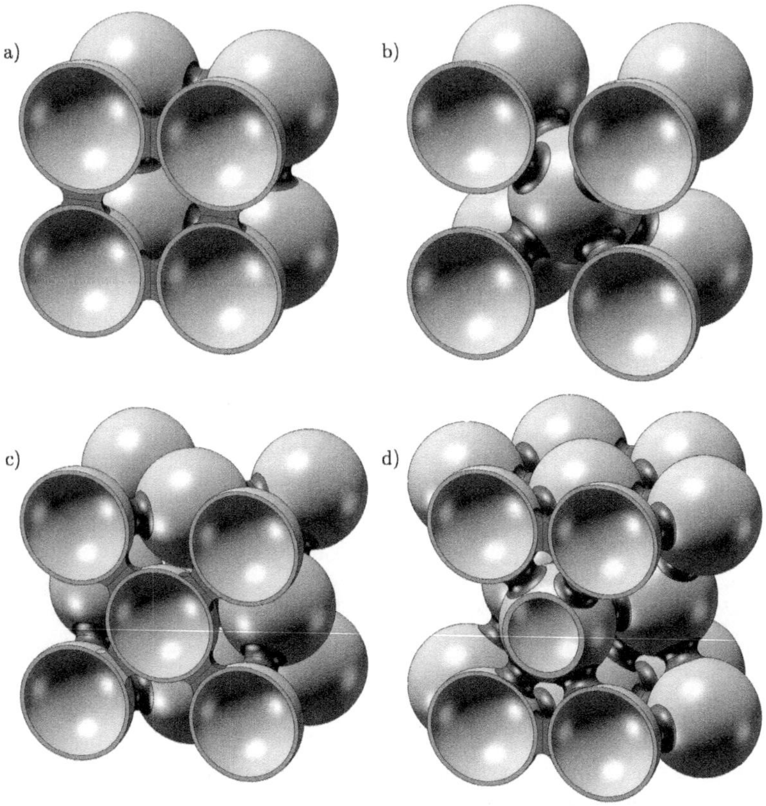

Fig. 3.9. Simple topologies of partial HSS: a) primitive cubic (PC); b) body-centered cubic (BCC); c) face-centered cubic (FCC); d) hexagonal (HCP)

Table 3.2. Periodic unit cell approach in the context of the finite element method

Packing	Reference
PC	[23, 7, 31, 11, 28, 8]
BCC	[31, 11, 29]
FCC	[23, 31, 12, 11, 29, 13]
HCP	[11]

Many numerical approaches based on the finite element method use the unit cell concept to model periodic structures and Tab. 3.2 collects a few references where such simplifications were applied.

Typical unit cells of the four configurations are shown in Fig. 3.10 where characteristic lengths of the cells are given.

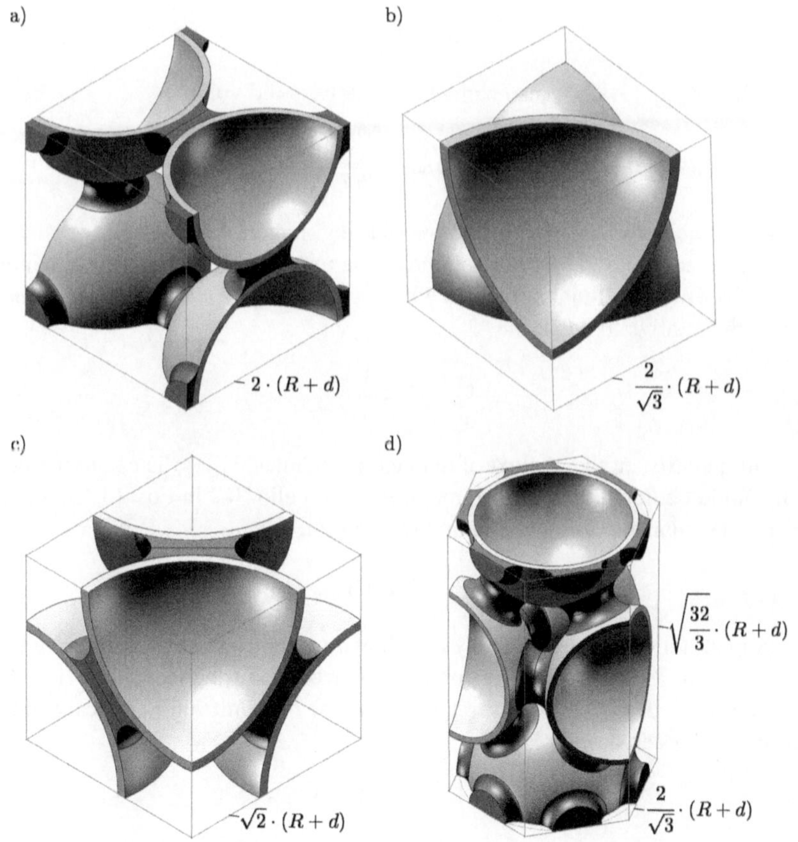

Fig. 3.10. Unit cell models under consideration of symmetry: a) primitive cubic (entire UC); b) body-centered cubic (one-eighth of UC); c) face-centered cubic (one-eighth of UC); d) hexagonal (entire UC)

3.3.2 Calculation of Volume Fractions, Densities and Further Geometrical Properties

In the case of cellular materials, physical properties are commonly described as a function of their relative density [14]

$$\rho_{\text{rel}} = \frac{\rho}{\rho_{\text{So}}}, \tag{3.11}$$

where ρ is the density of the cellular material and ρ_{So} is the density of the solid material from which the cells are made. Such scaling relations or power-laws which account for the density dependency can be found e.g. in [3]. Since the mass of the solid material and the mass of the cellular material itself is identical, the RHS of Eq. (3.11) can be divided by the mass to express the relative density in terms of volumes as:

$$\rho_{\text{rel}} = \frac{V_{\text{So}}}{V_{\text{UC}}}. \tag{3.12}$$

The value of Eq. (3.12) is also named the relative solid volume

$$V_{\text{rel,So}} = \frac{V_{\text{So}}}{V_{\text{UC}}}, \tag{3.13}$$

where V_{So} is the volume of the solid material (e.g. material of the cell walls, spheres, bond material etc.) and V_{UC} is the volume of the unit cell or of the entire specimen. In the case of open-cell cellular structures where a gas or liquid can flow through the open structure, the relative free volume

$$V_{\text{rel,free}} = \frac{V_{\text{free}}}{V_{\text{UC}}}, \tag{3.14}$$

is a useful quantity. In the context of damage mechanics, $V_{\text{rel,free}}$ is called the porosity [22]. It should be noted that the volume of the unit cell (UC) in Eqs. (3.13) and (3.14) can be expressed as the sum of solid and free volume as:

$$V_{\text{UC}} = V_{\text{So}} + V_{\text{free}}. \tag{3.15}$$

Looking at hollow sphere structures as shown in Figs. 3.4 and 3.5, one can see that the solid volume is composed of a contribution from the spherical shells (index 'Sp') and the matrix (index 'Ma'). Thus, the total volume of a unit cell in the case of hollow sphere structures can be expressed as

$$V_{\text{UC}} = V_{\text{Sp}} + V_{\text{Ma}} + V_{\text{free}}, \tag{3.16}$$

where the matrix volume in the case of Fig. 3.4 is equal to the bond volume V_{Bo} as given in Eq. (3.7).

Assuming that the densities of the matrix (bond) material and the sphere wall material are identical, i.e. $\rho_{\text{Ma}} = \rho_{\text{Sp}} = \rho_{\text{So}}$, the relative density is:

$$\rho_{\text{rel}} = \frac{\rho}{\rho_{\text{So}}} = \frac{V_{\text{Ma}}}{V_{\text{UC}}} + \frac{V_{\text{Sp}}}{V_{\text{UC}}}. \tag{3.17}$$

In the case that the sphere wall material and the matrix (or bond) material are different (e.g. aluminium and epoxy), the indication of a *relative* density is no longer possible and and an *average* density ρ can be used instead:

$$\rho = \frac{V_{\text{Ma}}}{V_{\text{UC}}} \cdot \rho_{\text{Ma}} + \frac{V_{\text{Sp}}}{V_{\text{UC}}} \cdot \rho_{\text{Sp}}. \tag{3.18}$$

The conditional equations for the unit cell volume V_{UC} of simple topologies in the form $a \cdot (R+d)^3$ and the number of spheres per unit cell N_{UC} are collected in Tab. 3.3.

The sphere wall volume of a single hollow sphere, V_{Sp}, is given for all configurations (cf. Fig. 3.3) as

$$V_{\text{Sp}} = V_{\text{Sp,o}} - V_{\text{Sp,i}} = \frac{4}{3}\pi \cdot \left(R^3 - (R-h)^3\right), \tag{3.19}$$

Table 3.3. Unit cell (UC) volumes $a \cdot (R+d)^3$ and number of spheres N_{UC} of simple topologies

Packing	UC volume	Spheres in UC	Figure
PC	$8 \cdot (R+d)^3$	1	3.10 a)
BCC	$\frac{64}{3\sqrt{3}} \cdot (R+d)^3$	2	3.10 b)
FCC	$16\sqrt{2} \cdot (R+d)^3$	4	3.10 c)
HCP	$8\sqrt{2} \cdot (R+d)^3$	2	3.10 d)

where R is the outer radius of the sphere and h is the sphere wall thickness. The total sphere wall volume within a unit cell can simply be obtained by multiplying the volume given in Eq. (3.19) with the number of spheres per unit cell as given in Tab. 3.3. It should be noted here that the internal volume of a sphere, $V_{\mathrm{Sp,i}} = \frac{4}{3}\pi \cdot (R-h)^3$, can be considered as a contribution to the free volume (cf. Eq. (3.16)) but does not allow any fluid flow through the cellular structure.

3.3.2.1 Spheres in Point Contact and Periodic Arrangement

In the case of hollow spheres with a neck region at a minimum distance of $2d$, cf. Fig. 3.4, Eq. (3.18) can be written as

$$\rho = \frac{N_{\mathrm{Bo}} \cdot V_{\mathrm{Bo}}}{a(R+d)^3} \cdot \rho_{\mathrm{Bo}} + \frac{N_{\mathrm{UC}} \cdot \frac{4}{3}\pi(R^3 - (R-h)^3)}{a(R+d)^3} \cdot \rho_{\mathrm{Sp}}, \tag{3.20}$$

where N_{Bo} is the number of bond volumes according to Eq. (3.7) per unit cell: PC: 3; BCC: 8; FCC: 12; HCP: 12. In the case that the density of the bond region and the sphere wall are the same ($\rho_{\mathrm{Sp}} = \rho_{\mathrm{Bo}} = \rho_{\mathrm{So}}$), e.g. sintered spheres, Eq. (3.20) reduces to

$$\frac{\rho}{\rho_{\mathrm{So}}} = \frac{N_{\mathrm{UC}} \cdot \frac{4}{3}\pi(R^3 - (R-h)^3)}{a(R+d)^3} + \frac{N_{\mathrm{Bo}} \cdot V_{\mathrm{Bo}}}{a(R+d)^3}. \tag{3.21}$$

Assuming that $d = 0$, i.e. that the spheres are in point contact, Eq. (3.21) can be further simplified to

$$\frac{\rho}{\rho_{\mathrm{So}}} = \frac{4N_{\mathrm{UC}}\pi}{3a} \cdot \left[3\left(\frac{h}{R}\right) - 3\left(\frac{h}{R}\right)^2 + \left(\frac{h}{R}\right)^3\right] + \frac{N_{\mathrm{Bo}} \cdot V_{\mathrm{Bo}}}{a(R+d)^3}. \tag{3.22}$$

In the case that the spheres are in point contact and that there is no neck region, cf. Fig. 3.2, the sphere packing factor (SPF) or packing fraction can be evaluated. This dimensionless number indicates the fraction of volume in a sphere structure (for the case of $h \to R$, i.e. a 'solid' sphere, the SPF is equal to the relative solid volume $V_{\mathrm{rel,So}}$) that is occupied by the spheres and is always less than unity[1].

[1] This definition is identical to the atomic packing factor (APF) in crystallography.

Table 3.4. Sphere packing factor (SPF) for simple topologies in the case of $d = 0 \wedge \theta = 0°$

Packing	Spheres in UC	SPF
PC	1	$\frac{\pi}{6} \approx 0.52$
BCC	2	$\frac{\sqrt{3}\pi}{8} \approx 0.68$
FCC	4	$\frac{\pi}{3\sqrt{2}} \approx 0.74$
HCP	2	$\frac{\pi}{3\sqrt{2}} \approx 0.74$

3.3.2.2 Periodic Syntactic Hollow Sphere Structures

In the case of syntactic hollow sphere structures, the hollow spheres are completely embedded in a matrix and Eq. (3.18) can be written for all configurations as (cf. Tab. 3.3)

$$\rho = \left(1 - \frac{N_{\text{UC}} \cdot \frac{4}{3}\pi R^3}{a \cdot (R+d)^3}\right) \cdot \rho_{\text{Ma}} + \frac{N_{\text{UC}} \cdot \frac{4}{3}\pi (R^3 - (R-h)^3)}{a \cdot (R+d)^3} \cdot \rho_{\text{Sp}}. \tag{3.23}$$

In the case that the density of the matrix and the sphere wall is the same, i.e. the case of spherical pores in a matrix, the thickness, h, can be set to zero since there is no more need to distinguish between the sphere and matrix material. In this case, Eq. (3.23) reduces to

$$\frac{\rho}{\rho_{\text{So}}} = 1 - \frac{4}{3} \cdot \frac{N_{\text{UC}} \cdot \pi}{a} \cdot \left(\frac{R}{R+d}\right)^3 . \tag{3.24}$$

Assuming in addition that $d = 0$, i.e. that the nearest spheres are in point contact, Eq. (3.24) reduces to

$$\frac{\rho}{\rho_{\text{So}}} = 1 - \frac{4N_{\text{UC}}}{3a} \cdot \pi , \tag{3.25}$$

which is equal to the maximum matrix fraction (max. relative solid volume, Eq. (3.13)) for a periodic arrangement of spherical pores. In addition, Eq. (3.25) can be expressed as $(1 - \text{SPF})$.

3.3.2.3 Statistically Random Dispersion of Spheres

The unit cell approach is no longer applicable in the case of randomly distributed spheres since there is no more symmetry. The geometry of such random structures can be accurately characterised by means of computer tomography. However, the discretisation in the scope of the finite element method is generally restricted to tetrahedron elements which often show poor performance for non-linear (e.g. plasticity) analysis. Looking at single spheres, a certain mesh density is required to obtain accurate results. In [24], a number of 80000 elements is reported for accurate results (less than 5% error) for two hemispheres welded together. Distributing such spheres in all space directions soon exceeds the capacity of actual computer hardware. In order to be able to consider a higher number of spheres, the mesh size must be reduced or the types of elements e.g. from solid to shell elements changed. Only a few references cover the finite element modelling of randomly distributed spheres. In [20] an algorithm was used to generate three-dimensional models of randomly distributed spherical particles in a matrix and the effective elastic mechanical properties were obtained using a numerical homogenization techniques. Application of a shell model allowed in [24] the simulation of a box with approx. 70 spheres. A new approach to characterise random hollow sphere structures based on a Lattice Monte Carlo has been introduced in [10] where the effective thermal conductivity of a random two-dimensional HSS structure has been investigated.

References

1. Anderson, O., Waag, U., Schneider, L., Stephani, G.: Novel metallic hollow sphere structures: processing and properties. In: Banhart, J., Ashby, M.F., Fleck, N.A. (eds.) Metal foams and porous metal structures, pp. 183–188. MIT Publishing, Bremen (1999)
2. Anderson, O., Waag, U., Schneider, L., Stephani, G., Kieback, B.: Novel Metallic Hollow Sphere Strcutures. Adv. Eng. Mater. 2, 192–195 (2000)
3. Ashby, M.F., Evans, A., Fleck, N.A., Gibson, L.J., Hutchinson, J.W., Wadley, H.N.G.: Metal foams: a design guide. Butterworth-Heinemann, Boston (2000)
4. Baumeister, E., Klaeger, S., Kaldos, A.: Lightweight, hollow-sphere-composite (HSC) materials for mechanical engineering applications. J. Mater. Process Tech., 155-156, 1839–1846 (2004)
5. Bonino, J.P., Rousset, A., Rossignol, C., Blottière, Y.: Composite material with hollow spherical modules made of microcrystallized or amorphus nickel alloys (in French). Matér. Tech. 78, 25–28 (1990)
6. De Graef, M., McHenry, M.: Structure of materials: an introduction to crystallography, diffraction and symmetry. Cambridge University Press, Cambridge (1997)
7. Fiedler, T., Öchsner, A., Grácio, J., Kuhn, G.: Modelling the mechanical behaviour of adhesively bonded and sintered hollow-sphere structure. Mech. Compos. Mater. 41, 559–570 (2005)
8. Fiedler, T., Öchsner, A., Grácio, J.: Uniaxial elasto-plastic behaviour of adhesively bonded hollow sphere structures (hss): numerical simulations and experiments. Mater. Sci. Forum 539-543, 1874–1879 (2007)
9. Fiedler, T., Öchsner, A.: On the anisotropy of adhesively bonded metallic hollow sphere structures. Scripta Mater. 58, 695–698 (2008)
10. Fiedler, T., Öchsner, A., Belova, I.V., Murch, G.E.: Recent Advances in the Prediction of the Thermal Properties of Syntactic Metallic Hollow Sphere Structures. Adv. Eng. Mater. 10, 361–365 (2008)
11. Gao, Z.Y., Yu, T.X., Karagiozova, D.: Finite element simulations on the mechanical properties of MHS materials. Acta Mech. Sin. 23, 65–75 (2007)
12. Gasser, S., Paun, F., Cayzeele, A., Bréchet, Y.: Uniaxial tensile properties of a regular stacking of brazed hollow spheres. Scripta Mater. 48, 1617–1623 (2003)
13. Gasser, S., Paun, F., Bréchet, Y.: Finite elements computation for the elastic properties of a regular stacking of hollow spheres. Mat. Sci. Eng. A-Struct. 379, 240–244 (2004)
14. Gibson, L.J., Ashby, M.F.: Cellular solids: structures & properties. Cambridge University Press, Cambridge (1997)
15. Green, R.J.: A plasticity theory for porous solids. Int. J. Mech. Sci. 14, 215–224 (1972)

16. Hartmann, M., Reindel, K., Singer, R.F.: Fabrication and properties of syntactic magnesium foams. In: Schwartz, D.S., Shih, D.S., Wadley, H.N.G., Evans, A.G. (eds.) Porous and cellular materials for structural applications, MRS symposium proceedings, vol. 521, pp. 211–216. MRS Publishing, Warrendale (1998)
17. Hartmann, M., Größmann, I., Reindel, K., Singer, R.F.: Microstructure and mechanical properties of cellular magnesium matrix composites. In: Banhart, J., Ashby, M.F., Fleck, N.A. (eds.) Metal foams and porous metal structures, pp. 331–336. MIT Publishing, Bremen (1998)
18. Hartmann, M.: Syntactic magnesium-based foams. Alum. 75, 154–156 (1999)
19. Hurysk, K.M., Clark, J.L., Nagel, A.R., Hardwicke, C.U., Lee, K.J., Cochran, J.K., Sanders, T.H.: Steel and titanium hollow sphere foams. In: Schwartz, D.S., Shih, D.S., Wadley, H.N.G., Evans, A.G. (eds.) Porous and cellular materials for structural applications, MRS symposium proceedinds, vol. 521, pp. 191–204. MRS Publishing, Warrendale (1998)
20. Kari, S., Berger, H., Rodriguez-Ramos, R., Gabbert, U.: Computational evaluation of effective material properties of composites reinforced by randomly distributed spherical particles. Compos. Struct. 77, 223–231 (2007)
21. Kohler, S., Fitzgerald, T.J., Singer, R.F.: Fabrication and properties of cellular magnesium. In: New materials in automotive engineering, VDI report No. 1235. VDI Verlag, Düsseldorf, pp. 305–319 (1995)
22. Lemaitre, J.: A Course on damage mechanics. Springer, Berlin (1998)
23. Luxmoore, A.R., Owen, D.R.J.: Syntactic foams. In: Hilyard, N.C. (ed.) Mechanics of cellular plastics, pp. 357–391. Macmillan Publishing Co., New York (1982)
24. Mamoud, F., Paun, F., Bréchet, Y.: Elasticity of hollow spheres based cellular material: non regular packing. In: Stephani, G., Kieback, B. (eds.) Proceedings of the international symposium on cellular metals for structural and functional applications, pp. 155–162. Fraunhofer IRB Verlag, Stuttgart (2005)
25. Palmer, R.A., Gao, K., Doan, T.M., Green, L., Cavallaro, G.: Pressure infiltrated syntactic foams - Process development and mechanical properties. Mat. Sci. Eng. A-Struct. 464, 85–92 (2007)
26. Rawal, S.P., Lanning, B.R., Misra, M.S.: Design and mechanical properties of microballon-aluminium composite-core laminates. In: Miravete, A. (ed.) Proceedings of the ninth conference on composite materials - ICCM/9. metal matrix composites, vol. I, pp. 203–210. Woodhead Publishing Limited, Abington (1993)
27. Rickles, S.A., Cochran, J.K., Sanders, T.H.: The production and compression characteristics of a low density syntactic metal/ceramic foam. Ceram. Eng. Sci. Proc. 10, 1472–1484 (1989)
28. Sanders, W.S., Gibson, L.J.: Mechanics of hollow spheres foams. Mat. Sci. Eng. A-Struct. 347, 70–85 (2003)
29. Sanders, W.S., Gibson, L.J.: Mechanics of BCC and FCC hollow-spheres foams. Mat. Sci. Eng. A-Struct. 352, 150–161 (2003)
30. Sypeck, D.J., Parrish, P.A., Wadley, H.N.G.: Novel hollow powder porous structures. In: Schwartz, D.S., Shih, D.S., Wadley, H.N.G., Evans, A.G. (eds.) Porous and cellular materials for structural applications, MRS symposium proceedings, vol. 521, pp. 205–210. MRS Publishing, Warrendale (1998)
31. Vesenjak, M., Fiedler, T., Ren, R., Öchsner, A.: Behaviour of syntactic and partial hollow sphere structures under dynamic loading. Adv. Eng. Mater. 10, 185–191 (2008)

4
Elastic-Plastic Properties: Simulation and Experiment

Thomas Fiedler[1] and Andreas Öchsner[1,2]

[1] The University of Newcastle, Callaghan NSW 2308, Australia
Thomas.Fiedler@newcastle.edu.au
[2] Technical University of Malaysia, Department of Applied Mechanics, 81310 UTM Skudai, Johor, Malaysia
Andreas.Oechsner@gmail.com

Abstract. This chapter addresses the elasto-plastic properties of novel metallic hollow sphere structures (MHSS). Numerical finite element analyses and experimental tests are conducted. The influence of the morphology, topology, joining technology and material composition on their mechanical properties is numerically investigated. Uni-axial compressive tests with adhesively bonded MHSS are performed in order to confirm the numerical findings. Young's modulus, Poisson's ratio and the initial plastic yield stress are determined. Furthermore, the elastic anisotropy of simple cubic MHSS is investigated.

Keywords: Metallic hollow sphere structures; Mechanical properties; Plasticity; Anisotropy; Computer simulation.

4.1 Introduction

Metallic hollow sphere structures (MHSS) are a new group of cellular metals characterised by easily reproducible geometry and therefore consistent mechanical and physical properties. Figure 4.1 shows examples of partially bonded MHSS (a) and syntactic MHSS (b). In the case of syntactic morphology, the hollow spheres are completely embedded within an adhesive matrix as in the case of classical fiber reinforced plastics. In contrast, the adhesive is concentrated at the contact points of neighboring spheres for partial MHSS.

Well-known advantages of cellular metals are their excellent ability for energy absorption (Evans et al. 1999, Nemat-Nasser et al. 2007), good damping behaviour (Golovin and Sinning 2003, Neugebauer et al. 2005), sound absorption (Hübelt and Bingel 2006), excellent heat insulation (Fiedler et al. 2008, Zhao et al. 2004) and a high specific stiffness (Ashby et al. 2000, Fiedler et al. 2005). The combination of these properties opens a wide field of potential multi-functional applications, i.e. as an ideal concept for structural design of lightweight transportation systems such as aircraft (Paik et al. 1999), high-speed trains (Kim et al. 2007) and fast ships (Knox et al. 1998).

However, despite more than 30 years of intensive scientific research on cellular materials few industrial applications of these technologies can be found. Essential limiting factors for the utilisation are unevenly distributed material parameters (Banhart et al. 1994, Ramamurty and Paul 2004) and relatively high production costs.

A. Öchsner, C. Augustin (Eds.): Multifunctional Metallic Hollow Sphere Struct., EM, pp. 47–70.
springerlink.com

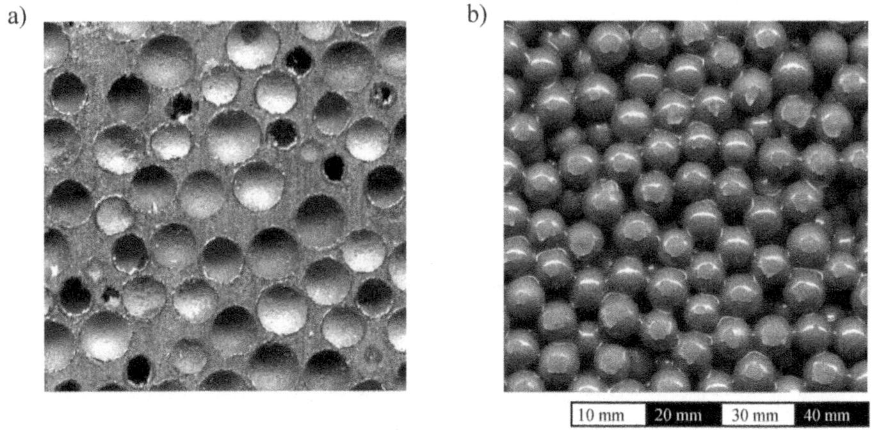

Fig. 4.1. Metallic hollow sphere structures: a) cross-section of a syntactic MHSS, b) Partial morphology

Metallic hollow sphere structures (MHSS) are a new group of cellular metals characterised by easily reproducible geometry and therefore consistent mechanical and physical properties. A powder metallurgy based manufacturing process enables the production of metallic hollow spheres of defined geometry (Jäckel 1987, Löthman et al. 2000). This technology brings a significant reduction in costs in comparison to earlier applied galvanic methods. In addition, all materials suitable for sintering can be used for the hollow spheres.

Initial research on hollow sphere structures focused on metallic matrices with embedded ceramic hollow spheres. Kohler et al. first investigated a syntactic magnesium matrix containing ceramic hollow spheres (Kohler et al. 1995). It was found that the specific mechanic properties were comparable to aluminium foams. However, the specimens were sensitive to fracture and could not absorb as much energy as typical aluminium foams. This work was continued (Hartmann et al. 1998) and the influence of the sphere wall thickness of alumina hollow spheres on the mechanical properties was investigated. An increase of strength in conjunction with a decrease of energy absorption efficiency was found for larger sphere wall thicknesses. Palmer and colleagues recently advanced this line of research by addressing aluminium syntactic foams containing ceramic microspheres (Palmer et al. 2007). Good energy absorption behaviour and repeatability of the results was observed.

An alternative approach is to substitute the metallic matrices by epoxies in order to improve energy absorption and decrease costs. The potential of these materials was shown by Baumeister who conducted experimental studies on their mechanical properties, damping behaviour and thermal characteristics (Baumeister 2002). In (Baumeister et al. 2004) the applicability of corundum based hollow spheres in an epoxy resin matrix was shown for machine tool design. Good mechanical properties and dimensional accuracy of manufactured machine building parts was found. Adrien and colleagues addressed hollow glass microspheres in polymer matrices and used in situ X-ray tomography to study their deformation behaviour (Adrien et al. 2006). Depending on the stiffness of the syntactic matrix, different failure modes could be identified:

in the case of a strong syntactic matrix, failure occurs at higher stress levels and is concentrated within deformation bands that form an angle of 45° with the loading direction. For soft matrices, the matrix acts as a fluid and the load is homogeneously distributed among all spheres.

A possibility to improve the mechanical properties of hollow sphere composites is the substitution of relatively brittle ceramic spheres by metallic ones. Various experimental and numeric analyses of such materials were conducted recently. In (Heine et al. 2002) cellular metals were evaluated in regard to application in automotives. Within these studies, hollow sphere structures showed good sound insulation and predictable isotropic mechanical properties. High manufacturing and processing costs were identified as the biggest problem. Lim et al. conducted experimental investigations on single and sintered bulk steel MHSS (Lim et al. 2002). A behaviour similar to open-celled metallic foams was observed. High strength was found for the compaction of single hollow spheres but could not be observed for bulk material. The likely explanation for this shortcoming is that the randomly interconnected network of spheres behaves more like a system of struts than a system of connected plates of shells. Friedl and colleagues (Friedl et al. 2005) conducted experimental tests on stainless steel MHSS and observed a distinct influence of the microstructure of the sintered sphere walls on the macroscopic properties. Apart from steel, metallic hollow sphere structures can also be manufactured from other metals. Koo and colleagues investigated copper hollow structures (Koo and Jung 2006). They discovered that increase in sintering temperature decreases the porosity of the base material and thereby improves mechanical properties.

In addition to these experimental analyses, several numerical investigations on partial MHSS were conducted. Gründer and Landgraf used cantilever experiments to obtain the elastic material properties of the sintered stainless steel base material (Gründer and Landgraf 2001). Using this material data they generated simple periodic 3D finite element models of single spheres and MHSS to simulate uni-axial compression. As a result of these analyses, typical cellular metal force-displacement curves were obtained. Sanders and Gibson conducted an extensive study on the elastic and plastic properties of sintered MHSS. In the scope of this study, three dimensional finite element analysis was used in order to investigate the orthotropic elasticity tensor and yield surface of simple cubic (Sanders and Gibson 2003), body-centred cubic and face-centred cubic (Sanders and Gibson 2003) unit cell models. A distinct dependence of Young's modulus and the initial yield stress on the loading direction was found. In contrast to the present studies, automatically generated meshes were used which allow for an easy variation of the geometry, but do not match the accuracy of regular meshes for non-linear material behaviour. Furthermore, only partially connected homogeneous steel structures were considered. Further finite element analyses were performed by Gasser and colleagues. They used relatively crude meshes as rough approximation of MHSS geometry to determine elastic (Gasser et al. 2004) and plastic (Gasser et al. 2004) properties. In a subsequent study using more accurate three dimensional models, face-centred cubic structures were analysed and a slight elastic anisotropy was observed. Most numerical analysis addresses cubic symmetric geometries in order to exploit symmetries and reduce the size of calculation models. In (Franeck and Landgraf 2005) MHSS with random arrangement of spheres were simulated using reduced bar-models. The disadvantage of this interesting approach is the necessary strong idealisation of the geometry.

The mechanical properties of MHSS can further be modified by embedding hollow spheres in a syntactic metallic matrix. It should be mentioned here that this approach also distinctly increases the density of such materials so that specific properties are not necessarily superior. Neville and colleagues (Neville and Rabiei 2007) investigated hollow steel spheres in a steel matrix produced by sintering. A very uniform deformation caused a higher strength to density ratio compared to any other steel foam reported in literature. Furthermore, very high energy absorption ability could be observed.

Instead of sintering, metallic hollow spheres can also be joined by adhesive bonding. Adhering is the most economic way of joining and therefore is attractive for a wide range of potential applications. Another important advantage is the possible utilisation of the mechanical behaviour and morphology of the adhesive layer as a further design parameter for the optimisation of the structure's mechanical properties for specific applications. Sukegawa et al. investigated the compression characteristics of sintered and adhesively bonded MHSS. The structures were inserted into pipes in order to simulate the strengthening of car parts (Sukegawa et al. 2005). An improvement of the energy absorption properties compared to empty pipes was found.

4.2 Continuum Mechanics

This chapter addresses the mechanical properties of MHSS with partial and syntactic morphologies. Thereby, the elastic properties and the initial yield stresses are determined. A geometry that exhibits three mutually perpendicular symmetry planes is orthotropic symmetric. In this case, the elasticity tensor is defined by nine independent coefficients:

$$C^{\text{orth}} = \begin{pmatrix} C_{11} & C_{12} & C_{13} & & & \\ & C_{22} & C_{23} & & 0 & \\ & & C_{33} & & & \\ & & & C_{44} & & \\ & \text{sym.} & & & C_{55} & \\ & & & & & C_{55} \end{pmatrix}. \tag{4.1}$$

In the scope of this work, the finite element models exhibit cubic symmetries and therefore the equalities $C_{11} = C_{22} = C_{33}$, $C_{12} = C_{13} = C_{23}$ and $C_{44} = C_{55} = C_{66}$ are valid. Consequently, the elastic behaviour can be described based on only three independent elastic parameters e.g. C_{11}, C_{12} and C_{44}. Two of these elastic constants can be expressed in terms of the elastic parameters Young's modulus E and Poisson's ratio ν:

$$\begin{aligned} C_{11} &= \frac{E^{0^\circ} \cdot (1-\nu^{0^\circ})}{(1+\nu^{0^\circ}) \cdot (1-2\nu^{0^\circ})}, \\ C_{12} &= \frac{E^{0^\circ} \cdot (1-\nu^{0^\circ})}{(1+\nu^{0^\circ}) \cdot (1-2\nu^{0^\circ})}. \end{aligned} \tag{4.2}$$

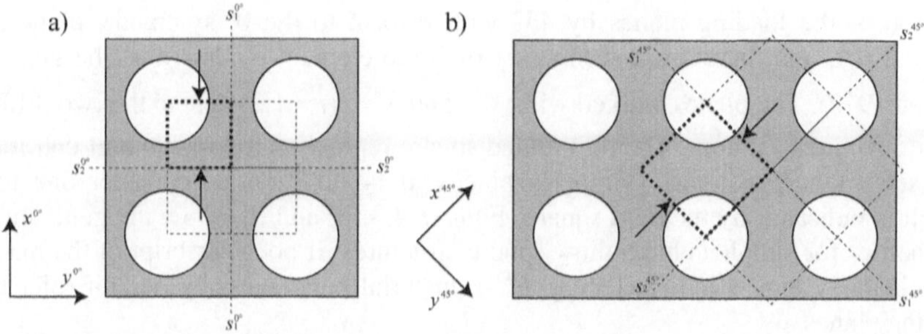

Fig. 4.2. Orientation of the loading planes for simple cubic topology: a) 0°-loading planes; b) 45°-loading planes

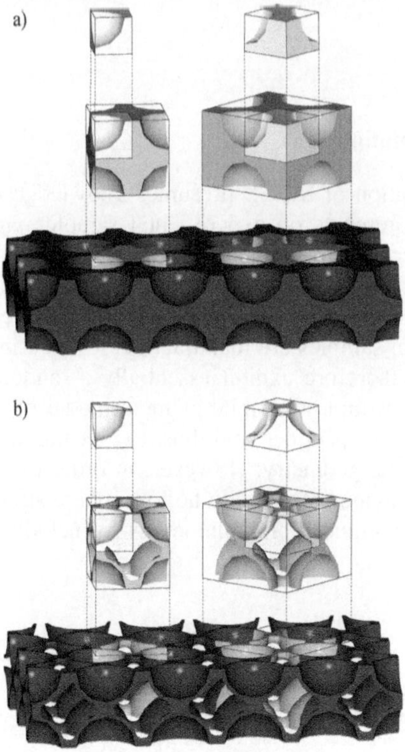

Fig. 4.3. Three dimensional representation of MHSS geometries: a) syntactic geometry, b) partial geometry

The orientation of the applied loads is thereby aligned with the axis of the cubic symmetry planes $s_i^{0^o}$ according to Fig. 4.2 (a). The third elastic constant C_{44} can be expressed in terms of the shear modulus ($C_{44} = 1/G^{45^o}$). This analysis requires the rotation of the loading planes by 45° with respect to the 0°-symmetry planes, (cf. Fig. 4.2 (b)) and therefore a second set of finite element models must be generated (Green 1972). The planes marked with $s_i^{0^\circ}$ and $s_i^{45^\circ}$ (i = 1,2) denote the two different sets of symmetry planes. The thin dotted square represents a possible unit cell for the two sets of perpendicular symmetry planes. It is sufficient to consider one fourth which is indicated by the bold square. Figures 4.3 (a) and (b) show the generation of geometries for simple cubic hollow sphere structures. If no superscript of the material coordinate system is indicated (e.g. E^{45°), material parameters by default refer to 0° loading planes.

In order to determine the initial macroscopic plastic compressive yield stress $\sigma_{Y,c}$ and tensile yield stress $\sigma_{Y,t}$, a method proposed in (König 1986) is applied where the macroscopic yield stress is recorded when the equivalent von Mises stress σ_{eff} of an integration point reaches the yield stress of the base material.

4.3 Methodology

4.3.1 Numerical Simulation

The numerical investigation of MHSS presumes knowledge of its geometric dimensions. Therefore, measurements on experimental samples are performed. The averaged value of the outer sphere radius R is 1.5 mm, the half minimum distance d between two neighbouring spheres is 0.18 mm and the sphere wall thickness h is 0.075 mm. Based on these measurements, volumetric finite element models are generated. The experimental samples are manufactured by pouring the hollow spheres in bulk into a mould and therefore exhibit essentially a random arrangement (cf. Fig. 4.1). The best approach would be a finite element model of such a random structure which comprises the entire specimen and thus, to generate a model without simplifications with respect to the geometry. However, in order to obtain significant results for the macroscopic behaviour of metallic hollow sphere structures, it is necessary to model a representative volume. This volume has to include approximately 10 to 15

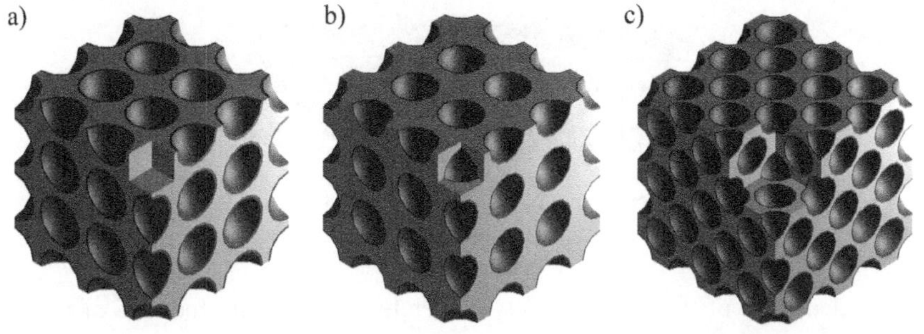

Fig. 4.4. Geometric approximation of random Hollow Sphere Structures: a) simple cubic (sc); b) body-centred cubic (bcc); c) face-centred cubic (fcc)

spheres in each direction (Lim et al. 2002). The required refinement of the mesh in conjunction with its dimensions would yield a huge number of unknowns which would exceed the capacity of available computer hardware. Therefore, a cubic symmetry of the arrangement of the spheres is assumed. This approach allows for the utilisation of symmetry boundary conditions and a distinct reduction of the required mesh size. The considered cubic symmetric arrangements of spheres, in the following also designated topologies, are shown in Fig. 4.4.

In addition, two different morphologies are considered. Figure 4.5 shows finite element meshes of syntactic and partial MHSS. The dark grey spherical shells are connected by a light grey matrix which completely embeds the spheres (syntactic, cf. Fig. 4.5 (a)-(c)) or is only concentrated in proximity to their contact points (partial, cf. Fig. 3.5 (d)-(e)).

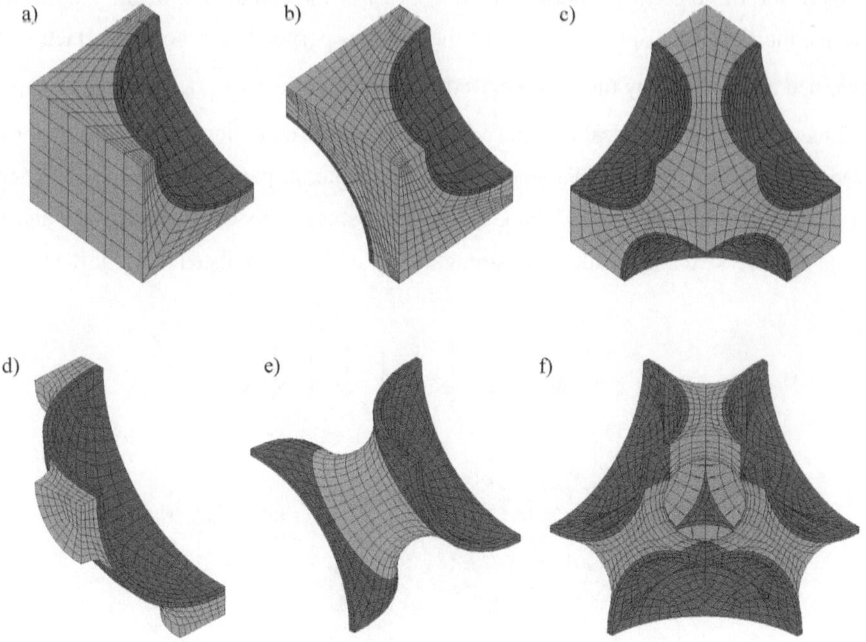

Fig. 4.5. Examples of finite element models: a) synt. simple cubic; b) synt. body-centred cubic; c) synt. face-centred cubic; d) part. simple cubic; e) part. body-centred cubic; f) part. face-centred cubic

Uni-axial tensile and compressive tests of MHSS are simulated in order to determine their mechanical properties. Figure 4.6 illustrates the mechanical boundary conditions of these analyses. Due to the cubic symmetry of the MHSS and the applied loads, only one eighth of a unit cell needs to be considered within the finite element models. The uni-axial loading is prescribed by the definition of a nodal displacement $u(t)$ on all nodes that lie inside the upper surface. This time dependent boundary condition $u(t)$ is a ramp function which displaces the nodes in the negative y-direction. In order to simulate the mechanical response of the whole structure repetitive

(Zienkiewicz and Scorft 1972) and reflective (Cook et al. 2002) boundary conditions are defined. The reflective symmetry conditions constrain the displacement perpendicular to the surface on which they are defined and simulate the influence of the remaining unit cell. Repetitive boundary conditions model the interaction with neighbouring unit cells. According to these constraints, all nodes exhibit the same displacement in the direction of the normal vector of these surfaces. As result of these symmetry boundary conditions, the finite element models simulate an infinite structure where the influence of a free boundary is disregarded. This assumption holds for MHSS which comprise at least 10 spheres in each direction (Lim et al. 2002). The macroscopic strains ε and stresses σ are obtained using a Fortran subroutine. The nodal forces F_n of all nodes within one of the three surfaces, where a repetitive boundary condition is prescribed, are summed up. The macroscopic engineering stresses σ are obtained by dividing this sum by the areas of the corresponding surfaces of the finite element models ($\sigma = F_\mathrm{n} /A$). For the calculation of the engineering strains, the average displacement $\overline{u_n}$ of these nodes perpendicular to the surface is determined and divided by the half length of the unit cell ($\varepsilon = 2\overline{u_n}/l_{\mathrm{cell}}$).

Linear elasticity and ideal plasticity (von Mises yield condition) are presumed for the base materials of the finite element models. The material parameters of the considered base materials, namely Young's modulus E_S, Poisson's ratio ν_S, the density ρ_S and the initial compressive and tensile yield strengths $\sigma_{\mathrm{Y,c}}$ and $\sigma_{\mathrm{Y,t}}$, are listed in Table 4.1.

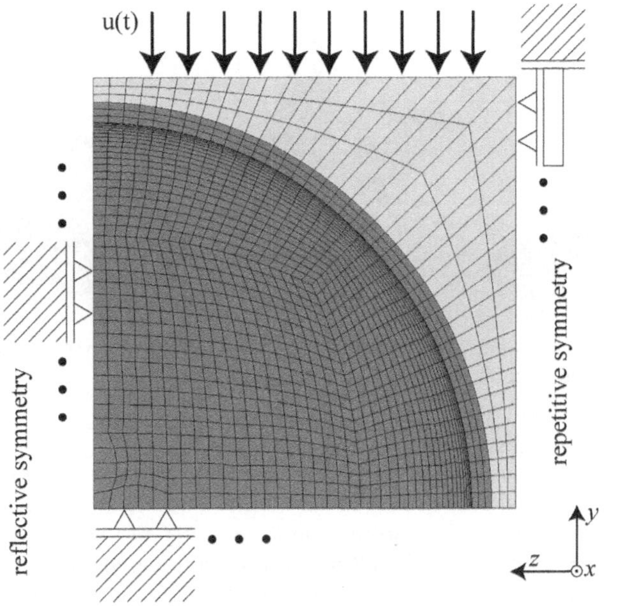

Fig. 4.6. Boundary conditions for uni-axial compression test simulation

Table 4.1. Mechanical properties of base materials

Mat. Parameter	Steel (St)	Adhesive (Ep) (Jeandrau 1991)
E_S [N/mm²]	210 000	4 500
ν_S [–]	0.3	0.36
$\sigma_{Y,t}$ [N/mm²]	300	61.5
$\sigma_{Y,c}$ [N/mm²]	300	195
ρ_S [kg/dm³]	6.95	1.13

For the three dimensional discretisation of the simple cubic model geometries, principally tetrahedral or hexahedral elements can be employed. Earlier investigations (Benzley et al. 1995, Fiedler et al. 2006) have shown that hexahedral elements yield superior performance, especially in the case of plastic material behaviour. Therefore, the geometry of the structures is discretised based on regular hexahedral elements. This approach is much more time-consuming, but it is important in order to achieve a more accurate simulation of the elastic and non-linear (i.e. plasticity) behaviour of such materials. Figure 4.7 shows Young's modulus convergence behaviour for partial finite element models with different mesh densities. Finer meshes containing higher number of nodes yield more accurate results. However, the calculation time increases exponentially with the number of nodes and therefore a compromise between calculation time and accuracy must be found.

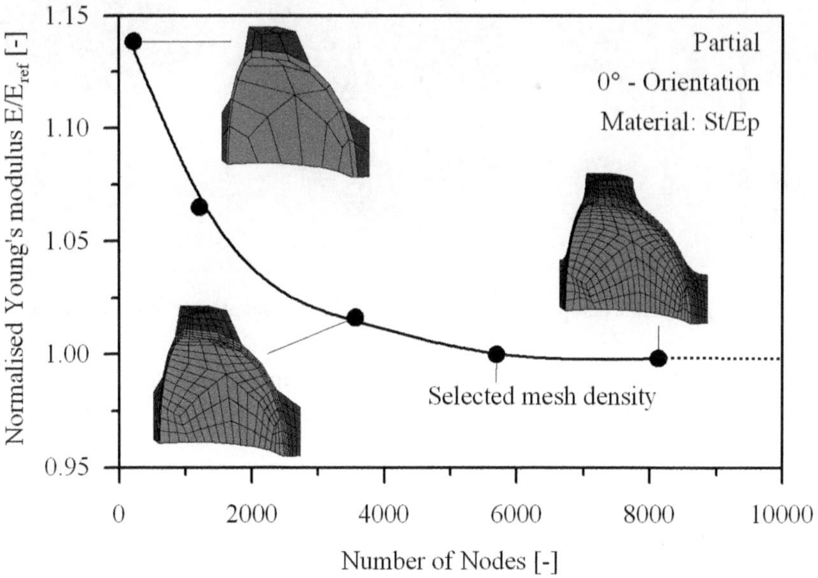

Fig. 4.7. Convergence analysis for unit-cell models under compressive load

4.3.2 Experimental Approach

Compressive testing is used to determine Young's modulus E, the 0.2% offset yield strength $\sigma_{0.2}$ and the hardening behaviour of MHSS. The experimental setup is schematically drawn in Fig. 4.8 (a). In order to minimise friction, the surfaces of both pressure stamps are lubricated. The upper pressure stamp moves downwards with a constant velocity $u(t)$ = 1 mm/min and thereby continuously compacts the specimen. The samples are of cubical shape and have a side length of 30 mm. Figure 4.8 also shows a specimen before (b) and after (c) compressive loading. The upper pressure stamp is attached to the load cell which records the reaction force of the sample. The displacement is measured with an inductive displacement transducer (LVDT) which is attached to both pressure stamps. Compared with the measurement of the displacement at the crosshead of the testing machine, this arrangement avoids perturbations of the measurements due to the finite stiffness of the machine frame. The force and displacements signals are recorded with an MGC+ measuring amplifier and processed with the software Catman®. For the evaluation of the compressive tests, the engineering stresses and strains are calculated. The engineering stress σ is defined by the measured force $F(t)$ divided by the initial cross section of the specimen A = 900 mm². The engineering strain ε is the measured displacement $u(t)$ divided by the initial height (30 mm) of the sample.

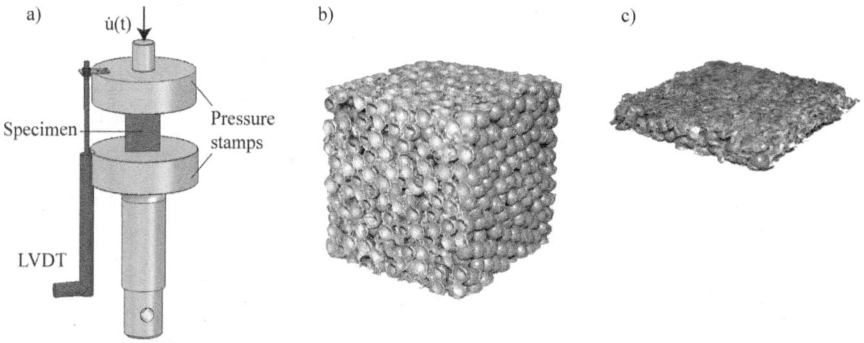

Fig. 4.8. Experimental setup (a) and specimens before (b) and after (c) compressive testing

Young's modulus can then be determined according to the well-known equation $E = \sigma/\varepsilon$. The 0.2 % offset yield strength $\sigma_{0.2}$ and the hardening behaviour are obtained by the evaluation of the stress-strain diagram.

4.4 Numerical Results

4.4.1 Adhesively Bonded MHSS

First, adhesively bonded structures are considered. Two morphologies can be distinguished: partial MHSS and syntactic MHSS (cf. Fig. 4.1). The stress-strain relations of these structures are plotted in Fig. 4.9. It can be observed that the syntactic MHSS exhibit higher stresses in comparison to the partial structures. Furthermore, the stresses increase with the density of the structure. This change in density is caused by different sphere wall thicknesses *h* of the spherical shells. The stress-strain curves can be divided in a linear-elastic area, an elastic-plastic transition zone and the typical stress plateau for cellular materials. It should be mentioned here that within the scope of the finite element analysis all stress-strain relations exhibit the same characteristics. Therefore, only the material parameters E, ν and σ_Y will be discussed beyond this point.

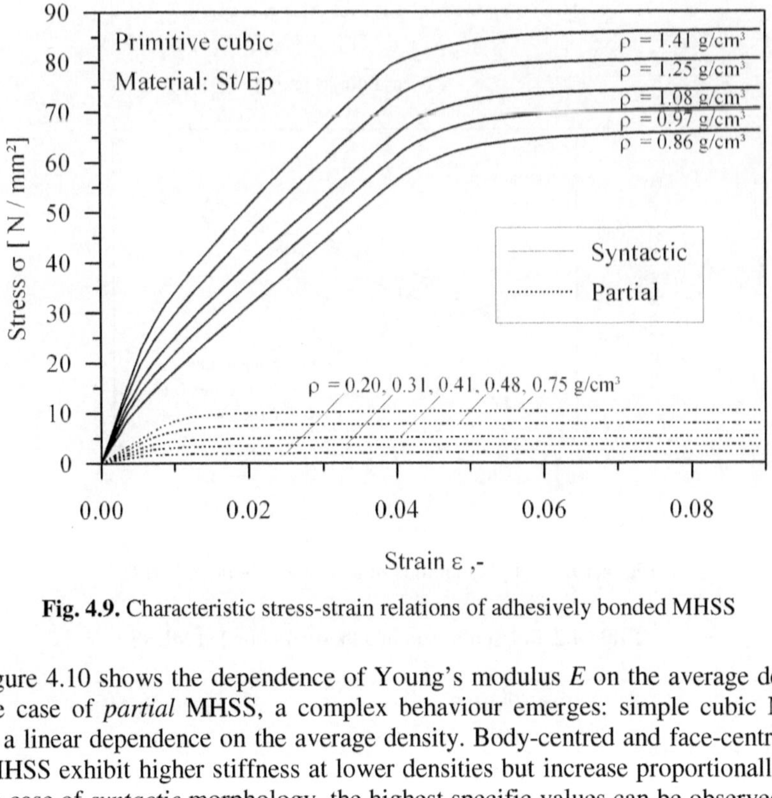

Fig. 4.9. Characteristic stress-strain relations of adhesively bonded MHSS

Figure 4.10 shows the dependence of Young's modulus E on the average density. In the case of *partial* MHSS, a complex behaviour emerges: simple cubic MHSS show a linear dependence on the average density. Body-centred and face-centred cubic MHSS exhibit higher stiffness at lower densities but increase proportionally less. In the case of *syntactic* morphology, the highest specific values can be observed for a face-centred cubic arrangement of the hollow spheres. However, the deviation in stiffness between the different topologies is small. As a consequence, Young's modulus E can be estimated by the linear function $E(\rho) = 3200 \cdot \rho$ [N/mm^2].

Poisson's ratio of adhesively bonded MHSS is shown in Table 4.2. In the case of *syntactic* morphology, this material parameter is widely independent of the topology and sphere wall thickness and equal to $\nu = 0.34 \ldots 0.35$. In contrast, Poisson's ratio of *partial* MHSS shows a strong dependence on the sphere wall thickness *h* of the hollow spheres and generally increases with this value, in particular for the body-centred cubic topology.

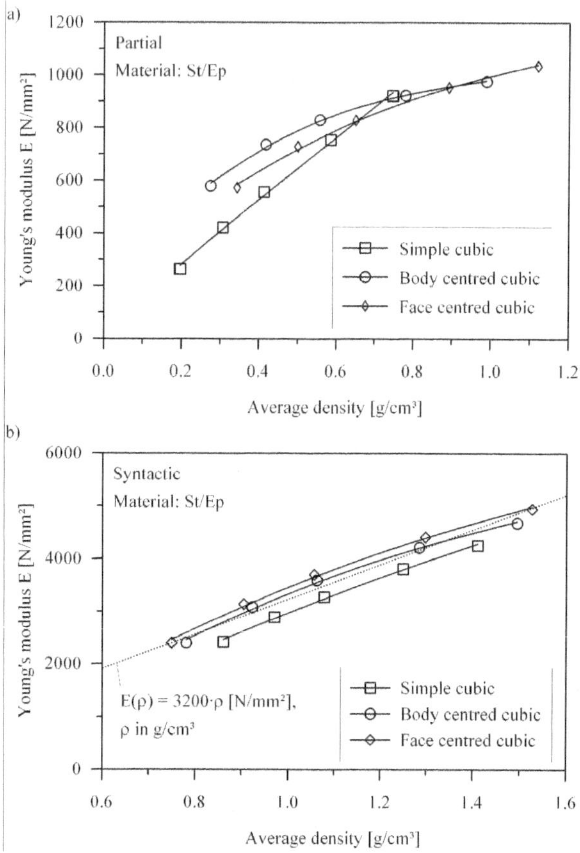

Fig. 4.10. Young's modulus of adhesively bonded MHSS

Table 4.2. Poisson's ratio of adhesively bonded MHSS

	Partial MHSS			Syntactic MHSS		
Thickness *h*	sc	bcc	fcc	sc	bcc	fcc
0.030 mm	0.11	0.07	0.14	0.35	0.34	0.34
0.052 mm	0.10	0.13	0.17	0.35	0.34	0.34
0.075 mm	0.09	0.18	0.18	0.35	0.35	0.34
0.113 mm	0.09	0.22	0.19	0.34	0.34	0.34
0.150 mm	0.08	0.25	0.20	0.34	0.34	0.34

Figure 4.11 shows the initial compressive $\sigma_{Y,c}$ and tensile $\sigma_{Y,t}$ yield stress dependencies on the average density ρ. For *partial* MHSS (cf. Fig. 4.11 (a)) the solid lines correspond to the compressive yield stress, whereas the dotted lines indicate the tensile yield stress.

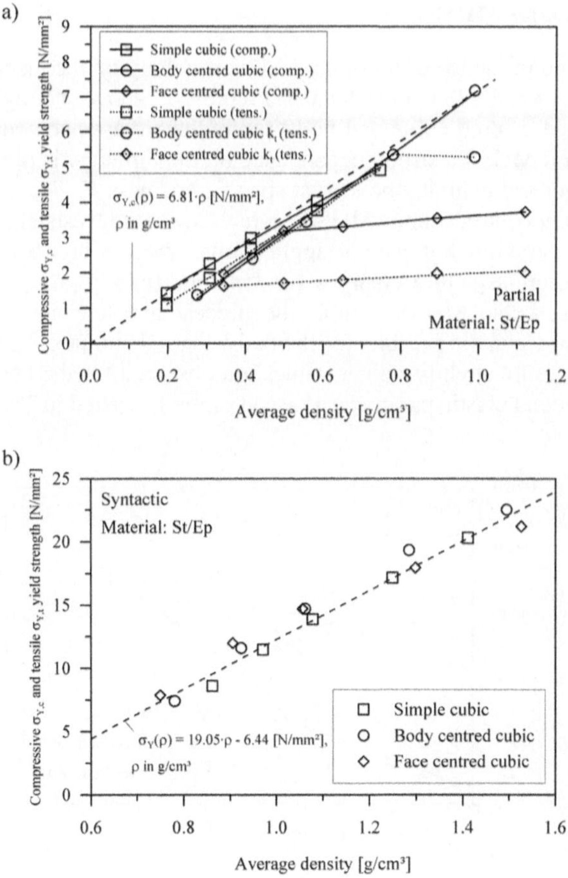

Fig. 4.11. Compressive and tensile yield strength of adhesively bonded MHSS

A linear fit can be employed in order to estimate the dependence of the initial compressive yield stress $\sigma_{Y,c}$ on the average density. By comparing the tensile and compressive yield stresses it becomes clear that for low densities the values coincide. However, with increasing density the curves of the tensile yield stresses exhibit a sharp edge. The explanation is the relocation of the initial region of plastification. With increase of the density and therefore sphere wall thickness h, the initial area of plastification changes from the isotropic metallic sphere to the anisotropic adhesive matrix and therefore affects the macroscopic response of the structure. The initial compressive and tensile yield stresses of *syntactic* MHSS (cf. Fig. 4.11 (b)) are not affected by the topology of the structure. The material parameter is well approximated by the linear function $\sigma_{Y,c}(\rho)$ as indicated in the figure. Furthermore, no dependency on the loading direction is observed for these *syntactic* MHSS.

4.4.2 Homogeneous MHSS

In this section, the influence of topology and morphology on the macroscopic properties of *homogeneous* MHSS is investigated. Figure 4.12 shows Young's modulus *E* for simple cubic, face-centred cubic and body-centred cubic arrangements of spheres. In the case of *partial* MHSS a strong dependence on the topology is observed. The simple cubic arrangement exhibits the lowest specific stiffness *E*. The highest values are obtained for body-centred cubic MHSS, a fact which suggests the use of this arrangement of spheres for lightweight applications, where a high specific stiffness is aimed for. A different picture emerges for *syntactic* (cf Fig. 4.12 (b)) morphology. Here, the simple cubic MHSS exhibit the highest absolute and specific Young's modulus. The corresponding results for the body-centred and face-centred cubic structures are similar, only slightly higher values are observed for the bcc structures. The values for the second elastic parameter Poisson's ratio are listed in Table 4.3.

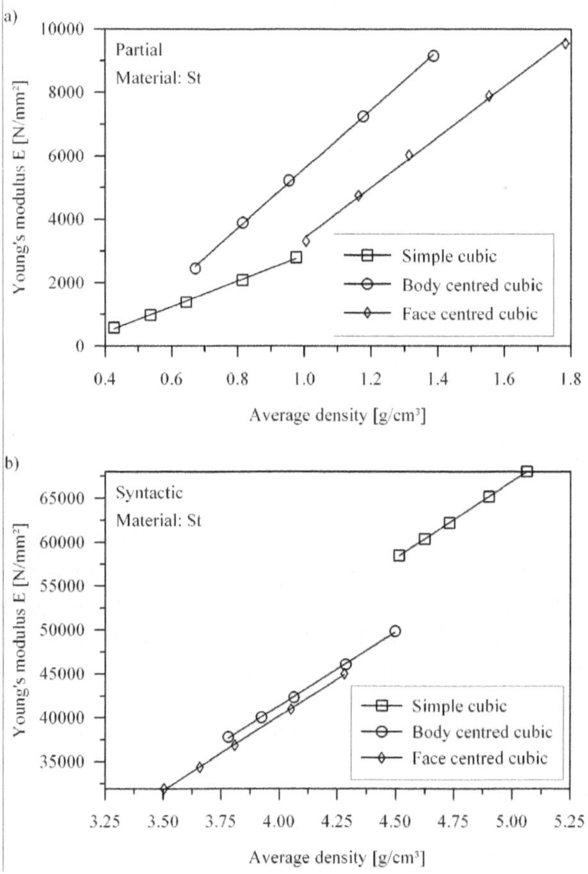

Fig. 4.12. Young's modulus of homogenous MHSS

Table 4.3. Poisson's ratio of homogeneous MHSS

	Partial MHSS			Syntactic MHSS		
Thickness *h*	sc	bcc	fcc	sc	bcc	fcc
0.030 mm	0.19	**-0.04**	0.17	0.24	0.28	0.29
0.052 mm	0.20	**-0.01**	0.17	0.24	0.28	0.29
0.075 mm	0.20	0.01	0.17	0.25	0.28	0.29
0.113 mm	0.22	0.04	0.18	0.25	0.28	0.29
0.150 mm	0.23	0.07	0.18	0.25	0.28	0.28

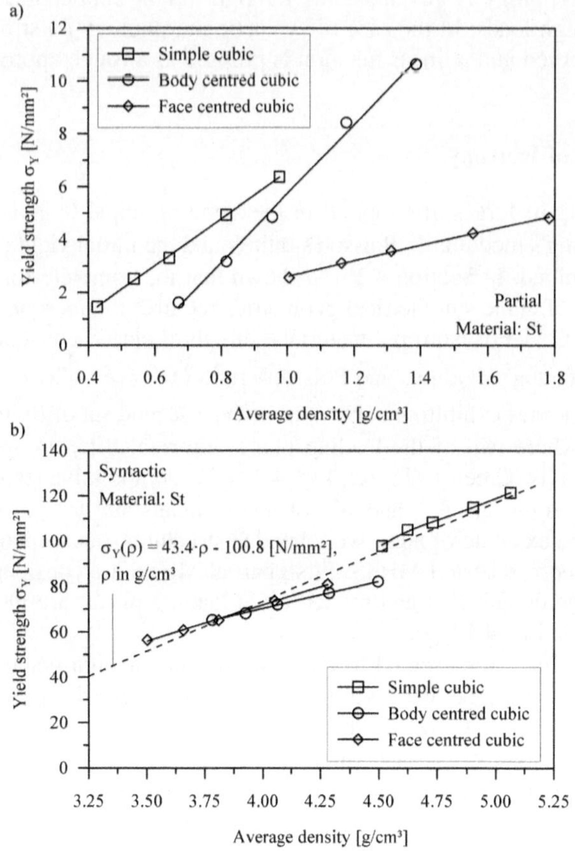

Fig. 4.13. Compressive and tensile yield strength of homogeneous MHSS

The Poisson's ratio for syntactic morphology varies slightly with the topology, but shows no significant dependence on the density. In general, Poisson's ratio is distinctly higher for syntactic morphology than for partial morphology. An anomaly is observed for partial morphology and body-centred cubic topology. Due to the elastic deformation mechanism inside the structure a negative Poisson number is obtained for the two smallest densities. The localisation of the deformation yields a positive transverse contraction for tensile and a negative transverse contraction for compressive loading. This phenomenon was already observed by Gasser et al. for particular face-centred cubic arrangements of hollow spheres (Gasser et al. 2004).

Figure 4.13 shows the initial yield stress plotted versus the average density. The tensile $\sigma_{Y,t}$ and compressive $\sigma_{Y,c}$ yield stresses coincide for *homogeneous* structures because of the isotropy of the base material steel (cf Table 4.1). The yield stresses of partial homogeneous structures exhibit a distinct dependence on the topology. The simple cubic arrangement of spheres yields the maximum specific yield strength. Accordingly, this topology is advantageous for lightweight applications which require the bearing of high loads. In the case of syntactic morphology no strong influence of topology is observed and a linear function is propsed as a rough approximation of the yield strength.

4.4.3 Analysis of Isotropy

Up to this point, all forces are applied in the same principal 0°-material coordinate system and Young's modulus *E*, Poisson's ratio ν and the initial yield stresses $\sigma_{Y,c}$ and $\sigma_{Y,t}$ were determined. In Section 4.2 it is shown that the complete formulation of the elasticity tensor of cubic symmetrical geometries requires the incorporation of a third elastic constant $G^{45°}$. For isotropic materials, this third elastic constant can be calculated based on Young's modulus and Poisson's ratio ($G^{\alpha} = E^{\alpha} / \left[2(1+\nu^{\alpha})\right]$, $\alpha = 0°$). In the case of geometries exhibiting cubic symmetry, a second set of finite element models is required where two of the loading planes are rotated by an angle of $\alpha = 45°$ about e.g. the *y*-axis (Green 1972) (cf. Fig. 4.2, 4.3). Compressive testing is simulated and the elastic parameters $E^{45°}$ and $\nu^{45°}$ of these models are determined. The results are shown for the example of adhesively bonded structures, but similar results are obtained for sintered and casted MHSS. First, partial MHSS are considered. The elastic parameters of the default (0°) and rotated (45°) loading planes are plotted versus the average density in Fig. 4.14 (a).

Young's modulus is represented by the solid line and no significant dependence on the loading direction can be observed. In the case of low densities a small deviation of Poisson's ratio is visible. However, this deviation disappears quickly with increase of density. Based on these numerical findings, the shear moduli *G* and $G^{45°}$ are determined. The ratio $G/G^{45°}$ can be interpreted as an indicator for the anisotropy of the material. A completely isotropic material will have a ratio of $G/G^{45°} = 1$, any deviation from this value signals anisotropic behaviour. According to the results shown in Table 4.4 it can be concluded that simple cubic and partial MHSS can be regarded as isotropic within the linear-elastic range.

Figure 4.15 (a) shows the initial compressive yield stresses $\sigma_{Y,c}$ for the two different loading planes. Slightly higher results are obtained for the default 0° loading planes, but again the material can be considered as isotropic to a good approximation.

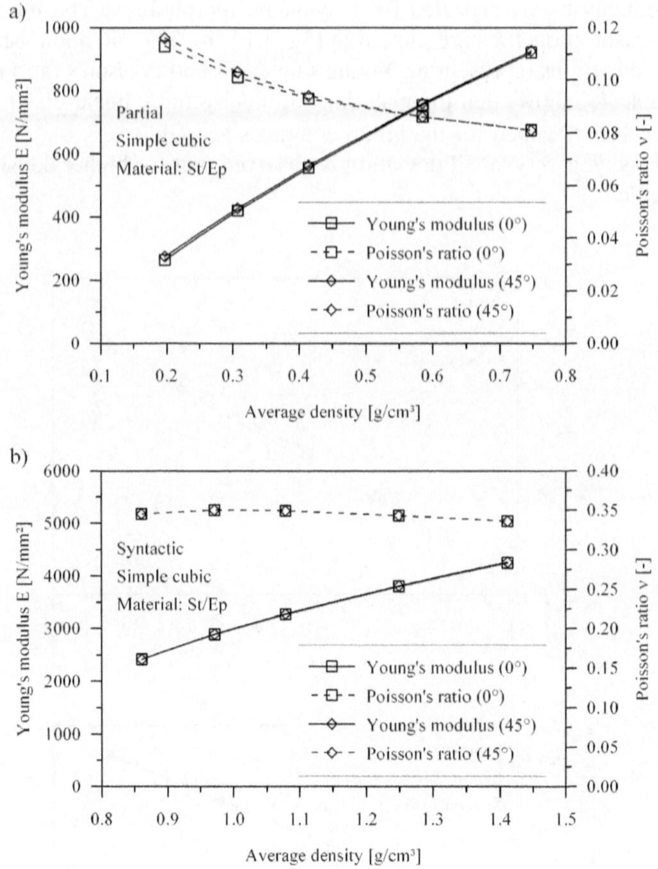

Fig. 4.14. Young's modulus dependence on the loading direction: a) partial MHSS; b) syntactic MHSS

Table 4.4. Elastic anisotropy of MHSS

Partial MHSS		Syntactic MHSS	
ρ	$G^{0°}/G^{45°}$	ρ	$G^{0°}/G^{45°}$
[g/cm³]	[-]	[g/cm³]	[-]
0.20	0.967	0.86	1.00
0.31	0.985	0.97	1.00
0.41	0.992	1.08	1.00
0.58	0.991	1.25	1.00
0.75	0.995	1.41	1.00

The same analyses are repeated for a syntactic morphology. The numerical findings for the elastic properties are plotted in Fig. 4.14 (b). The full and dashed lines for both loading directions representing Young's modulus and Poisson's ratio respectively coincide. Table 4.4 shows that the material exhibits isotropic linear-elastic behaviour. The same result is obtained for the initial compressive yield stress $\sigma_{Y,c}$. In the case of the lowest density, a very small deviation is observed, but for higher densities the two curves coincide.

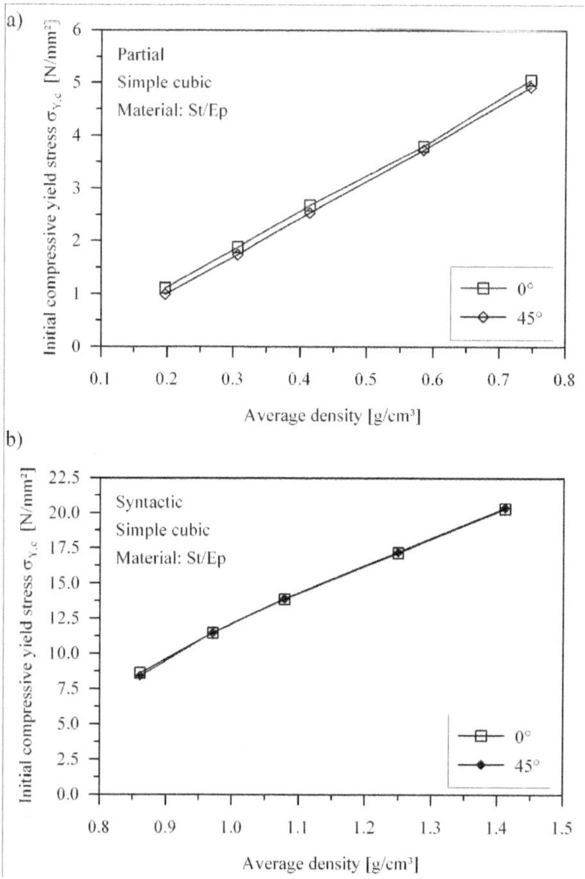

Fig. 4.15. Initial yield stress dependence on the loading direction: a) partial MHSS; b) syntactic MHSS

4.5 Experimental Findings

In addition to the numerical analyses, experimental tests on *adhesively bonded* MHSS are performed. Compressive testing is used to determine Young's modulus E, the plateau stress σ_{pl} and the hardening behaviour of these structures. The experimental samples investigated were provided by the German enterprise Glatt GmbH, Dresden.

Figure 4.16 shows the stress-strain relation of MHSS with partial and syntactic morphology. The curves exhibit the typical characteristics of a porous material. A linear elastic area is followed by a stress plateau, before the stress rises exponentially during the densification of the structure. The gradient of the curve inside the linear elastic area is equal to Young's modulus E and the elastic zone is confined by the stress peak σ_{max}. Within the stress plateau, the stress level is approximately constant and indicated by the plateau stress σ_{pl}. After reaching the densification strain ε_D the porosity in the MHSS decreases rapidly and the stress increases. In contrast to the partial specimens, the observed Young's modulae and stress levels of syntactic MHSS are distinctly higher. Furthermore, because of the lower porosity of syntactic MHSS the densification starts at slightly lower strains ε_D.

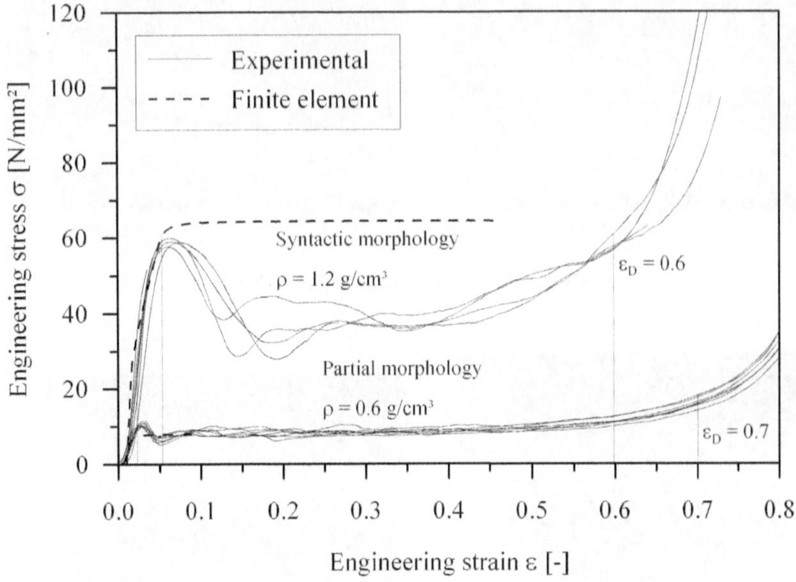

Fig. 4.16. Experimental and numerical stress-strain data for uni-axial compression

In addition to the experimental data, Fig. 4.16 contains results of the finite element analysis. Good agreement between numerical and experimental data is found for partial morphology. However, the numerical analysis requires several simplifications which do not allow for the incorporation of all physical effects that occur in the experimental tests at certain strain levels. In contrast to the results of the finite element analysis, the measured stresses inside the plateau are not constant and oscillate around the plateau stress. Furthermore, the experimental curves exhibit an initial stress peak $\sigma_{max} > \sigma_{pl}$ at the end of the linear elastic area. This behaviour is caused by a 'macroscopic' failure mechanism in partial MHSS. The plastic deformation is concentrated inside bands which collapse after a critical strain is exceeded, each time coinciding with a macroscopic stress oscillation. In this, the weakest hollow spheres collapse first and initiate the collapse band. The majority of these bands form an angle of approx.

30-40° with the loading direction as shown in Fig. 4.17. The first collapse band weakens the structure by creating 'seeds' for subsequent deformation bands so that the plateau stress σ_{pl} is smaller than the initial stress peak σ_{max}. This effect is not observed in the numerical analysis, since the symmetry boundary conditions do not allow for the formation of collapse bands.

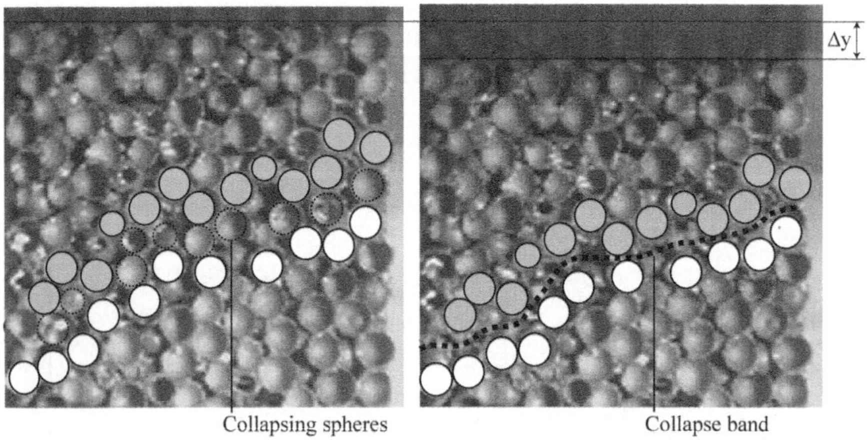

Fig. 4.17. Collapse band in partial MHSS

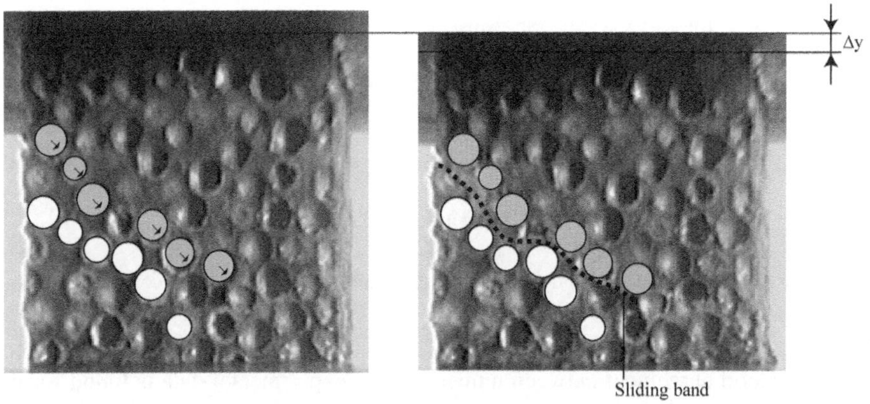

Fig. 4.18. Shear band in syntactic MHSS

In the case of syntactic morphology, the comparison with numerical results only yields good agreement at small strains. At higher strains, the stress of the numerical solution remains constant whereas the real stresses drop towards the plateau stress σ_{pl}. Analogous to partial MHSS, the explanation for this deviation is a macroscopic failure mechanism. In Fig. 4.18, a syntactic specimen is shown under compressive loading at two different strains. After reaching a critical strain, the geometry shears along a sliding band which forms an angle of approximately 45° with the loading direction.

A similar failure mode was already observed for syntactic glass microsphere foams with epoxy matrix (Adrien et al. 2006). The simulation of this phenomenon in the numerical investigation requires the modelling of the whole geometry and exceeds the capacity of the available computer hardware.

4.6 Summary

In the scope of this chapter elasto-plastic properties of metallic hollow sphere structures were investigated. The results of the finite element analysis indicate a strong dependence of MHSS material properties on the morphology and joining technology.

Syntactic morphology increases the stiffness and strength in comparison to partial structures. This finding is independent of the considered topology or material combination. However, due to lower porosity, syntactic MHSS also exhibit a distinctly higher average density.

Adhesively bonded structures generally exhibit lower stiffness and strength than homogeneous ones. Young's modulus and the initial yield stress of *syntactic* adhesively bonded structures can be estimated based on linear fits and do not depend on the topology. In contrast, the material properties of *partial* adhesively bonded structures change with the topology. The highest Young's moduli are obtained for a body-centred cubic arrangement of spheres. A linear fit was proposed to estimate the compressive yield stress in dependence on the average density. Under tensile loading, the initial plastification occurs within the adhesive matrix for structures with higher sphere wall thickness. In these cases, the anisotropic material behaviour of the adhesive matrix is reflected in the macroscopic properties of the MHSS. This anisotropy should not be confused with the later addressed geometric anisotropy.

Homogeneous structures show a distinct dependence on the topology. The highest Young's modulae are observed for a body-centred cubic arrangement in the case of *partial* MHSS and simple cubic topology in the case of *syntactic* structures. The initial yield stresses are independent of the loading direction (compressive or tensile) and can be approximated by a linear fit for *syntactic* MHSS. The highest specific yield stress of *partial* homogeneous hollow spheres structures is observed for simple cubic topology.

Based on the analysis of geometric anisotropy it can be concluded that MHSS with a simple cubic arrangement of spheres can be considered as isotropic in relation to their linear-elastic behaviour and uni-axial compressive yield stresses. This result is independent of the morphology or material composition of the structure. It should be mentioned here that this conclusion cannot be transferred to other cubic symmetrical topologies. However, the simple cubic topology exhibits the highest deviation of the distance of spheres in the dependence on the loading direction. Accordingly, it is likely that body-centred cubic and face-centred cubic structures behave at least approximately isotropic. Another indicator for this assumption are finite element results by Gasser et al. that indicate only slight elastic anisotropy of face-centred cubic MHSS (Gasser et al. 2004).

The comparison of the numerical and experimental results obtained for adhesively bonded structures shows good agreement within the elastic range and the elastic-plastic transition zone. However, at higher strains deviations can be observed. The explanation are macroscopic deformation mechanisms which in the numerical calculation models are suppressed by symmetric boundary conditions. Uni-axial compressive tests on *partial* MHSS showed a localisation of the deformation inside collapse bands. In the case of *syntactic* hollow sphere structures shear deformation along sliding bands was predominant. The numerical analysis of these deformation mechanisms will require accurate calculation models with a detailed representation of real geometries, e.g. obtained by computer tomography.

References

Adrien, J., Maire, E., Gimenez, N., et al.: Experimental study of the compression behaviour of syntactic foams by in situ X-ray tomography. Acta Mater. 55, 1667–1679 (2007)

Ashby, M.F., Evans, A., Fleck, N.A., Gibson, L.J., Hutchinson, J.W., Wadley, H.N.G.: Metal Foams: A Design Guide. Butterworth-Heinemann, Boston (2000)

Banhart, J., Baumeister, E., Weber, M.: Metal foam - a material with future (in German). Aluminium 70(3/4), 209–212 (1994)

Baumeister, E.: Hollow-spheres-composites - as new light-weight materials for mechanical engineering. In: Proceedings Materials Week 2002, the European Congress on Advanced Materials, their Processes and Applications (2002) (Published on CD-ROM)

Baumeister, E., Klaeger, S., Kaldos, A.: Lightweight, hollow-sphere-composite (HSC) materials for mechanical engineering applications. J. Mater. Process Tech., 155–156, 1839–1846 (2004)

Benzley, S.E., Perry, E., Merkle, K., Clark, B., Sjaardema, G.F.: A Comparison of All Hexagonal and All Tetrahedral Finite Element Meshes for Elastic and Elasto-plastic Analysis. In: Fourth International Meshing Roundtable, Albuquerque, New Mexico, pp. 179–191 (1995)

Cook, R.D., Malkeis, D.S., Plesha, M.E., Witt, R.J.: Concepts and Applications of Finite Element Analysis. John Wiley & Sons, New York (2002)

Evans, A., Hutchinson, J.W., Ashby, M.F.: Multifunctionality of cellular metal systems. Prog. Mater. Sci. 43, 171–221 (1999)

Fiedler, T., Öchsner, A., Grácio, J., Kuhn, G.: Structural Modeling of the Mechanical Behavior of Periodic Cellular Solids: Open-Cell Structures. Mech. Compos. Mater. 3, 277–290 (2005)

Fiedler, T., Sturm, B., Öchsner, A., Grácio, J., Kuhn, G.: Modelling the mechanical behaviour of adhesively bonded and sintered hollow sphere structures. Mech. Compos. Mater. 42, 559–570 (2006)

Fiedler, T., Solórzano, E., Öchsner, A.: Numerical and Experimental Analysis of the Thermal Conductivity of Metallic Hollow Sphere Structures. Mater. Lett. 62, 1204–1207 (2008)

Franeck, J., Landgraf, G.: Determination of linear and non-linear mechanical properties of sintered hollow-sphere structures. In: Singer, R.F., Körner, C., Altstädt, V., Münstedt, H. (eds.) Cellular Metals and Polymers 2004, pp. 139–142. Trans Tech Publications, Zürich (2005)

Friedl, O., Motz, C., Färber, J., Stoiber, M., Pippan, R.: Tension and compression behaviour of stainless steel (316L) hollow sphere structures. In: Singer, R.F., Körner, C., Altstädt, V., Münstedt, H. (eds.) Cellular Metals and Polymers 2004, pp. 135–138. Trans Tech Publications, Zürich (2005)

Gasser, S., Paun, F., Bréchet, Y.: Finite elements computation for the elastic properties of a regular stacking of hollow spheres. Mat. Sci. Eng. A-Struct. 379, 240–244 (2004)

Gasser, S., Paun, F., Riffard, L., Bréchet, Y.: Microplastic yield condition for a periodic stacking of hollow spheres. Scripta Mater. 50, 401–405 (2004)

Golovin, I., Sinning, H.: Damping in some cellular metallic materials. J. Alloy. Compd. 355, 2–9 (2003)

Green, R.J.: A plasticity theory for porous solids. Int. J. Mech. Sci. 14, 215–224 (1972)

Gründer, J., Landgraf, G.: Modelling and simulation of the meso- and macro-mechanical properties of hollow sphere structures. In: Banhart, J., Ashby, M.F., Fleck, N.A. (eds.) Cellular Metals and Metal Foaming Technology, pp. 397–402 (2001)

Hartmann, M., Reindel, K., Singer, R.F.: Fabrication and properties of syntactic magnesium foams. Mat. Res. Soc. Symp. Proc. 521, 211–216 (1998)

Heine, B., Andersson, K., Baur, H., Lempenauer, K.: Porous Metallic Structures for Automotive Application – Overview and Perspectives. In: Proceedings Materials Week 2002, the European Congress on Advanced Materials, their Processes and Application (2002) (Published on CD-ROM)

Hübelt, J., Bingel, G.: Excellent sound absorption by metallic hollow sphere structures. Cellmet-News 1, 1–2 (2006)

Jäckel, M.: German Patent 3 724 156 (1987)

Jeandrau, J.P.: Analysis and design data for adhesively bonded joints. Int. J. Adhes. Adhes. 11, 71–79 (1991)

Kim, J., Lee, S., Shin, K.: Manufacturing and structural safety evaluation of a composite train carbody. Compos. Struct. 78, 468–476 (2007)

Knox, E., Cowling, M., Winkle, I.: Adhesively bonded steel corrugated core sandwich construction for marine applications. Mar. Struct. 11, 185–204 (1998)

Kohler, S., Fitzgerald, T.J., Singer, R.F.: Manufacturing and properties of celluar magnesium (in German) In: VDI Berichte Nr. 1235:305-319 (1995)

König, M.: Yield surface for perforated sheets. Res. Mech. 19, 61–90 (1986)

Koo, J.M., Jung, S.B.: Mechanical Properties of Copper Hollow Sphere Manufactured by Sintering Process. Mater. Sci. Forum 510–511, 730–733 (2006)

Lim, T.J., Smith, B., McDowell, D.L.: Behavior of a random hollow sphere metal foam. Acta Mech. 50, 2867–2879 (2002)

Löthman, P.A., Andersen, O., Gründer, J., Landgraf, G.: Mechanical properties of metallic hollow sphere bodies and characterisation of their cell walls (in German). Materialwiss. Werkst. 31, 529–530 (2000)

Nemat-Nasser, S., Kang, W., McGee, J., Guo, W., Isaacs, J.: Experimental investigation of energy-absorption characteristics of components of sandwich structures. Int. J. Impact Eng. 34, 1119–1146 (2007)

Neugebauer, R., Hipke, T., Hohlfeld, J.: Metal Foam as a Combination of Lightweight Engineering and Damping. In: Singer, R.F., Körner, C., Altstädt, V., Münstedt, H. (eds.) Cellular Metals and Polymers 2004, pp. 13–18. Trans Tech Publications, Zürich (2005)

Neville, B.P., Rabiei, A.: Composite metal foams processed through powder metallurgy. Mater. Design 27, 388–396 (2007)

Paik, J., Thayamballi, A., Kim, G.: The strength characteristics of aluminum honeycomb sandwich panels. Thin Wall. Struct. 35, 205–231 (1999)

Palmer, R.A., Gao, K., Doan, T.M., Green, L., Cavallaro, G.: Pressure infiltrated syntactic foams-Process development and mechanical properties. Mat. Sci. Eng. A-Struct. 464, 85–92 (2007)

Ramamurty, U., Paul, A.: Variability in mechanical properties of a metal foam. Acta Mater. 52, 869–876 (2004)

Sanders, W.S., Gibson, L.J.: Mechanics of hollow sphere foams. Mat. Sci. Eng. A-Struct. 347, 70–85 (2003)

Sanders, W.S., Gibson, L.J.: Mechanics of BCC and FCC hollow-sphere foams. Mat. Sci. Eng. A-Struct. 352, 150–161 (2003)

Sukegawa, Y., Waag, U., Murakami, K., Yoshimura, H., Shinagawa, K., Mihara, Y.: Compression Characteristics of Metallic Hollow Spheres with Pipe Overcoat. In: Nakajima, N., Kanetake, N. (eds.) Porous Metals and Metal Foaming Technology, pp. 571–576. The Japan Institute of Metals, Sendai (2005)

Sukegawa, Y., Waag, U., Murakami, K., Yoshimura, H., Shinagawa, K., Mihara, Y.: Metallic Hollow Sphere Structures Bonded by Adhesion. In: Nakajima, H., Kanetake, N. (eds.) Porous Metals and Metal Foaming Technology, The Japan Institute of Metals, Sendai, Japan, pp. 381–386 (2005)

Zhao, C., Lu, T., Hodson, H., Jackson, J.: The temperature dependence of effective thermal conductivity of open-celled steel alloy foams. Mat. Sci. Eng. A-Struct. 367, 123–131 (2004)

Zienkiewicz, O.C., Scorft, F.C.: On the principle of repeatability and its application in analysis of turbine and pump impellers. Int. J. Numer. Meth. Eng. 4, 445–452 (1972)

5

On the Vibroacoustic Behaviour of HSS

Markus Merkel, Wolfram Pannert, and Rolf Winkler

University of Applied Sciences Aalen, Beethovenstrasse 1
73430 Aalen, Germany
Markus.Merkel@htw-aalen.de, Wolfram.Pannert@htw-aalen.de,
Rolf.Winkler@htw-aalen.de

Abstract. In contrast to classic engineering materials hollow sphere structures show special properties regarding vibroacoustic behaviour. Due to their macro- and microscopic structure these materials seem to provide properties engineers require, but previously could not be achieved with standard materials. In this chapter the vibroacoustic behaviour is discussed in detail. The chapter is roughly split into two parts. While the first part deals with the vibrational behaviour, the second focuses on acoustic behaviour. For each part we provide a section with theoretical background, numerical simulation and experiments.

Keywords: Metal Hollow Sphere Structures, Vibration Analysis, Acoustical Analysis, Impedance Tube.

5.1 The Vibrational Behaviour of Hollow Sphere Structures

Engineers always inspect systems, subsystems, components or parts loaded by vibrations. Generally just in interest to see in which pattern and with which magnitude structures are vibrating. In worse case scenarios strength and stiffness of structures are weakened by vibrations depending on the amplitude, frequency or duration of the vibrational load. Fatigue or damage is a typical consequencee.

5.1.1 Theoretical Background / Motivation

From an engineer's point of view the macroscopic behaviour of a structure is what is of interest. For any standard engineering material with homogenous structure, it is known how to describe the vibrating behaviour by a set of parameters like density, Young's modulus, Poisson's ratio.

As the HSS represent an inhomogeneous material more information is necessary. The understanding of microscopic mechanisms, the kind and intensity of interaction between basic elements will help to find a correlation between microscopic and macroscopic parameters.

The basic units of HSS are spheres, which are interconnected by other spheres. Spheres can either be closed or partially open. All spheres can be embedded in an

A. Öchsner, C. Augustin (Eds.): Multifunctional Metallic Hollow Sphere Struct., EM, pp. 71–88.
springerlink.com

overall matrix without touching each other or can be interconnected by special joining elements. Figure 5.1 shows a sphere with six interconnections. This sphere represents a typical unit in a regular structure.

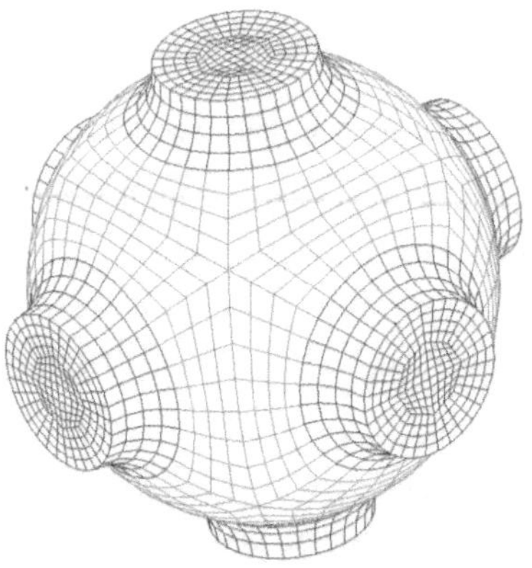

Fig. 5.1. FE-model for a single sphere with joining sections

The model represents a FEM mesh for a standard analysis of strength. More than 1,000 of these units build a real structure. In some computational runs as a preliminary investigation the computing time for models with some hundred units already exceeded standard resources. The need for a reduced model resulting in less computing time is obvious.

5.1.2 Experiments on HSS

The vibrational behaviour is analysed for several test specimens. Figure 5.2 shows a set of samples with different geometry. The microscopic parameters vary in each specimen. These are geometric parameters, like the diameter of a single sphere, thickness of a sphere, and material parameter like Young's modulus, Poisson's ratio.

Fig. 5.2. A set of test specimens based on Hollow Sphere Structures

The pattern of the structure is not unique. It is neither regular, nor body or face centred nor hexagonal closest-packing.

The plates being analysed for vibration behaviour have the dimensions 300 mm x 110 mm x 30 mm. The diameter of the hollow spheres is 2 mm and the weight of a complete specimen is in average 442 g.

The vibrational behaviour is measured by a standard test. Figure 5.3 shows the test bench and a hollow sphere structure specimen.

Fig. 5.3. Test bench and HSS specimen

The specimen is layed on a weak ground to approximate free-free condition of the vibration analysis. The vibrations are introduced by a hammer. The movement of the specimen is measured by a laser scanning vibrometer. The test set up is shown in Fig. 5.4.

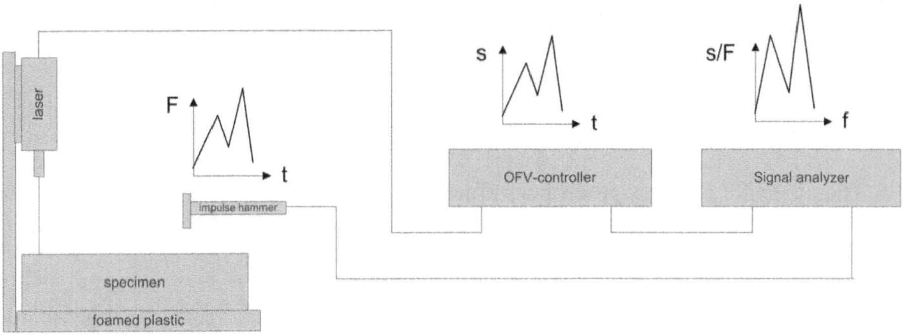

Fig. 5.4. Test set up

Figure 5.5 shows the measured mode shapes of the test specimen.

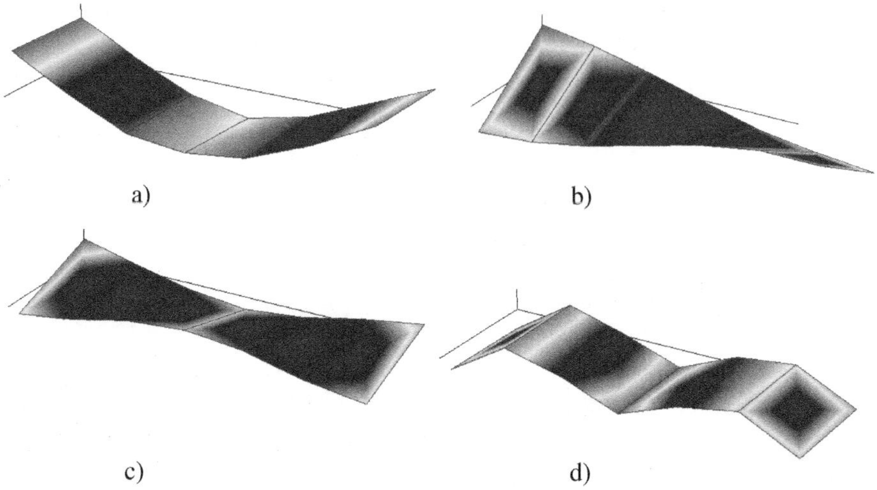

Fig. 5.5. Mode shapes gained by measurement a) 1st Eigenmode b) 2nd Eigenmode c) 3rd Eigenmode d) 6th Eigenmode

5.1.3 Vibration Analysis by Numerical Simulation

In general the quality of a FE analysis strongly depends on the details taken into account. A good approximation is to be expected when sufficient details are reflected in the FE model. In Fig. 5.1 the basic unit of a FEM model was already presented. Up to now it has only been used for static analysis of specimens with a few spheres. The FE-model for the vibration analysis is built up with thousands of single spheres including joints. It is expected that this model would exceed available computer resources due to the fine mesh.

An approximation of this detailed model will help to reduce computing time. A coarse approach is to replace the hollow sphere by a mass point and joint between spheres by beams. This idea stems from classical mass-spring-damper models used to idealise vibrating systems. The approximation has the following variables:

1. Mass
2. Stiffness (bending & torsional)
3. Moment of inertia (geometrical & mass)

The objective is to assign proper values to those microscopic variables, so that the macroscopic vibration behaviour can be described accurately. We defined two types of FE models for our investigations.

Model type 1

The specimen is modelled as homogenised mass with constant density. The FE model consists of 3D solid elements.

Model type 2

Spheres are replaced by mass points and joints are replaced by beam elements. The stiffness and the geometrical moment of inertia are assumed to be constant along the beam. The structure of the FE models is defined as a primitive cubic arrangement. Other models where spheres are ordered in face or body centred cubic scheme are considered later.

5.1.4 Validation Process

The validation process has to consider mode shapes as well as eigenfrequencies. The FE model of type 1 with homogenized mass shows the same mode shapes as already depicted in Fig. 5.5. The FE model of type 2 with mass points and beams also shows equivalent mode shapes. In Figs. 6a–6d the mode shapes are plotted. They correspond exactly to those gained by the experiment and previously shown in figures 5.5a – 5.5d.

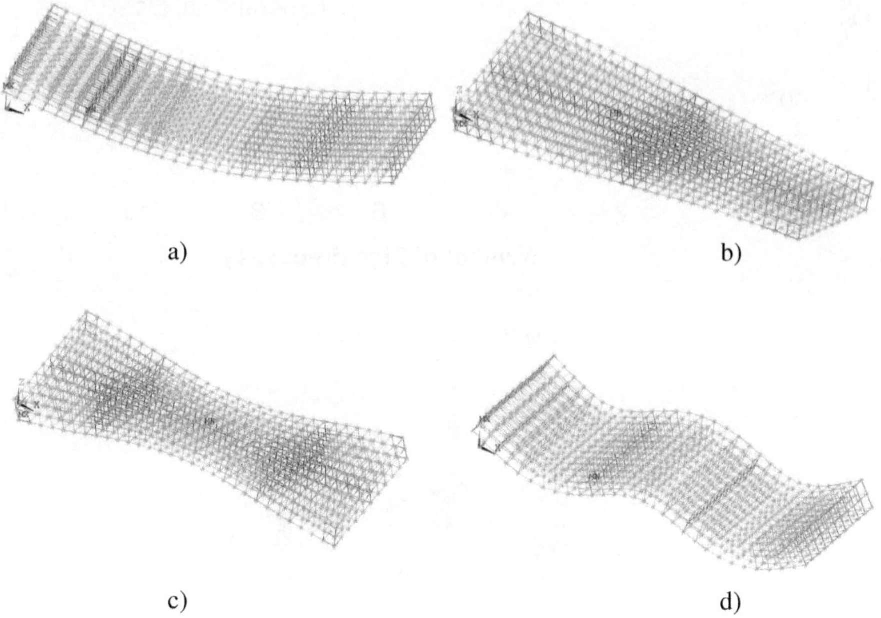

Fig. 5.6. Mode shapes gained by numerical simulation

Regarding the FE model the overall mass can be assigned either to the mass points or to the beams. In order to simplify the model, i.e. to clearly separate different effects, all mass is put to the mass points and the beams are defined without mass.

All simulations were carried out with the professional software system "ANSYS". Beams are represented as beam4 elements, mass points as mass21 elements. Solid186 elements are taken for the homogenised mass model. The computation time for a single run takes a few minutes on a standard office machine.

In Fig. 5.7 the eigenfrequencies are plotted for three different types of results: FE model type 1 and type 2 as well as experimental data. For model type 2 the Young's modulus was assigned 1,100 Pa and Poisson's ratio as 0.15.

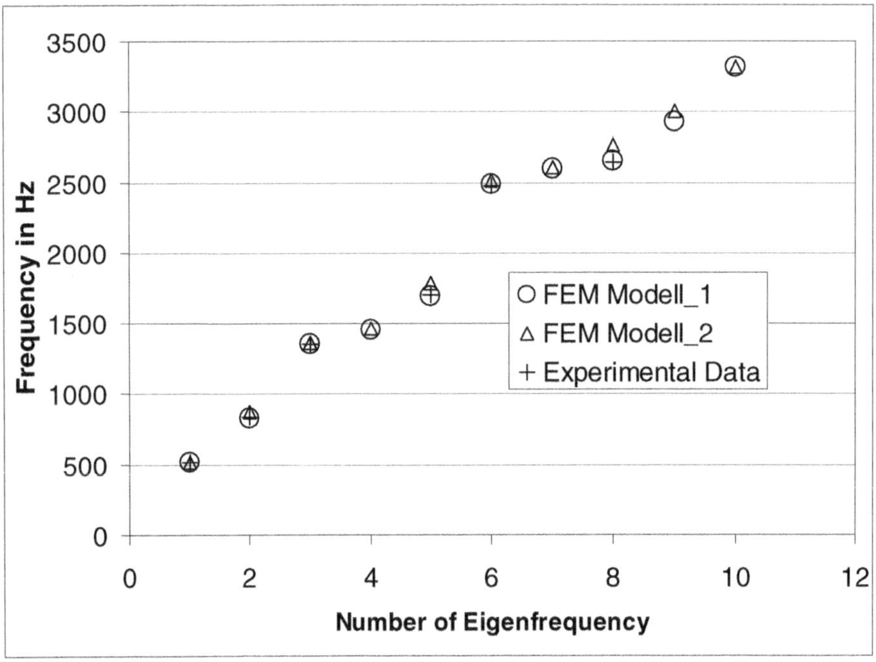

Fig. 5.7. Eigenfrequencies

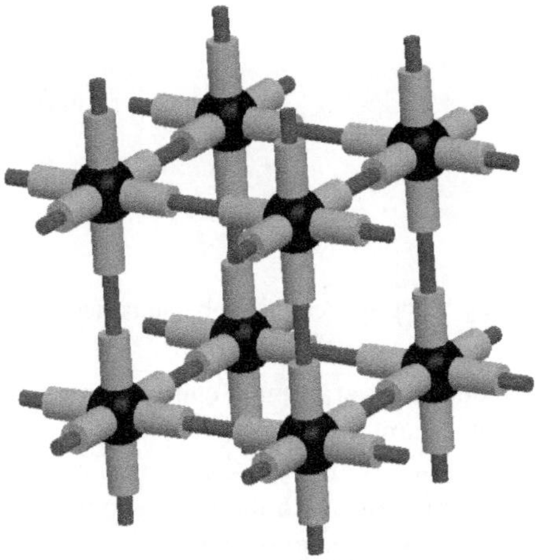

Fig. 5.8. Extended FE Model

A very good correlation for all results has to be stated. Experimental results for the 4th ad 7th eigenfrequency are missing. These are inplane motions that cannot be detected by the laser scanning method.

Despite the fact that a simple FE-model with constant stiffness along the beam delivers quite good results, a more detailed model is required when additional components are added to the hollow sphere structure, e.g. thin ceramic layers around the hollow spheres. These are expected to influence the vibration behaviour. Therefore an extended FE model is defined, where the stiffness as well as the geometrical moment of inertia are formulated as variables along the principal beam axis. In Fig. 5.8 a

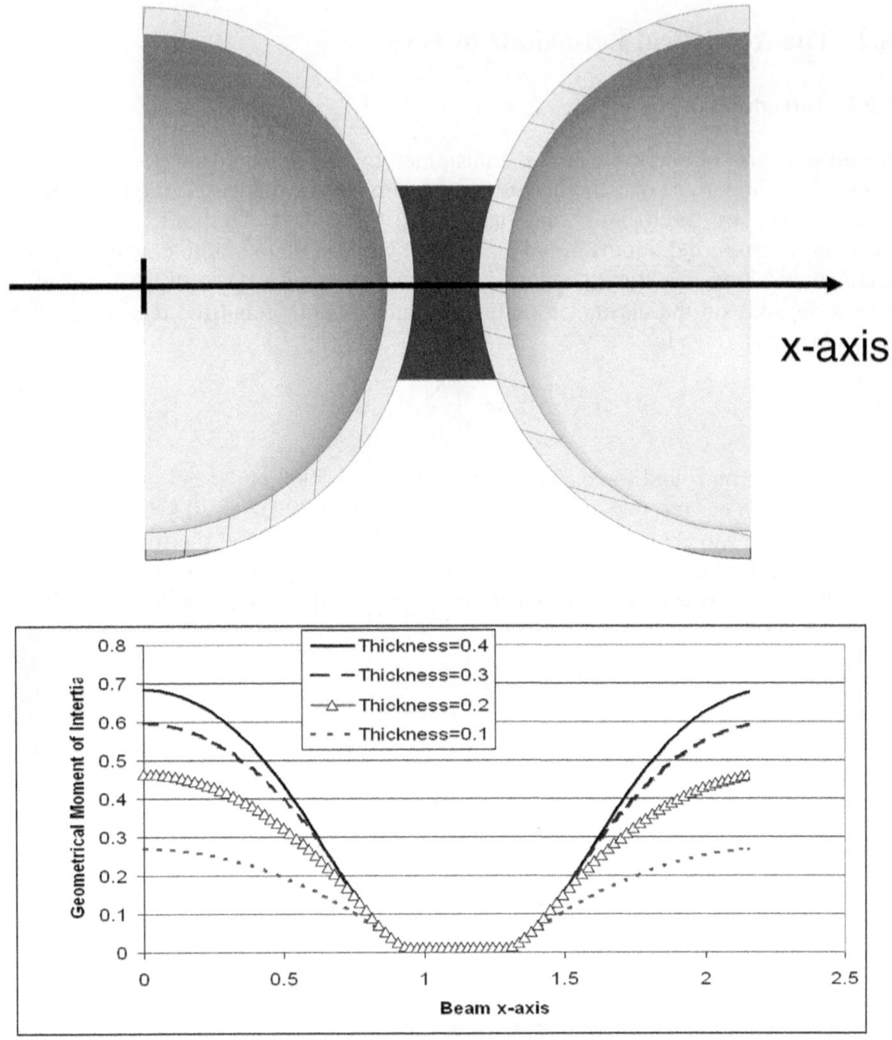

Fig. 5.9. Extended simulation model. a) spheres with joint b) geometrical moment of inertia along a beam axis

configuration with a regular pattern of the structure is depicted: a mass point (enlarged), a beam element represents the stiffness of the sphere while the joining element is represented by another beam.

In Fig. 5.9a the basic configuration is seen along the axis between two spheres. The two halves of spheres are connected by a joining element. In Fig. 5.9b the resulting moment of inertia is plotted along the beam axis.

The hollow spheres thickness reach from 0.1 mm to 0.4 mm. This non-constant curve for the moment in combination with different Young's modulae results in a non-constant bending stiffness. The FE-model has to reflect this by several subsections. Beams have to be split into shorter sub-beams.

5.2 The Acoustical Behaviour of HSS

5.2.1 Introduction

Sound is a form of energy, which is transmitted by the collisions of atoms and molecules. It is important that there is no transfer matter in this case. The spreading of sound occurs in the form of longitudinal and transversal vibrations (waves) of the medium. In gases and liquids only longitudinal vibrations are possible, while in solids there are also transversal modes due to the shear forces. The speed of the longitudinal waves depends on the elastic properties (K-bulk modulus) and the density ρ of the medium and is given by

$$c \propto \sqrt{\frac{K}{\rho}}. \tag{5.1}$$

For this reason sound travels at velocities between 5000 m/s (steel) and 340 m/s in air. The human ear responds to frequencies from about 20 Hz to 20 kHz, which in air corresponds to wavelengths of 17 m to 17 mm. The amplitude of the sound waves determines the loudness sensation. The amplitude of sound pressure is measured in Pascals (Pa), it is however more convenient to use a logarithmic scale in decibels (dB).

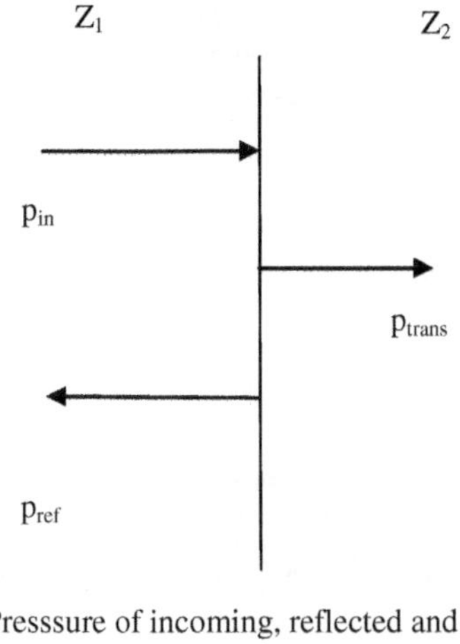

Fig. 5.10. Presssure of incoming, reflected and transmitted wave

At the boundary between two different materials an incoming plane wave is reflected due to the difference in the acoustic impedances Z of the material (see Fig. 5.10).

With

$$p_{ref} = r \cdot p_{in},$$

it follows

$$r = \frac{Z_2 - Z_1}{Z_2 + Z_1}, \tag{5.2}$$

where p_{in} is the amplitude of the incoming wave, p_{ref} is the amplitude of the reflected wave and r is the sound reflection factor. For the intensity I, this relation reads

$$I_{ref} = |r|^2 \cdot I_{in} \quad \text{and} \quad I_{trans} = (1-|r|^2) \cdot I_{in} = \alpha \cdot I_{in}\,. \tag{5.3}$$

I_{ref} is the reflected intensity and I_{trans} is the transmitted intensity, which is directly related to the energy transport across the boundary. The coefficient $(1-|r|^2) = \alpha$ is also called absorption coefficient and is a very important functional property of materials.

The absorption coefficient depends on the frequency and angle of the incident wave and for a given material it is a function of the material thickness, density and flow resistance (pore size). On small samples (up to 100 mm in diameter) the absorption coefficient of normally incident waves can easily be measured by an impedance tube. If the absorption coefficient for all angles of incidence is required, the measurement in a reverberation room is usually required. Much larger samples are needed in this case (up to 10 m^2).

For sound absorption purposes materials like open-cell polymer foam and glass or mineral wool fibers are used. Both types of absorbers usually require various facing materials in order to improve durability or to protect the absorber from contamination.

As opposed to closed-cell metal foams, which are poor sound absorbers, we study metal hollow sphere structures, which show up an open porosity. So it is expected, that these materials show similar behaviour to open-celled metal foams, which have good absorption behaviour [7].

Fig. 5.11. MHSS 86 and MHSS 113 Samples with thickness varying from 10 mm to 60 mm

5.2.2 Experimental Method

5.2.2.1 Sample Preparation

Two different samples of MHSS were investigated (MHSS 113 with 2.5 mm sphere-diameter and MHSS 86 with 1.5 mm sphere-diameter). The material was cut into pieces, which fit perfectly in the impedance tube (see Fig. 5.11).

The thickness of the samples varies from 10 mm to 60 mm. In other experiments, the porosity was determined to be $\sigma = 0.4$ for MHSS 113 and $\sigma = 0.6$ for MHSS 86.

5.2.2.2 Measurement

The measuring equipment included a signal generator for the generation of white noise, a loudspeaker, two impedance tubes (for different frequency regions) with two high quality microphones and a frontend for data acquisition. The analysis of the data was carried out using the program MatLab and the included signal processing toolbox.

Fig. 5.12. MHSS sample in the impedance tube with steel back plate

A number of preliminary tests with different materials were performed to calibrate the test. The sound absorption coefficient of the steel back plate of the impedance tube was also measured and found to be less than a value of 0.1 at all relevant frequencies.

The maximum frequency of the impedance tube is limited by its inner diameter. Therefore we used two different tubes with diameters of 100 mm and 40 mm. The larger tube is used for the frequency range from 150 Hz to 1000 Hz and the smaller one for frequencies from 400 Hz to 3500 Hz.

5.2.3 Theoretical Background of Measurement

As mentioned above, there are two main methods for the determination of the acoustic absorption coefficient of materials – the reverberation room method and the impedance tube method. In this investigation we used the second method, since it is faster and generally reproducible and because it requires only relatively small samples. Two different impedance tube methods are available. The transfer function method is a relatively recent development. In this method a broadband noise signal is used as sound source. A detailed description of this method is given in DIN EN ISO 10534-2. The sound absorption coefficient and the impedance ratio of the tested materials can be measured much faster and easier, compared to other methods. Therefore the transfer technique was adopted in this study.

The transfer function method is based on the fact that the sound reflection factor r at normal incidence, can be determined from the measured transfer function H_{12} between two microphones, which are positioned in front of the material under investigation. The complex transfer function H_{12} is defined as

$$H_{12} = \frac{p_2}{p_1} = \frac{e^{jk_0x_2} + r \cdot e^{-jk_0x_2}}{e^{jk_0x_1} + r \cdot e^{-jk_0x_1}}. \tag{5.4}$$

Where p_1 and p_2 are the complex sound pressure amplitudes at the two microphone positions x_1 and x_2, which are measured from the reference plane $x = 0$. k_0 is the wave vector and is defined by

$$k_0 = \frac{2\pi}{\lambda}. \tag{5.5}$$

The transfer function for the incident wave H_{in} and for the reflected wave H_{ref} can be calculated by

$$\begin{aligned} H_{in} &= e^{-jk_0(x_1-x_2)}, \\ H_{ref} &= e^{jk_0(x_1-x_2)}. \end{aligned} \tag{5.6}$$

Combining these equations, the complex reflection factor r can be calculated

$$r = \frac{H_{12} - H_{in}}{H_{ref} - H_{12}} \cdot e^{j2k_ox_1}. \tag{5.7}$$

The sound absorption coefficient α can be determined in terms of r

$$\alpha = 1 - |r|^2. \tag{5.8}$$

5.2.4 Comparison with Theory of Sound in Porous Media

According to the simple Rayleigh-model of a porous absorber, the porous medium can be considered as an ensemble of many small tubes and the sound propagates in these small tubes. The flow in each tube can be calculated exactly and gives the known result in form of the "Hagen-Poiseuille-Law". It is possible to introduce the flow-resistance Ξ, which is responsible for the absorption of energy in the porous material.

Solving the wave-equation with the appropriate boundary conditions, it is possible to calculate the complex frequency dependent impedance Z(ω) of an absorber with thickness d [6].

$$Z(\omega) = -j\rho\, c\frac{\sqrt{\chi}}{\sigma}\sqrt{1 - j\frac{\Xi\sigma}{\omega\rho\chi}} \cdot ctg(k_a d). \tag{5.9}$$

Definition of the parameters (typical values in brackets)

d: thickness of the absorber

k_a: wave number in the absorber

ρ: density of air

c: speed of sound in air

χ: structure form factor (χ≥1)

ω: frequency

Ξ: flow resistance (5000 Ns/m^4 < Ξ < 100000 Ns/m^4).

From this model one can calculate the coefficient of absorption α equal to:

$$\alpha = \frac{4\,\mathrm{Re}(\frac{Z}{\rho c})}{[\mathrm{Re}(\frac{Z}{\rho c}) + 1]^2 + [\mathrm{Im}(\frac{Z}{\rho c})]^2}. \tag{5.10}$$

In Figure 5.13 the absorption coefficient is plotted over a wide frequency range.

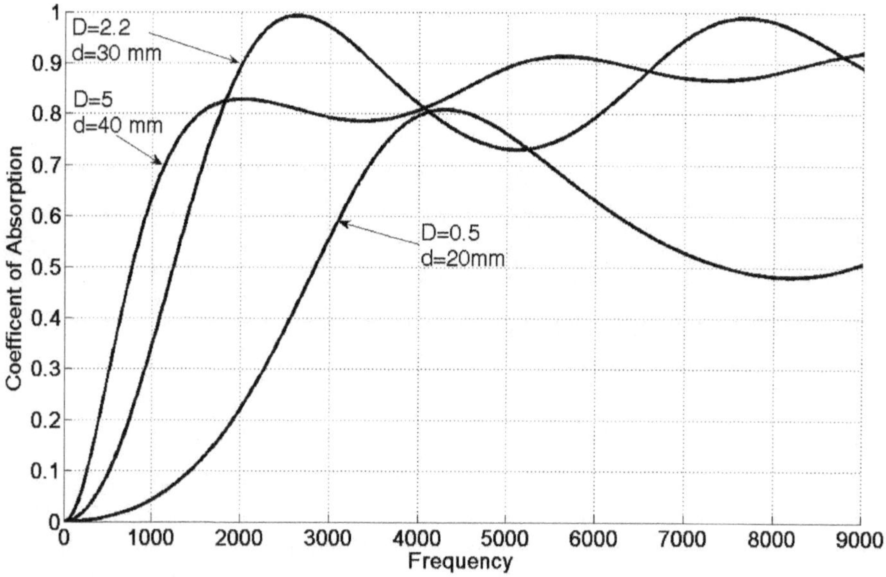

Fig. 5.13. Examples of theoretical calculations based on the simple Rayleigh-model for different values of the parameter D and thickness d

The shape of the α (ω) curve is governed by a dimensionless parameter D.

$$D = \frac{\Xi\ d}{\rho\ c}. \tag{5.11}$$

This parameter controls the damping of the sound wave in the porous medium. Small values of D < 0.5 result in low and strongly fluctuating absorption, while high values of D result in a smooth shape with low absorption at low frequencies. An optimal value of around D ≈ 2 leads to a curve-shape with α ≈ 1 at λ/4 and little variation for higher frequencies. With this simple theory we are not able to fit all the measurements. A more detailed treatment of this problem is found in [1] and [5], where Johnson introduced the concept of dynamic tortuosity and effective density.

Compared to the simple theory, thermal effects due to the adiabatic compression and thermal exchange processes are taken into account. In addition to the parameters defined above, the tortuosity is used to account for the fact that the tubes are not straight lines and have complicated shapes. The exact formulas to calculate the impedance Z(ω) and the absorption coefficient α of the porous material are given in the appendix.

The theoretical description neglects all effects due to an interaction of the sound waves with the rigid frame of the porous absorber. These effects are covered by the theory of Biot [2].

5.2.5 Results

In the present work the absorption coefficients for two different types of MHSS with varying thickness were measured in the frequency range from 500 Hz to 3500 Hz. The two types of MHSS consisted of spheres with diameters of 1.5 mm and 2.5 mm.

Table 5.1. Optimal values for parameters used for adjusting the theory in the appendix to measured data

Parameter		MHSS 86 (exp. Value)	MHSS 113 (exp. Value)
σ:	porosity	0.84 (0.6)	0.37 (0.4)
Ξ:	flow resistance	17600 [Ns/m^4]	13500 [Ns/m^4]
χ:	structure form factor	3.25	2.2
c_v:	shape factor of the tubes; viscous effects	1.7	0.16
c_t:	shape factor of the tubes; thermal effects	2.0	2.7

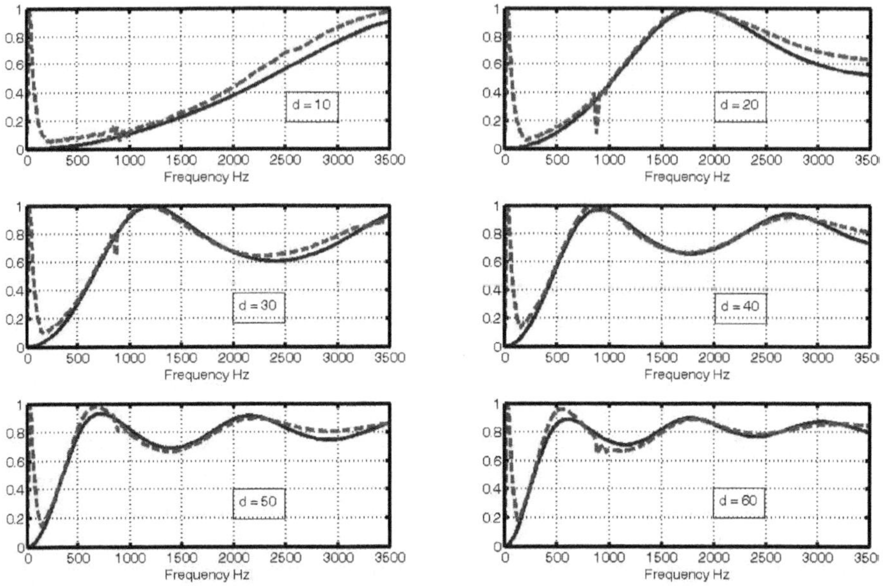

Fig. 5.14. Sample MHSS 86; Coefficient of Absorption - (- - -measurement; — theory)

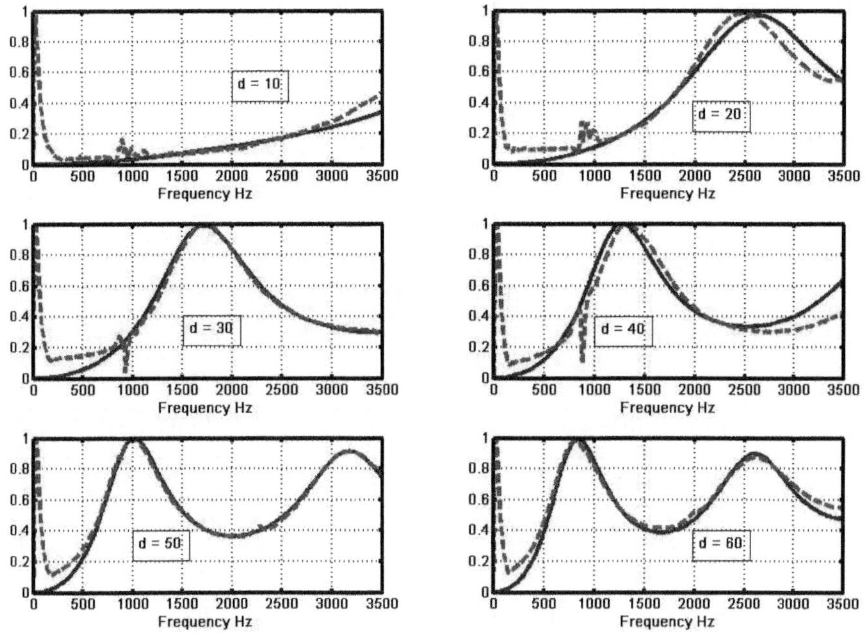

Fig. 5.15. Sample MHSS 113; Coefficient of Absorption - (- - - measurement; — theory)

For both types of MHSS samples, we fit the measured data with the theory of Champoux and Allard[1] mentioned above. An optimal set of parameters is shown in Tab 5.1.

From figure 5.14 and 5.15 one can clearly see the effect of the increasing thickness. The first maximum for the absorption is reached at lower frequencies for increasing thickness. This is the case if $\lambda/4$ is equal to the sample thickness d. The following maxima are found at odd multiples of $\lambda/4$.

Comparing the two samples with d = 60 mm, the sample MHSS 86 clearly shows a broadband absorption above f = 500 Hz with little variation in frequency. The absorption is always greater than $\alpha = 0.7$. Sample MHSS 113 shows the first maximum around f = 800 Hz, but in contrast it shows large variation and a decrease to $\alpha = 0.4$. One parameter – the porosity σ - was determined separately by experiments and can be compared with the values, which are used to fit the data by theory. The porosity for specimen MHSS 113 is $\sigma = 0.4$ from experiments and $\sigma = 0.37$ in the calculation. For MHSS 86 we find $\sigma = 0.6$ from experiments and $\sigma = 0.84$ in the calculation.

A possible explanation of this discrepancy between theory and experiment could be deduced from a microscopic analysis of the sphere walls. In Fig. 5.16 one can clearly see holes in the wall (red arrow) of the small spheres of MHSS 86. This could account for an additional porosity which is not measured by the usual experimental methods, which use low viscosity fluids which fill the space between the spheres.

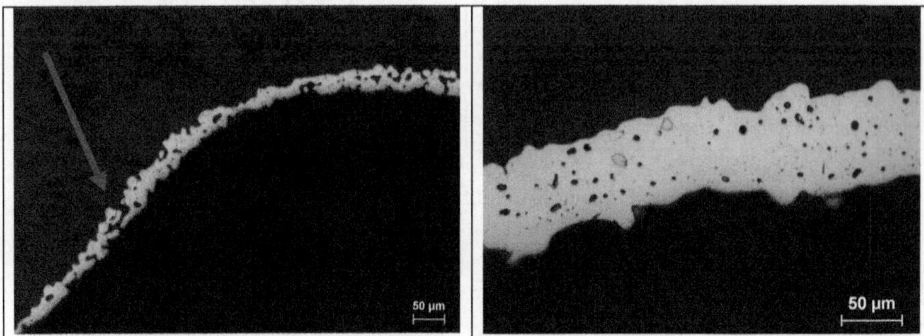

MHSS 86; Sphere Diameter = 1.6mm ; Wall-Thickness = 29.409µm	MHSS 113; Sphere Diameter = 2.7mm; Wall-Thickness = 67.827µm

Fig. 5.16. Microscopic view of the sphere-wall

Comparing the results with absorption measurements of metal foams we present some results from the work of several authors (see Fig. 5.17).

For similar sample thickness of 20 mm, the first maximum is also found around 2 kHz for two examples and at slightly higher values in the other cases.

Therefore, we can conclude that for the acoustical absorption, open celled metal foams behave very similar to MHSS.

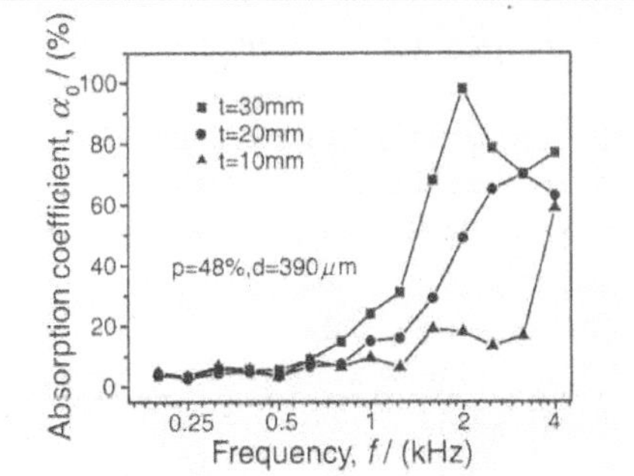

Effect of sample thickness for Lotus Type porous Magnesium (porosity 48%, pore size 390 μm); from Xie et al. [7];

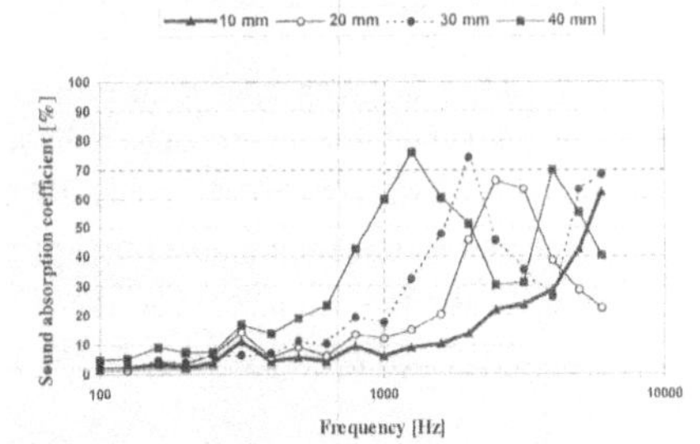

Effect of sample thickness for AlSi sponge(pore size 4 – 5mm); from [3];

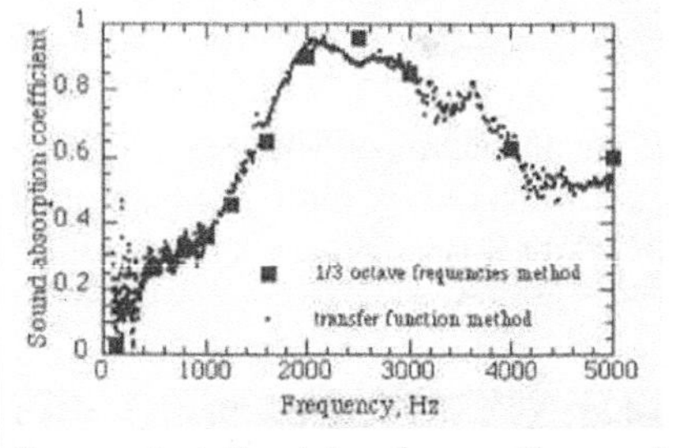

Open celled Aluminium foam with sample thickness of 20 mm (porosity 57% - 61%, pore size 0.5 mm); from Han et al. [4];

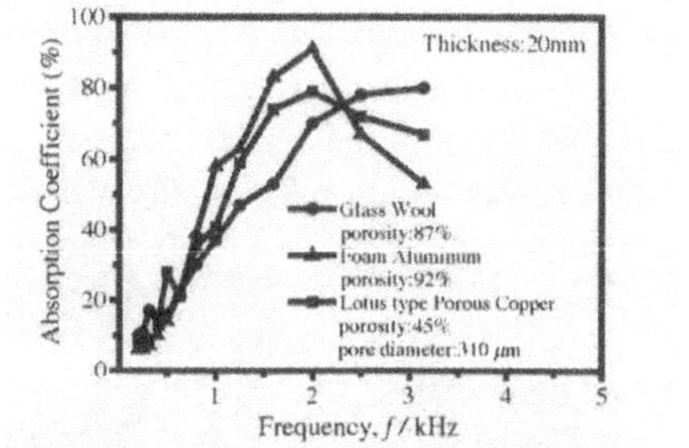

Comparison of various materials; from Xie et al. [8]

Fig. 5.17. Comparison of different results of open celled metal foams

5.3 Summary

The absorption coefficient α of MHSS was measured with the impedance tube in the range from 500 Hz to 3500 Hz. The absorption coefficient shows a very typical dependence on frequency with maxima and minima that can be explained by theory as an effect of an energy loss due to frictional and thermal effects. The detailed shape and frequency dependence is influenced strongly by material parameters such as sample thickness, pore size, porosity and sphere diameter, which influence the flow resistance Ξ. It is possible to reach high absorption over a wide range of frequencies.

5.4 Conclusion

MHSS are broad-band sound absorbing materials. By varying different technological parameters, the ability for sound absorption can be adapted to various noise spectra. These parameters include thickness, sphere diameter, packing density, mixture of different sphere diameters and others. Thus the material can be used for an efficient and new design of noise control systems.

Acknowledgments

We thank Glatt GmbH / hollomet GmbH (Germany) for providing us with material for the vibrational as well as for the acoustical measurements.

The microscopic investigations at the computer tomograph were done at Aalen University in the Institute of Material Science headed by Prof. Dr. Schneider.

References

[1] Allard, J.F.: Propagation of sound in porous media. Elsevier Applied Science, Amsterdam (1993)
[2] Biot, M.A.: The theory of propagation of elastic waves in a fluid-saturated porous solid. J. Appl. Phys. 26, 182–185 (1955)
[3] Degischer, H.P., Krizst, B.: Handbook of Cellular metals. Wiley-VCH, Weinheim (2002)
[4] Han, F., Seiffert, G., Zhao, Y., Gibbs, B.: Acoustic absorption behavior of an open-celled aluminium foam. J. Phys. D Appl. Phys. 36, 294–302 (2003)
[5] Johnson: Theory of dynamic permeability and tortuosity in fluid saturated porous media. J. Fluid Mech. 176, 379–402 (1987)
[6] Möser, M., Müller, H.A.: Technische Akustik, 3rd edn. Springer, Berlin (2004)
[7] Xie, Z., Teruyuki, I., Yoshiyuki, O., Hideo, Y.: Characteristics of sound absorption in Lotus Type Porous Magnesium. Japanese J. Appl. Phys. 43, 7315–7319 (2004)
[8] Xie, Z., Teruyuki, I., Yoshiyuki, O., Hideo, Y.: Sound Absorption characteristics of lotus-type porous copper fabricated by unidirectional solidification. Mat. Sci. Eng. A-Struct. 386, 390–395 (2004)

Appendix

According to Champoux and Allard, who introduced the concept of characteristic dimensions, it is possible to calculate the bulk modulus and the effective density for porous media.

The meaning of the symbols in the following formulas are:

$\rho_{\tilde{0}}$: density of air
P_0: air pressure
σ : porosity
χ : structure form factor
ω : frequency
Ξ : flow resistance
μ : dynamic viscosity of air
c_v : shape factor of the tubes; viscous effects
c_t : shape factor of the tubes; thermal effects
Pr: Prandtl number
γ : adiabatic exponent for air

Viscous effects:

$$\Lambda_v = \frac{1}{c_v}\sqrt{\frac{8\chi\mu}{\sigma\,\Xi}} \qquad G_v(\omega) = \sqrt{1+\frac{j4\chi^2\mu\rho_0\omega}{\sigma^2\Lambda_v^2\Xi^2}} \qquad \lambda_v = c_v\sqrt{\frac{8\chi\rho_0\omega}{\sigma\,\Xi}}$$

Thermal effects:

$$\Lambda_t = \frac{1}{c_t}\sqrt{\frac{8\chi\mu}{\sigma\,\Xi}} \qquad G_t(\omega) = \sqrt{1+\frac{j4\chi^2\mu\rho_0\omega\Pr}{c_t^4\sigma^2\Lambda_t^2\Xi^2}} \qquad \lambda_t = \frac{1}{c_v}\sqrt{\frac{8\chi\rho_0\omega}{\sigma\,\Xi}}$$

Bulk-modulus:

$$K(\omega) = \frac{\gamma\, p_0}{\gamma-(\gamma-1)\left[1+\frac{8}{j\Pr\lambda_t^2}G_t(\omega)\right]^{-1}}$$

Effectiv density:

$$\rho_{eff}(\omega) = \rho_0\chi\left[1+\frac{8c_v^2}{j\lambda_v^2}G_v(\omega)\right]$$

Impedance of the medium:

$$Z_c = \sqrt{\rho_{eff}(\omega)K(\omega)}$$

Wave vector in the porous medium:

$$k_a = \omega\sqrt{\frac{\rho_{eff}(\omega)}{K(\omega)}}$$

Impedance of sample with thickness d:

$$Z = \frac{-jZ_c ctg(k_a d)}{\sigma}$$

Coefficient of absorption:

$$\alpha = \frac{4\operatorname{Re}\left(\frac{Z}{\rho c}\right)}{\left[\operatorname{Re}\left(\frac{Z}{\rho c}\right)+1\right]^2+\left[\operatorname{Im}\left(\frac{Z}{\rho c}\right)\right]^2}$$

6

Thermal Properties of Hollow Spheres

Eusebio Solórzano, Miguel Angel Rodriguez-Perez, and José Antonio de Saja

CellMat Group, Condensed Matter Physics Department, Faculty of Science,
University of Valladolid 47011, Valladolid, Spain
esolo@fmc.uva.es, marrod@fmc.uva.es, sajasaez@fmc.uva.es

Abstract. A good knowledge of the thermal properties, and more specifically of the thermal conductivity, is determinant for a proper materials selection and technical design of real applications using metallic hollow spheres structures (MHSS). As a consequence, it is necessary to know with precision which are the MHSS parameters that influence on these properties. The modelling presented in this chapter is carried out taking into account the different processing possibilities for these materials. Sphere size, sphere thickness, sphere packing, and metallic/adhesive bonding are the main parameters considered in this exhaustive study. To this end analytic models are combined to obtain a final expression that considers all the above-mentioned processing parameters. Additionally, thermal conductivity of the MHSS is compared with other cellular metallic materials produced with different technologies and based on several alloys such aluminium, copper, magnesium and zinc.

Keywords: Thermal conductivity; thermal properties; sphere packing; cellular metals; foams.

6.1 Introduction

6.1.1 Thermal Properties

We can cite three main thermal properties of interest in materials science: thermal conductivity (λ), thermal diffusivity (κ) and specific heat (C_p). Nevertheless, these properties are not independent and are related, considering also the mass density, ρ, through the following equation:

$$\kappa = \frac{\lambda}{\rho \cdot C_p} \tag{6.1}$$

On the other hand, it is admitted that specific heat of a composite can be exactly determined by using the rule of phases [18]:

$$C_p = C_p^{\text{phase-1}} \cdot \%mass_{\text{phase-1}} + C_p^{\text{phase-2}} \cdot \%mass_{\text{phase-2}} + \ldots \tag{6.2}$$

In case of cellular metals, only two phases are present (metal and gas). Then, if we consider that the $\%mass_{\text{gas}}$ in a cellular metal is negligible (compared to $\%mass_{\text{met}}$) then, the specific heat (per mass unit) of a cellular metal is approximately the same of the constituent alloy itself.

A. Öchsner, C. Augustin (Eds.): Multifunctional Metallic Hollow Sphere Struct., EM, pp. 89–107.
springerlink.com

Under these conditions it is only necessary to determine the thermal conductivity or the diffusivity to know these three important thermal properties. Typically, thermal conductivity is the property determined since measurement of thermal diffusivity involves more complexity in terms of experimental set up.

We can define the thermal conductivity as the material's intrinsic characteristic that expresses its ability to conduct the heat. Its SI units are W/m·K and, as shown previously, commonly it is denoted with the Greek symbol λ.

6.1.2 Experimental Methods to Determine Thermal Conductivity

It is important to know the heat transfer properties for a wide variety of scientific and industrial applications. Due to this reason, a number of different experimental techniques have been developed to measure the thermal conductivity for different experimental conditions and for different materials [44,45]. The process of measurement of this property is complicated by the fact that, in several practical situations, different heat transfer mechanisms (conduction, convection and radiation) are present.

The overall range of thermal conductivity for cellular metals is approximately two orders of magnitude above approximately 1 W/m·K. In practice, there is no "ideal" method of measurement for this range of λ. It is necessary to choose a procedure depending on factors such as the expected conductivity, the shape of the material, its density, the available sample size (needing that sample is representative of the bulk material or the application).

In order to measure thermal conductivity or a related property by a steady-state or a transient method, the experimental arrangement must simulate a solution to the basic heat conduction equation for a homogeneous isotropic solid:

$$\frac{1}{k}\frac{\partial T}{\partial t} = \nabla^2 T + \frac{A(x,y,z,t)}{\lambda}, \tag{6.3}$$

Where T is the temperature, t is the time, k is the thermal diffusivity and $A(x,y,z,t)$ is the heat generation per unit volume and unit time in the medium. Note that Eq. (6.1) may induce a change in Eq. (6.3).

For longitudinal unidirectional heat flow, no radial losses and without power supplied into the solid, analysis of the steady-state term leads to a linear dependence between temperature and dimension in the heat flow direction. Then the heat per unit time and unit area through a sample can be determined by Fourier's law:

$$Q_a = G\lambda\frac{\Delta T}{d}, \tag{6.4}$$

where Q_a is the heat flow generated by the application of a temperature difference ΔT in the two sides of the sample material, separated over a distance d (the material thickness), and G is a constant, evaluated by calibration, for a given apparatus. Figure 6.1 shows a schematic diagram of the typical equipment based on this equation.

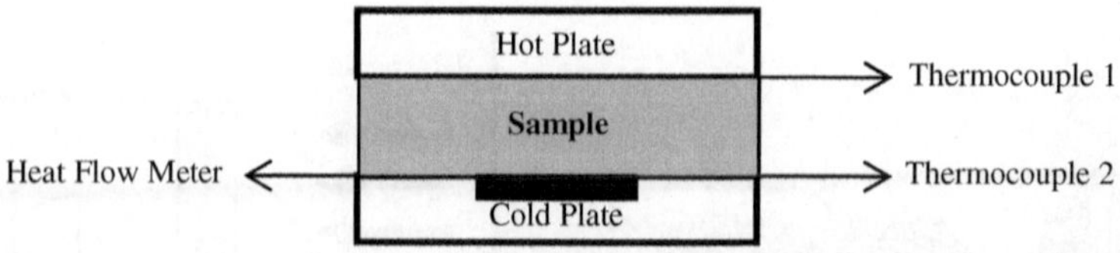

Fig. 6.1. Schematic representation of typical standard equipments based on Fourier's law

The steady-state techniques are commonly used and, on the other hand, they are supported by standard methods (ISO & ASTM) based on this procedure [4, 25, 26].

In general, the heat generated in the upper plate is not all conducted to the lower plate. Thus, it is necessary in all cases to account for heat losses. Moreover, not always the heat flow is normal to the heat surfaces and there is a small gap between the two heater surfaces and those of the sample. This gap contributes to the reduction of the heat transferred (interfacial heat transfer resistance). On the other hand, although the current equipments are relatively fast in operation, there is still a need to reduce such times, particularly for quality control applications.

Transitory methods are based on the analysis of the transient term solution of Eq. (4.3), which relates change in temperature with time. Transient hot wire, transient hot strip, transient plane source and laser flash methods are probably the most known techniques based on measuring the sample behaviour in the transient regime of heat flow. Nevertheless, laser flash method differs from the others since it is a non contact method and determines the thermal diffusivity of the sample instead of the thermal conductivity. Transient plane source (TPS) can be considered as the evolution of both transient hot wire and transient hot strip (by combining some aspects of the transient hot probe method, not mentioned before). Therefore we can consider the TPS method as the most representative of these transient contact methods.

In the TPS method a round and plane heat source is used (see Fig. 6.2). It behaves as a transient plane source working simultaneously as a temperature sensor. This element consists of an electrical conducting pattern of thin nickel foil in the form of double spiral, inserted between two insulating layers made of Kapton. The total thickness of the ensemble is only a few tens of microns. The TPS element is located between two samples with both sensor faces in contact with the two samples surfaces (see Fig. 6.2-left-). Two samples of similar characteristics are required for this purpose. During the testing a constant electric power is supplied to the sensor and the temperature increase is recorded. To relate the change in temperature with time, the equation for the heat conduction, assuming the conditions reported by Log et al. and Gustavsson et al., is applied [30, 23].

Finally, the laser flash system is based on a laser beam pulse (typically $< 1\,\text{ms}$) focused on one of the faces of a relatively thin sample while the temperature increase in the opposite parallel face is recorded by a contactless method (IR pyrometer). From this temperature increase it is possible to determine the thermal diffusivity of the sample. The schematic description of the measurement procedure is shown in Fig. 6.2-right- [7].

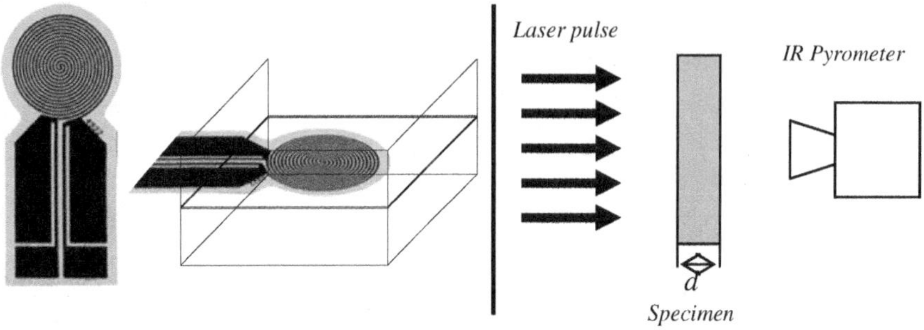

Fig. 6.2. Schematic draws for transient plane source (left) and laser flash method (right)

Transitory methods present several advantages compared to steady state ones, for example, it is possible to obtain values of thermal conductivity, thermal diffusivity and specific heat, simultaneously. The range of measurement comprises five orders of magnitude (0.01 to 400 W/m·K). Among other secondary benefits, these methods are faster, can be used to determine properties of inhomogeneous and/or anisotropic materials and offer the ability to measure in small samples compared to big plates conventionally needed for steady-state standard methods.

On the other hand, it is important to remark that transient methods are not well standardized (testing procedure). Nevertheless, in the last years some standard procedures have been approved for determining the thermal properties by the laser flash method [5, 6] and several efforts are being developed to standardise the TPS method [46]. It is also interesting to comment that thermal properties of non-conventional new materials have been determined by using transient methods in the last years [2, 11, 21, 32, 36, 39-42].

In addition, several works have showed that the steady-state technique and the transient techniques give comparable results [3, 22].

6.1.3 Thermal Conductivity in Cellular Materials

Thermal properties of cellular metals differ from those of dense solids; it is known that the thermal conductivity, thermal diffusivity and specific heat per unit volume of cellular metals are smaller than those of conventional solid materials [1]. These differences have important consequences on several aspects of the processing, post-processing and applications of cellular metals as it will be described at the end of this chapter.

It is well-known that λ changes extensively depending on the density, cellular structure and morphology of the matrix, etc. [1, 12, 19, 20, 28]. Moreover, thermal conductivity also exhibits a strong dependence on temperature and pressure.

In order to analyse the thermal properties of cellular solids it is necessary to evaluate the three mechanisms whereby, under the influence of a temperature gradient, energy can be transported from one region of space to another. These mechanisms are radiation, conduction through the solid and gaseous phase and convection, as expressed in Eq. (6.5):

$$\lambda = \lambda_{\text{conduction}} + \lambda_{\text{radiation}} + \lambda_{\text{convection}} \tag{6.5}$$

Nevertheless, the weights of these three mechanisms are far from being similar and in many situations only conduction needs to be taken into account. This is the case of most of cellular metals since it has been demonstrated that radiation in cellular metals does not contribute significantly at ambient temperatures [31]. On the other hand, the convection mechanism is assumed to be negligible considering that: part of the air is occluded inside the hollow sphere, the temperature is relatively low and the interconnected fraction of air is low, with narrow interconnections [24].

As a conclusion only the conduction through solid and gaseous phases is considered in this chapter. This term can be predicted by analytic/empirical models as it will be described in next section.

6.2 Thermal Conductivity as a Function of MHSS Parameters

Metallic hollow sphere structures (MHSS) are a group of cellular metals characterised by easily reproducible geometry and therefore consistent physical properties. They are not foams neither lattice materials since they present a regular geometry (sphere) but, usually, a random arrangement of these individual entities. This apparent paradox (i.e. regularity and randomness) conditions the modelling of their thermal conductivity.

Thermal properties of hollow spheres materials depend on many factors. Their defining characteristics can be divided into two different groups, consistently with the previous-mentioned characteristics. The first group contains all the characteristics related to the structure regularity, i.e. each single sphere: external diameter and thickness of the metallic shell, metal matrix of the sphere and microstructure of the constituting alloy.

The second group includes those characteristics associated to the way each individual hollow sphere is connected to the neighbouring spheres forming the structure, i.e. topology. Sphere packing and the joining technology used to assemble the single hollow spheres to interdependent structures (sintering, brazing, and adhering) are the main characteristics in this second group. In case of joining with adhesives, contact angle and the kind of adhesive used need also to be considered to model thermal properties.

Macroscopic properties as well as the relative density (or porosity) are connected with both the individual sphere characteristics but also depends on the second group. Up to now, all the models employed to calculate properties of hollow spheres were based in mono-sized hollow spheres and conventional crystallographic lattice topologies, i.e. cubic primitive (CP), body centred cubic (BCC), face centred cubic (FCC), etc. This means in practice that random topology was not considered in order to simplify the problem (to obtain symmetry simplifications in later numerical analysis). The use of lattice arrangements presents certain incoherencies if the results are compared to real HS structures. As an example the density does not match exactly with real density of these materials and properties present preferred directions [14, 35]. The approach considered here uses the packing density and the coordination number for random arrangements in order to obtain more representative results. Nevertheless, as the approach is valid for latticed arrangements results will be compared to those, always using simple analytic modelling.

It is important to mention that the different MHSS parameters (metal matrix, sphere geometries, type of joining technology, etc.) would give as a result too many combinations to consider in this chapter and only the most representative types of MHSS are included in the discussion.

6.2.1 MHSS Main Characteristics

Shell Characteristics

Base material and processing: Although HS can be produced from many diverse materials they are mostly fabricated from *carbon steel* (cheaper, no specific sintering conditions) and *stainless steel* with thermal conductivities of 50 and 13.8 W/m·K, respectively. These two materials will be taken into account in the later analysis presented in this chapter. We have selected the powder metallurgical –binder assisted– process [17] to produce steel sintered spheres since it is the most typical process. Although shells produced by using this process may present a certain micro-porosity we will assume that shells are fully dense.

Thickness and diameter: Elemental entities forming the structure have been chosen to be monosized with diameters ranging from 1.5 to 10 mm. On the other hand, shell thickness could be produced independent from the shell diameter but in practice the thickness is increased with increasing diameter (large spheres with thin shells are weak). In our model it will be considered a linear dependency of thickness (t) with sphere radius (R) [34]. To express the independency of thickness with diameters different constants of proportionality between these two parameters have been selected (t/R = 0.02, 0.033, 0.054, 0.1013). It is important to mention that HS typically produced present a t/R ratio in the lower range modeled (i.e. 0.02-0.054) which means that producers manufacture light-weight materials more than stiff ones.

Topological Characteristics

Type of structures: Three types of structures are studied. Probably the most characteristic structure types are *sintered hollow spheres structures* (SHSS) that present only two phases (steel+air). On the other hand, adhering is an economic way of joining and also is attractive for a wide range of potential applications. Other interesting advantage is the possibility of using the adhesive as an additional parameter to modify the mechanical properties and the thermal insulation. There are two different possibilities of obtaining MHSS using an adhesive as third phase: in *syntactic adhesive* morphology (SAHSS) the spheres are completely embedded within the adhesive layer. In case of *partial-adhesive* (PAHSS) the adhesive is concentrated at the contact points of spheres forming adhesive necks so this HS type present interconnected porosity. In the models we assume that the bonding phase is a polymeric adhesive and therefore it presents a low thermal conductivity (0.22 W/m·K).

Spatial arrangement: We will consider four different spatial arrangements: CP, BCC, FCC and random. Through their characteristic packing densities and coordination numbers we will obtain the thermal conductivity as described in next section. Although syntactic structures do not necessarily reach high filling factors, packings considered for these structures are those corresponding to the spatial arrangements previously mentioned. On the other hand, the distance between spheres is supposed to be zero, i.e. spheres are in contact. This is obvious in case of SHSS and close from experimental findings in case of PAHSS [16] (spheres very near from each other).

The random sphere arrangement considered in this work is, actually, a dense random packing, (DRP) with a sphere-filling efficiency of 0.64 [27]. This value is different from packing factors of 0.52, 0.68 and 0.74 for CP, BCC and FCC sphere arrangements, respectively. A filling factor of 0.64 can be considered as the maximum efficiency when spheres are subjected to mechanical vibration during processing and under other conditions probably this value tends to be lower (as well as the average coordination number of 6.2 [29]). Nevertheless, this assumption allows for comparison with other arrangements and can be further easily modified by only slightly reducing its value for the packing factor (and the corresponding coordination number) in case it is necessary. Table 6.1 shows the characteristic packing factors and coordination numbers of the different arrangements considered in this chapter. Please note the differences in factors and coordination numbers that make, in practice, impossible predicting the real density of the materials by using lattice arrangements, if we consider that random packing is the closest to experimental values of porosity. Only the BCC structure presents similar values to the DRP which explains that BCC better predicts real properties of HS [35].

Table 6.1. Packing factors and coordination number of diverse spatial arrangements

Spatial arrangement	Packing factor(*PF*)	Coordination number(*CN*)
CP	0.52	4
BCC	0.68	6
FCC	0.74	8
DRP	0.64	6.2

6.2.2 Models to Estimate Thermal Conductivity

Analytical models utilised to estimate thermal conductivity use volume fractions and conductivities of each phase. Therefore first are presented the volume fractions calculations to later implement models for thermal conductivity considering the calculated volume fractions.

Volume fractions

Metal volume fraction (V_{met})

The procedure to estimate the fraction of metal in the whole volume is relatively simple. Having selected a packing efficiency its value should be multiplied by the factor $C=[1-((R-t)/R)^3]$, which represents the fraction of metal in each single sphere –sphere shell– and it is calculated by subtracting the internal to the external sphere volumes.

$$V_{met} = PF \cdot C \quad \text{where} \quad C = 1 - \left((R-t)/R\right)^3 . \qquad (6.6)$$

The above-mentioned assumptions for the relation t/R could be applied over the calculated solid fraction to simplify it.

This model allows comparing the experimental with the theoretical relative density by introducing the R and t values found in the real structure. If results does not match exactly, differences could be explained in terms of slightly lower sphere filling factors and coordination numbers. Moreover, the small influence of shell densification factor –possible presence of microporosity– and shell deformation in the contact points – characteristic for this kind of processing technique [34]- could be corrected [43].

Adhesive volume fraction (V_{adh})

As told in previous sections two different structures from those studied in this chapter include an adhesive phase. The calculation for SAHSS is relatively simple and can be obtained by considering that the external free volume (1-*PF*) is fully occupied by the embedding adhesive material. Of course, the presence of microporosity in the adhesive phase is supposed to be zero but could be taken into account by using a correcting factor.

$$\left[V_{\text{adh}}\right]^{\text{SAHSS}} = 1 - PF \, . \tag{6.7}$$

On the other hand, to calculate the total amount of adhesive for PAHSS, the quantity of adhesive phase in a *single neck* needs first to be calculated. This value can be estimated by using the following model developed by Sanders and Gibson [34] which depends only on sphere radius (R) and contact angle (θ):

$$V^{1}_{\text{adh}} = 2R^{3} \cdot 2.74 \cdot 10^{-7} \cdot \theta^{3.41} \, . \tag{6.8}$$

To calculate the *total amount of adhesive in a single sphere*, the previous value needs to be multiplied by *CN,* characteristic of each spatial arrangement. On the other hand, we can assign an average value for the contact angle between sphere and adhesive. This way, considering a contact angle adhesive-sphere of 25° and the mentioned average number of connections, the total amount of adhesive in a single sphere is:

$$\overset{\circ}{V}_{\text{adh}} = CN \cdot 0.032 \cdot R^{3} \, . \tag{6.9}$$

Finally, to calculate the total fraction of adhesive it is necessary to normalize this volume to the total of the representative unit cells (with side lengths –*SL*– of $2R$, $4/\sqrt{3}R$, $2\sqrt{2}R$, respectively for CP, BCC and FCC) and multiply by two times the total number of spheres ($2n$) contained in the representative unit cell (1, 2 and 4 for CP, BCC and FCC, respectively). Factor 2 express that junctions are actually constituted by two necks according to Eq. (6.8). Equation (6.10) shows the final expression for the *total volume fraction of adhesive* for PAHSS:

$$\left[V_{\text{adh}}\right]^{\text{PAHSS}} = \frac{NC \cdot 2n \cdot 0.032R^{3}}{(SL)^{3}} \, . \tag{6.10}$$

Above is described the case for lattice arrangements since representative unit cells are well-known but in the particular case of DRP there is not a clear representative unit cell; therefore we will use an *average* representative unit cell. In a simplistic approach this cell could be assumed to include only a single hollow sphere and 6.2 adhesive necks in concordance with the *CN* for this arrangement. Nevertheless, volume fraction of a single sphere in a cubic unit cell is 0.52 but the real volume fraction for DRP is 0.64; this approach seems to be not valid and it is necessary to apply a more precise correction. Therefore we use the topology continuity conditions to predict the average number of spheres contained in the average unit cell. Figure 6.3 a) shows relative *PF* (normalized to primitive arrangement -*PF/0.52*-) versus the number of spheres contained in the unit cell to calculate the average number of spheres in the unit cell for a random packing. After getting this value, it is possible to calculate the side length of the average unit cell (see Fig. 6.3 b). Although this procedure is not exact, it allows obtaining an estimative approach to predict the amount of adhesive in the DRP unit cell.

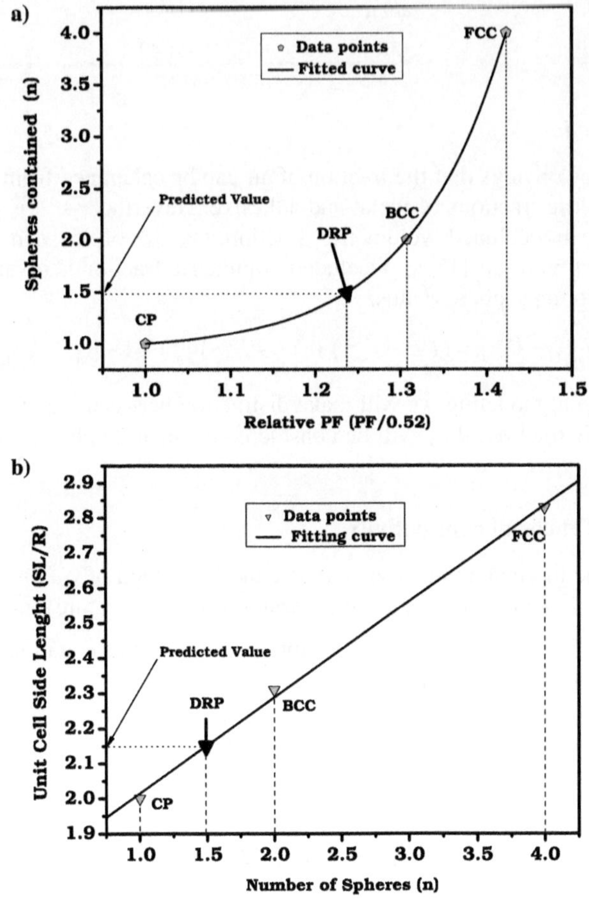

Fig. 4.3. a) Prediction of spheres contained in the unit cell for DRP b) Prediction of average unit cell side length for DRP

Table 6.2 summarizes values to be used in the calculations of adhesive volume fraction for PAHSS. Equation (6.10) shows that sphere arrangement largely conditions the volumetric fraction of adhesive. Actually, in the particular model developed (considering a constant contact angle of 25°) the amount of adhesive is independent from the sphere radius and only depends on the spatial arrangement and its derived *CN*, *n* and *SL* values.

Table 6.2. Side length and number of spheres within unit cells of different spatial arrangements

Spatial arrangement	Unit Cell Side Length (*SL*)	Number of Spheres (*n*)
CP	2R	1
BCC	$4/\sqrt{3}R$	2
FCC	$2\sqrt{2}R$	4
DRP	2.15R	1.49

Air volume fraction (V_{air})

Finally, it seems obvious that the fraction of air can be calculated from the subtraction of both volumetric fractions of metal and adhesive. Nevertheless, it is possible to distinguish between occluded volumetric fraction ($V_{\text{air\_O}}$) of air and interconnected volumetric fraction of air ($V_{\text{air\_I}}$). Occluded volumetric fraction is invariant in relation to packing factor and sphere radius:

$$V_{\text{air}} = V_{\text{air\_O}} + V_{\text{air\_I}} = (PF - V_{\text{met}}) + (1 - PF - V_{\text{adh}}) = 1 - V_{\text{met}} - V_{\text{adh}} \,. \tag{6.11}$$

In later thermal modeling we will make distinction between the interconnected and the occluded air fraction; they will be considered as separate phases to improve models accuracy.

Calculation of thermal conductivity

To calculate the thermal conductivity we use the formalism of *volume averaging* (expressed with $\langle\ \rangle$ notation) for the field variables *T* (temperature) and *Q* (heat flow) [37, 47]. Equations (6.12) and (6.13) express the working equations to be solved, based on volume fractions (ϕ_i).

$$\nabla\langle T\rangle = \sum_{i=1}^{n} \phi_i \langle \nabla T_i \rangle^i \,, \tag{6.12}$$

$$\lambda\nabla\langle T\rangle = \sum_{i=1}^{n} \phi_i \lambda_i \langle \nabla T_i \rangle^i \,. \tag{6.13}$$

Nevertheless, these equations require other extra equations to solve the problem since for *n* phases we have two equations and *n*+1 unknowns.

Two-phase systems

In the particular case of two-phase systems, Eqs. (6.14) and (6.15) can be rewritten as:

$$\nabla\langle T\rangle = V_{air}\langle\nabla T_{air}\rangle^{air} + (1-V_{air})\langle\nabla T_{met}\rangle^{met}, \tag{6.14}$$

$$\frac{\lambda}{\lambda_{air}}\nabla\langle T\rangle = V_{air}\langle\nabla T_{air}\rangle^{air} + (1-V_{air})\frac{\lambda_{met}}{\lambda_{air}}\langle\nabla T_{mat}\rangle^{met}. \tag{6.15}$$

Different closing equations lead to well known models. As an example, if closing equations are (6.16) and (6.17) the system is reduced to the simplest *series* and *parallel* models, respectively.

$$\langle\nabla T_{air}\rangle^{air} = f\langle\nabla T_{met}\rangle^{met} \quad / \; f = \frac{\lambda_{met}}{\lambda_{air}}, \tag{6.16}$$

$$\langle\nabla T_{air}\rangle^{air} = f\langle\nabla T_{met}\rangle^{met} \quad / \; f = 1. \tag{6.17}$$

On the other hand if f (Eqs. (6.16) and (6.17)) is assigned to have a value of 2/3, then the final equation is the well-known Maxwell model:

$$\lambda = \frac{2V_{air}\lambda_{air} + \lambda_{met}(3-2V_{air})}{3-V_{air}+V_{air}(\lambda_{met}/\lambda_{air})}. \tag{6.18}$$

This equation will be used to predict values for thermal conductivity in multiphase structures. It is important to mention at this point that, in the next section air occluded and air interconnected are considered as two different phases in all cases. Therefore, under this criterion, SHSS present three phases (V_{air_O}, V_{met}, V_{air_I}), SAHSS present three phases again (V_{air_O}, V_{met}, V_{adh}) and PAHSS is constituted by four phases (V_{air_O}, V_{met}, V_{air_I}, V_{adh}).

Multiphase systems

To estimate the effective thermal conductivity of multiphase systems we could start again from the volume averaging equations considering 3 and 4 phases respectively. Nevertheless, in order to shorten the equations development we will go one step forward and then we start directly with the Maxwell equation applying it recursively.

This way to calculate the thermal conductivity of three-phase systems (SHSS and SAHSS) we only need two equations. The first one -Eq. (6.19)- is used to predict the thermal conductivity of shells and their internal air fraction combined together. Later on, Eq. (6.20) is applied to combine both external phase (λ_{ext}, V_{ext} for air or adhesive) and the spheres (shell + internal air).

$$\lambda_{sph} = \frac{2V^*_{met}\lambda_{met} + \lambda_{air}(3-2V^*_{met})}{3-V^*_{met}+V^*_{met}(\lambda_{air}/\lambda_{met})}, \tag{6.19}$$

$$\lambda = \frac{2V_{ext}\lambda_{ext} + \lambda_{sph}(3-2V_{ext})}{3-V_{ext} + V_{ext}(\lambda_{sph}/\lambda_{ext})}. \tag{6.20}$$

Note that metal volume fraction (V^*_{met}) in Eq. (6.19) is normalized to relative volume fraction ($V_{air\_O} + V_{met}$) of phases considered in that equation. V_{sph} and V_{ext} are respectively *PF* and *1-PF*.

On the other hand, the calculation of thermal conductivity for four-phase systems (PAHSS) is based in the same formalism: first are combined phases two by two ($V_{met} + V_{air\_O}$; $V_{adh} + V_{air\_I}$) and subsequently both two-phase systems are combined together to obtain final effective thermal conductivity. Equations (6.19) and (6.21) are used to calculate intermediate conductivities and Eq. (6.20) allows for calculating the final value.

$$\lambda_{ext} = \frac{2V^*_{adh}\lambda_{adh} + \lambda_{air}(3-2V^*_{adh})}{3-V^*_{adh} + V^*_{adh}(\lambda_{air}/\lambda_{adh})}. \tag{6.21}$$

Again, in Eq. (6.21), adhesive volume fraction (V^*_{adh}) is normalized to relative volume fraction for both phases considered ($V_{air\_I} + V_{adh}$) in that equation.

It is also important to mention that intermediate equations, (6.19) and (6.20), are referred to high conductive phase whereas final equation has as a reference the low conductive one.

6.2.3 Comparative Results within Maxwell Model

The result of applying above-described models is a complex multivariable study since we have considered four parameters in our investigation (kind of alloy, t/R ratio, type of structure and spatial arrangement) and most of them present 3-4 levels. Moreover, *t/R* ratio influences on the fraction of occluded porosity in combination with the topology (CP, DRP, etc.) which means in practice there is an additional intrinsic variable that is interesting to study. Therefore, at this moment the reader can understand better the reasons of having selected simple models to study these kind of materials. In practice, authors could have avoided the variable "topology" considering that the most representative of a real structure is DRP. Nevertheless, most of previous studies have been carried out by using as a basis latticed topologies [13, 15], and this latticed topologies have been included here to provide further discussions.

In order to facilitate comprehension the results have been divided trying to show the separate influence of each variable considered. To this end, results have been firstly normalized to the high conductive phase conductivity (i.e. steel) to suppress this variable and facilitate understanding.

Figure 6.4 shows the influence of shell thickness on the effective normalized thermal conductivity. It can be appreciated that the sphere thickness largely influences the conductivity in case of SHSS whereas this influence is pretty lower for structures incorporating adhesive. The thinner is the shell the nearer are the conductivities for different structures and topologies since the amount of metal tend to a minimum in case of the thinners shells. The effect of topology has an additional effect on the overall metal fraction since higher sphere packing gives as result a higher amount of metal.

Finally, it is important to remark that, SHSS apparently exhibit a linear trend in the studied range but this is not observed in case of SAHSS and PAHSS which show curved trends, so this effect can be assigned to the incorporation of an adhesive phase.

On the other hand, in Fig. 6.5, when results are plotted versus the occluded air fraction (this parameter -and metal fraction- are the only phases in common for the three types of structures) it is possible to appreciate a clear modification in the plotting pattern since curves are shifted in the x-axis for the different topologies (grey arrow is not vertical but tilted). Again, SHSS present a linear trend in the studied range as well as those structures incorporating adhesive keep them curved shape with occluded

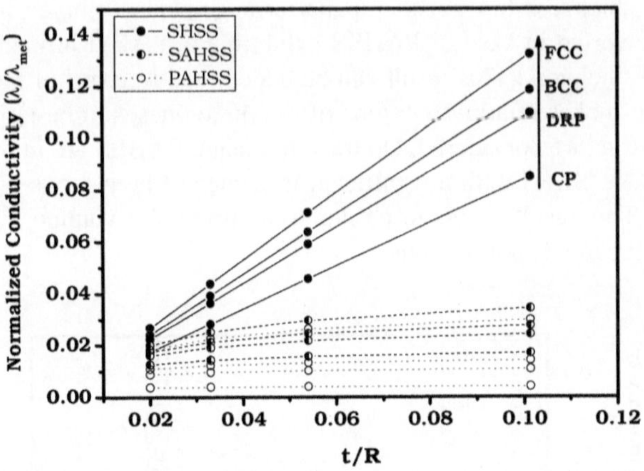

Fig. 6.4. Normalized thermal conductivity for different structures and different spatial arrangements plotted versus *t/R* ratio

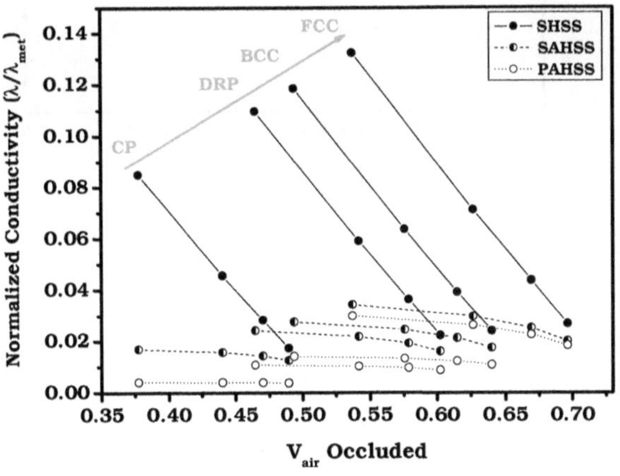

Fig. 6.5. Normalized thermal conductivity plotted versus occluded porosity (structure type and spatial arrangements are again parameters)

porosity. In addition, we can see how curves are intersected in the low thermal conductivity range. As an example, similar values of thermal conductivity can be reached for SHSS of minimum sphere thickness and SAHSS with intermediate shell thicknesses. Moreover, it is observed how topology influences the differences between SAHSS and PAHSS since curves for FCC are near for these structures and largely separated in case of CP. This effect can be assigned without doubt to the high packing factor making adhesive fraction to be not very different for SAHSS and PAHSS considering for FCC topologies.

The effect of structure on the thermal conductivity brings, in Fig. 6.6, interesting findings. It can be clearly observed that in the case of SHSS the main influencing factor is the shell thickness but not the spatial arrangement attributed to the structure. This effect is inverted in case of PAHSS exhibiting a marked influence of topology instead of shell thickness. This result can be understood in terms of the incorporation of adhesive that makes relative fractions of the three phases to be largely influenced by the type of structure considered. On the other hand, SAHSS are in the mid-way between PAHSS and SHSS with a significant influence of both factors (shell thickness and topology). This result seems to be due to an important volume fraction of adhesive that make the trends not obvious.

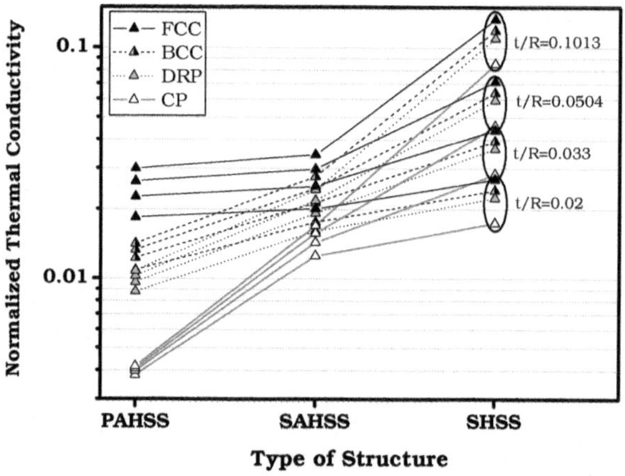

Fig. 6.6. Effect of type of structure on the thermal conductivity considering spatial arrangements and *t/R* ratio as parameters

We lastly analyze the effect of topology on the normalized thermal conductivity in Fig. 6.7. In this particular case we can observe that curves are completely parallel for the different shell thicknesses and keep a similar parallel trend for different types of structures. For a specific packing, only a couple of structures exhibit similar conductivities. Only in the case of FCC arrangement, results seem to be merged together, influenced again by the high packing factor of this topology.

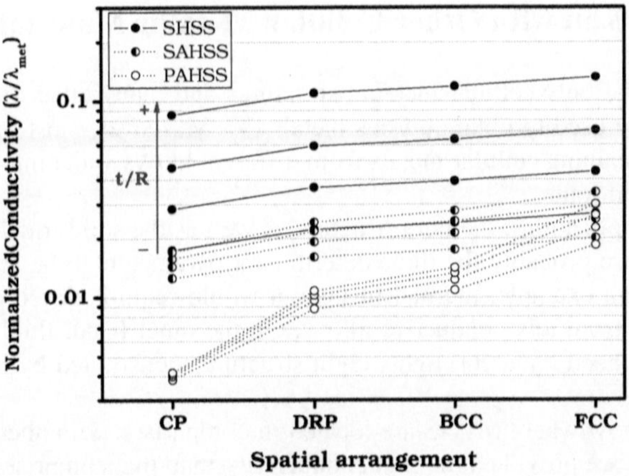

Fig. 6.7. Influence of spatial arrangement on the thermal conductivity

Once the normalized thermal conductivity results has been presented, attention is focused on the effect of the constituting alloy on the effective thermal conductivities, authors have suppress in this section the effect of possible topologies to provide a clear understanding. The selected topology has been chosen to be DRP and the results are provided in Fig. 6.8. Clearly, the kind of alloy shows capital influence in case of SHSS. Differences are also important for SAHSS and PAHSS for the different alloys but allow us to find other important result: it is possible to obtain structures of similar thermal conductivity by modifying the structure and/or the kind of alloy. It is important to mention that this graph represents the fraction of occluded porosity and probably it is more difficult to find structures with similar densities but with different metallic alloys and structures.

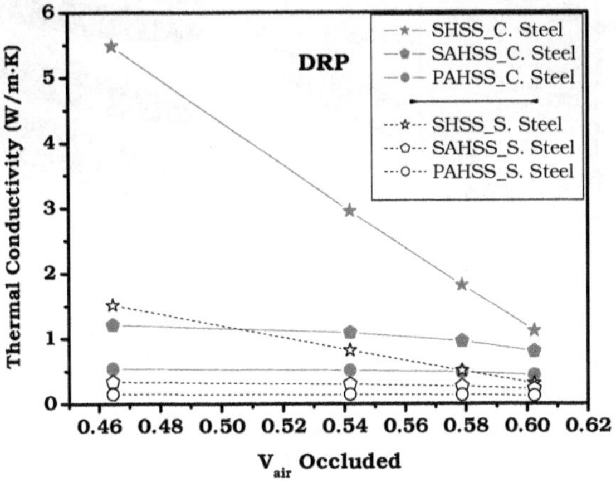

Fig. 6.8. Effective thermal conductivity versus occluded air fraction for a fixed DRP topology

6.3 Comparison with Other Cellular Metallic Materials

MHSS are not the only cellular metallic structures and many other different processing routes can be used to obtain porous metals [9]. Moreover, most of these methodologies allow obtaining cellular metals from different alloys widening the possibilities of each individual route.

In terms of applications, the pore structure as well as the solid fraction and the constituting alloys are probably the three determinant parameters to be adjusted for each application. As an example, thermal insulation would require closed cell lightweight structures made from low conductive alloys. On the other hand, fluid-based heat exchanging would need open cell lightweight structures constituted by high conductive alloys.

In order to know where MHSS are located in comparison with alternative processing technologies we provide a thermal conductivity map that comprises solid fraction, thermal conductivity, processing route and kind of alloys used for each processing route. To this end experimental values from literature as well as scaled values for other alloys are shown in Fig. 6.9. In that figure data for Lotus process [33], vacuum infiltration techniques (ERG, Recemat, etc.) [10], high pressure die casting –HPDC– on PS space holders [42], injection moulding [39], direct foaming with foaming agent (Alporas) [8] and powder metallurgical –PM– [40,42] processing routes are included. Alloys or metals shown in the graph are Zn, Al, Cu, Mg, Mb, stainless steel (316) and carbon steel. As most of the processing routes cannot produce cellular metals from all the mentioned alloys we show the typical alloys produced by each method.

We can observe how each different processing route presents a typical range of solid fraction. Lotus and injection moulding produce dense cellular solid whereas

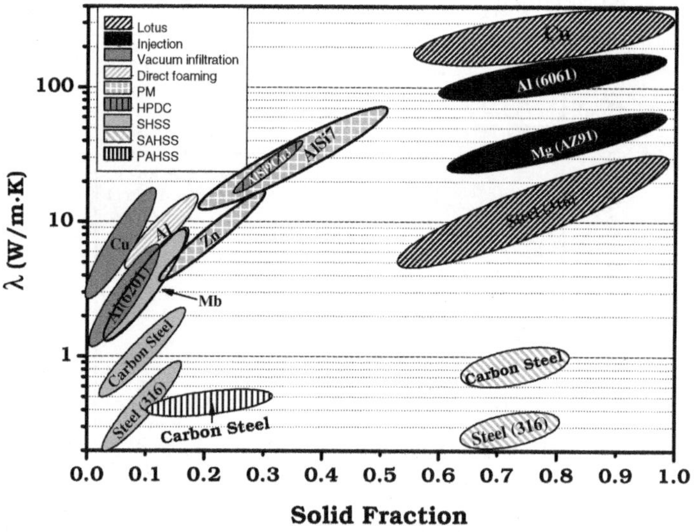

Fig. 6.9. Compared thermal conductivity of different cellular metallic materials

vacuum infiltration produce the lightest materials in this compared graph. In this way, SHSS and PAHSS are located in the lower range of porosities but SAHSS are located in the higher range. In most of cases HS present lower thermal conductivities than most of the cellular metals produced by alternative techniques which is mainly associated to both the thermal conductivity of the metallic phase and the presence of adhesive phase with much lower thermal conductivity of most of these metals.

6.4 The Role of Thermal Conductivity in Processing, Post-processing and Applications

To complete this chapter we indicate the importance of knowing the thermal conductivity of MHSS in different situations that comprise from the processing step to the end applications.

In many cases during production of metallic hollow sphere structures (MHSS) it is necessary to increase the metal temperature to a level that allows sintering (or even melting in case of brazed spheres) to create the HS metallic junctions. Due to low thermal conductivity and thermal diffusivity of the structure, time involved in this heating need to be estimated and should be higher than those needed for a dense solid. In addition it is expected the existence of temperature profiles at early times within the structures that may create an in-homogeneous response of the piece just in case the junctions were not enough time inside the furnace. Both factors have to be taken into account in order to asses, during processing, a homogeneous structure with reproducible physical properties and to this end the thermal conductivity of these materials should be considered (although it is expected that at those temperatures not only the conduction mechanism is present).

Another interesting case is related to the conventional method to improve mechanical properties of alloys: i.e. thermal treatments. To this end, the most common approach is quenching the material needing high cooling rates above a critical value. Due to the low thermal conductivity of cellular metals, reaching values upper the critical cooling rate could be difficult if the porosity and/or the size of the part to be treated is high [38]. In addition, it may occur that cooling rates higher than the critical value are only reached in areas close to the surface of the piece. This would result in a material with an in-homogeneous metal-matrix microstructure and as consequence mechanical properties varying through the part thickness. To know of thermal conductivity can be useful to simulate the temperature increase when these materials are subjected to secondary operations such as cutting, machining, etc.

Lastly, talking strictly about the application of MHSS, it is well-known that hollow spheres are multifunctional materials: they have excellent characteristics as energy absorbers, present good damping of structural vibrations and good acoustic absorption, own high specific stiffness and strength and are much better thermal insulators than dense solids. Therefore, the best potential for these materials lies in multifunctional applications in which mostly the previous characteristics are in use.

As an example, MHSS present, in comparison with other cellular metals, lower thermal conductivity and at the same time better mechanical properties, so they became very attractive for structural applications in which heat insulation is an additional requirement. On the other hand, their lower thermal conductivity can be exploited in applications in which an improved thermal insulation is required combined with an excellent fire resistance. Fire barriers produced from cellular metals, require from this combination of properties. In the first case the low thermal conductivity and diffusivity increases the time needed for rising high temperatures at the other side of the barrier, they do not smoke any toxic fumes and, of course, present a better performance than dense solids in these circumstances.

References

[1] Almanza, O., Rodríguez-Pérez, M.A., de Saja, J.A.: J. Polym. Sci. Part B: Polym. Phys. 38, 993–1004 (2000)
[2] Almanza, O., Rodríguez-Pérez, M.A., de Saja, J.A.: Polym. Int. 53, 2038–2044 (2004)
[3] Almanza, O., Rodríguez-Perez, M.A., de Saja, J.A.: J. Polym. Sci., Part B: Polym. Phys. 42, 1226–1234 (2004)
[4] ASTM E1225-04: Standard Test Method for Thermal Conductivity of Solids by Means of the Guarded-Comparative-Longitudinal Heat Flow Technique
[5] ASTM E1461-07: Standard Test Method for Thermal Diffusivity by the Flash Method
[6] ASTM E1461-92: Standard Test Method for Thermal Diffusivity of Solids by the Flash Method
[7] Baba, T., Ono, A.: Meas. Sci. Technol. 12, 2046–2057 (2001)
[8] Babcsán, N., Mészáros, I., Herman, N.: Materialwiss. Werkst. 34, 391–394 (2003)
[9] Banhart, J.: Prog. Mater. Sci. 46, 559–632 (2001)
[10] Bhattacharya, A., Calmidi, V.V., Mahajan, R.L.: Int. J. Heat Mass Tran. 45, 1017–1031 (2002)
[11] Bouguerra, A., Aït-Mokhtar, A., Amiri, O., Diop, M.B.: Int. Commun. Heat Mass Transfer 28, 1065–1078 (2001)
[12] Collishaw, P.G., Evans, J.R.G.: J. Mater. Sci. 29, 486–498 (1983)
[13] Fiedler, T., Öchsner, A., Belova, I.V., Murch, G.E.: Adv. Eng. Mater. 10, 361–365 (2008)
[14] Fiedler, T., Oschner, A.: Scr. Mater. 58, 695–698 (2008)
[15] Fiedler, T., Solórzano, E., Öchsner, A.: Mater. Lett. 63, 1204–1207 (2008)
[16] Fielder, T.: Numerical and Experimental Investigation of Hollow Sphere Structures in Sandwich Panels. Trans. Tech Publications, Switzerland (2007)
[17] German Patent DE 3902032 (1989)
[18] Gibson, L.J., Ashby, M.F.: Cellular solids: Structure and properties. Cambridge University Press, Cambridge (1997)
[19] Glicsksman, L.R.: Heat transfer in foams. In: Hilyard, N.C., Cunningham, A. (eds.) Low density Cellular Plastics: Physical Basis of Behaviour, ch. 5, Chapman and Hall, London (1994)
[20] Glicksman, L.R., Schuetz, M., Sinofsky, M.: Int. J. Heat Mass Tran. 30, 187–197 (1987)
[21] Grujicic, M., Zhao, C.L., Biggers, S.B., Morgan, D.R.: J. Mater. Sci. 41, 2309–2317 (2006)
[22] Gustafsson, S.E.: Rev. Sci. Instrum. 62, 797–804 (1991)
[23] Gustavsson, M., Karawacki, E., Gustafsson, S.E.: Rev. Sci. Instrum. 65, 3856–3859 (1994)
[24] Holman, J.P.: Heat Transfer. McGraw-Hill, New Cork (1981)
[25] ISO 13787:2003: Thermal insulation products for building equipment and industrial installations (2003)

[26] ISO 8302:1991: Thermal insulation -Determination of steady-state thermal resistance and related properties- Guarded hot plate apparatus (1991)
[27] Jaeger, H.M., Nagel, S.R.: Science 255, 1524–1531 (1992)
[28] Leach, A.G.: J. Phys. D: Appl. Phys. 26, 733–739 (1993)
[29] Lochmann, K., Oger, L., Stoyan, D.: Solid State Sci. 8, 1397–1413 (2006)
[30] Log, T., Gustafsson, S.E.: Fire Mater. 19, 43–49 (1995)
[31] Lu, T.J., Chen, C.: Acta Mater. 47, 1469–1485 (1999)
[32] Nishi, T., Shibata, H., Waseda, Y., Ohta, H.: Metall. Mater. Trans. A 34, 2801–2807 (2003)
[33] Ogushi, T., Chiba, H., Nakajima, H., Ikeda, T.: J. Appl. Phys. 95, 5843–5847 (2004)
[34] Sanders, W.S., Gibson, L.J.: Mater. Sci. Eng. A 347, 70–85 (2003)
[35] Sanders, W.S., Gibson, L.J.: Mater. Sci. Eng. A 352, 150–161 (2003)
[36] Saxena, N.S.G., Pradeep, P., Mathew, G., Thomas, S., Gustafsson, M., Gustafsson, S.E.: Eur. Polym. J. 35, 1687–1693 (1999)
[37] Slattery, J.C.: AlChE J. 16, 345–352 (1970)
[38] Solórzano, E., Escudero, J., Lázaro, J., Rodriguez Perez, M.A., de Saja, J.A.: MetFoam 2007: Proceedings of the Fifth International Conference on Porous Metals and Metallic Foams, pp. 75–78. Destech Publications, Lancaster (2007)
[39] Solórzano, E., Hirschmann, M., Rodriguez-Perez, M.A., de Saja, J.A.: Mater. Lett. 62, 3960–3962 (2008)
[40] Solórzano, E., Reglero, J.A., Rodríguez-Pérez, M.A., Lehmhus, D., Wichmann, M., de Saja, J.A.: Int. J. Heat Mass Tran. 51, 6259–6267 (2008)
[41] Solórzano, E., Rodriguez-Perez, M.A., Reglero, J.A., de Saja, J.A.: J. Mater. Sci. 42, 2557–2564 (2007)
[42] Solórzano, E., Rodriguez-Perez, M.A., de Saja, J.A.: Adv. Eng. Mater. 10, 596–602 (2008)
[43] Solórzano, E., Rodriguez-Perez, M.A., de Saja, J.A.: Mater. Lett. (publication pendant) (2009)
[44] Tye, R.P.: Proceedings of the Cellular polymers: An international conference. Rapra Technology, London (1991)
[45] Tye, R.P., Coumou, K.G.: Temp. High Press. 13, 695–704 (1981)
[46] Tye, R.P., Kubicàr, L., Lockmuller, N.: Int. J. Thermophys. 26, 1917–1938 (2005)
[47] Whitaker, S.: AlChE J. 13, 420–427 (1967)

7
Hollow Spheres in Composite Materials and Metallic Hollow Sphere Composites (MHSC)

Erika Baumeister and Martin Molitor

Otto-von-Guericke-University Magdeburg
Institute of Manufacturing Technology and Quality Management
Postfach 4120, D-39016 Magdeburg, Germany
erika.baumeister@t-online.de,
martin.molitor@ovgu.de

Abstract. The newly developed metallic hollow spheres are used in combination with a polymeric matrix for producing metallic hollow-sphere-composites (MSHC), which have been developed for mechanical engineering applications in the "InnoZellMet" project.

This project was supported by the Federal Ministry of Education and Research under reference No. 03WK001E. The responsibility for the contents lies with the author.

Keywords: Hollow Spheres in Cast, Polymeric Materials, Metallic Hollow Sphere Composites.

7.1 Introduction

The aim of employing the metallic hollow spheres in polymeric materials (MHSC) was to reduce the mass of polymer concrete (PC), which is a widely used material in mechanical engineering to produce support structures, jigs, casings, tables etc. Typical advantages of PC are the very good damping behaviour in comparison to steel or cast iron, low-cost production, lower densities than that of steel or cast iron and excellent corrosive properties [1-4].

These advantageous properties of PC can be chiefly attributed to the hollow-sphere-composites (HSC). Because of the lower densities of HSC in comparison to PC, lighter constructions can be achieved, which can be rapidly moved and so the dimensioning of the motors can be reduced. This has a positive influence on the energy demand and on the ecology. Light machine parts can be transported by the operator with ease and smaller cranes can be used for material moving with less fuel consumption.

Hollow-sphere-composites can be used not only in mechanical engineering but also in the automotive and aircraft industry, for example as filler of hollow profiles to absorb the energy similarly to other cellular materials. For different applications appropriate material properties are required, which are in correlation with the structure and composition of hollow-sphere-composites.

Hollow-sphere-composites (HSC) consist of hollow spheres in high volume fraction (up to 80 vol. - %) and a polymer binder material, mostly epoxy resin. In the last years non-metallic hollow spheres HSC were developed and their thermal and mechanical properties investigated [5-8]. The recently developed metallic hollow spheres

A. Öchsner, C. Augustin (Eds.): Multifunctional Metallic Hollow Sphere Struct., EM, pp. 109–136.
springerlink.com

(MHS) have been used to produce metallic hollow-sphere-composites (MHSC). In these structures the heavy ceramic macro spheres were substituted by metallic hollow spheres having smaller bulk densities. Epoxy resins are very good adhesives and can be combined easily with fibrous materials causing the MHSC to be bonded with either short or long fibres (cf. Fig. 7.1). This is called a Metallic-Hollow-Sphere-Composite-Structure (MHSCS).

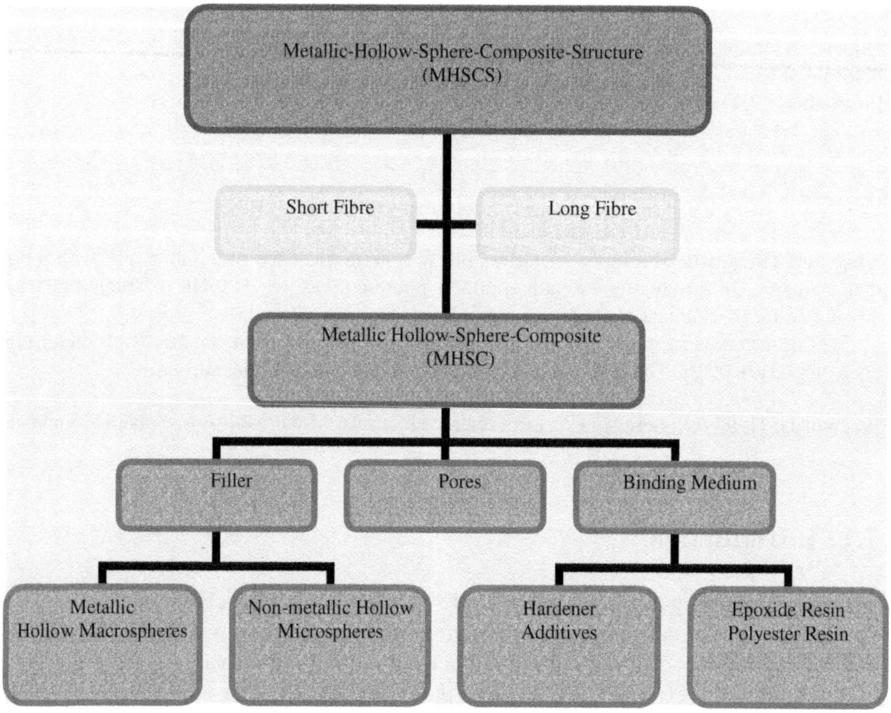

Fig. 7.1. Composition of Metallic-Hollow-Sphere-Composites (MHSC) and Metallic-Hollow-Sphere-Composite-Structures (MHSCS)

Because of the encapsulated gases in the spheres the MHSC consists of pores and this structure forms a cellular material. Due to the metallic character of the spheres, the structure can be considered as metal foam. The scope of material selection (spheres and resins), i.e. selecting the most appropriate materials, e.g., for mechanical engineering applications, is to have a minimal pore concentration between the spheres in the matrix material.

7.2 Raw Materials

When developing composite-materials and selecting raw materials, it is important to know the demands on these materials relating to the requirements of the machine components. For the design of a mechanical structure information is needed on the optimal production conditions, e.g. size and shape of components, batch size, process parameters, cost of production as well as sufficient stiffness, strength, functionality and reliability of assembled component junctions. These requirements can be met by using thermosetting materials with good mechanical and thermal properties, good handling, and rapid processing as matrix material. The thermosets are very often used to produce high performance composite materials with fibre reinforcing [9]. But also the PC includes thermoset materials, mostly epoxy resins, because of their better mechanical and thermal properties, adhesion, handling and smaller shrinkage in comparison to other resins. The process of producing MHSC will be carried out similarly to the production of PC when the use of epoxy resin (EP-resin) as matrix material in the production and development of MHSC is highly recommended.

7.2.1 Bonding Medium

In the project two different cold hardening epoxide resin systems were used in preparation of MHSC.

The EP-resin Ebalta AH 120 with hardener TL was used in the production of HSC based only on non-metallic hollow spheres. To compare the results of material testing of MHSC with those of HSC the same epoxy resin was used for producing MHSC. This resin was selected for their high thermal and stiffness properties. The technical data of processing for EP-resin Ebalta AH 120 with hardener TL is given in Table 7.1. The mechanical and thermal properties provided by Ebalta are given together with the measured properties in Table 7.2.

Table 7.1. Processing data from epoxy resin (EP-resin) Ebalta AH with hardener TL, epoxy resin (EP-resin) Araldit and unsaturated polyester resin (UP-resin) Palatal [10]

Resin system	*EP-resin Ebalta*			*EP-resin Araldit*		*UP-resin Palatal*
	Mixture AH 120/ TL	*Resin AH 120*	*Hardener TL*	*Resin Araldit BY 3739*	*Hardener HY 3741*	*Resin Palatal 80-02*
Aspect	White transparent	Milking cloudy	Yellow transparent	clear fluid	yellow clear	Clear
Viscosity [mPa s]	800±100	850±100	300±50	200-400 (ISO 12058-1)	5-25 (dynamisch)	300-360
Density [g/cm³]	1.11±0.02	1.15±0.02	0.96±0.02	1.15	0.91-0.92	1.1
Price [€/kg]	9.9	7.3-9.4	12.3-14.5	6	7	4.7

Table 7.2. Experimental determined values (1) and parameters from producer (2) and of used resin systems

Parameters	*EP-resin* *Ebalta[1)]*	*EP-resin* *Ebalta[2)]*	*EP-resin* *Araldit[1)]*	*UP-resin* *Palatal[2)]*
Density [g/cm³]	1.15	1.11±0.2	1.14	-
Tensile strength [MPa]	10.8	-	21.2	80
E-Modulus (Tensile) [GPa]	3.4	-	3.3	3.7
Bending strength [MPa]	47.1	125±10	80.4	145
Compression strength [MPa]	>100	107±10	94.1	-

The DSC investigations were monitored and the glass transition temperature (Tg) was equal to 70°C. The thermal degradation temperature is over 220°C. This means that the resin can resist a thermal load at this temperature only for a short time. Similar thermal effects were observed by the DSC-graph in the case of HSC [4] and it results that the MHSC can also support this temperature.

On the basis of convenient price/achievements relations the epoxy resin Araldit 3739 with the hardener HY 3741 (cf. Table 7.1) were well-tried by the PC manufacture companies, so they were selected for the production of MHSC [11]. The properties of this resin system were unknown, that is why they must be determined and serve as reference values (cf. Table 7.2).

In this investigation program, additionally to EP-resins, a polyester resin (UP-resin) Palatal 80-02 was also selected from the DSM company (cf. Tables 7.1 and 7.2) [12]. The polyester resins are more cost effective than EP-resins and they have a very short hardening time. However, the process reaction is extremely exothermic and it may cause the components with relatively thick cross-sections to be distorted due to internal stresses. The high reaction shrinkage of 5-10% is a notable disadvantage along with the lower mechanical properties of UP-resin in comparison to EP-resins. Therefore, the UP-resins were inserted in parts with modest requirements [13].

7.2.2 Hollow Spheres

The specific difference of hollow spheres in comparison to all other filler types exists in their spherical forms, so that their length/thickness ratio is 1. As they reveal a minimal contact surface with the matrix, the resin consumption can be reduced. As a result of their form, the spheres in the HSC show isotropic properties. The smaller the dimensions of the spheres are the smoother surfaces can be obtained. Fillers as hollow spheres are generally added to the plastic-matrix to improve elastic modulus, the compressive and bending strength, the hardness, and the shape stability and to reduce the shrinkage and percentage elongation at failure.

Hollow spheres have lower bulk densities than solid spheres and they induce the cellular structure of HSC and the lighter weight of these materials.

7.2.2.1 Metallic Hollow Spheres

For filler selection the German Standard DIN 51290 was considered, which prescribes that the minimum dimensions of the sample should not be smaller than three times the maximum grain size of the used filler. In compliance with the standard the grain-size of metallic hollow-spheres between 1.5 and 3.3 mm were used.

The metallic hollow spheres (MHS) made of steel alloy are delivered by the Glatt Company [14]. The technical data from Glatt Company about density and grain size of the spheres were completed with the photos of the spheres of the second delivery. The bulk densities in Table 7.3 varied between 0.42 and 0.77 g/cm³ depending on grain size and porosity. At present the size of the spheres cannot go under 1.5 mm.

In addition to metallic macro hollow spheres, non-metallic hollow microspheres are necessary to improve the package density by fillers in MHSC.

Table 7.3. Used metallic hollow spheres [14]

Notation	HK-420-15-2005 „E3261-3“	HK-440-27-2005 „E2261/2“	HK-770-32-2005 „HF1“	HK-430-32-2005 „E2152“
Grain size [mm]	1.5	2.7	3.2	3.2
Bulk densities [g/cm³]	0.42	0.44	0.77	0.43
Photos				

7.2.2.2 Non-metallic Hollow Spheres

The best non-metallic micro hollow spheres types from the earlier research of the thermal and mechanical properties of HSC spheres are selected to produce MHSC.

The hollow microspheres Fillite PG improves best properties of micro non-metallic hollow spheres filled HSC. Fillite PGs are fabricated of alumina-silicate. They are grey with a very high package density of 60-65%. The grain size of Fillite ranges between 5 and 355 μm and is distributed after a defined particle size distribution curve. Table 7.4 includes some properties of Fillite from the manufacturers.

These are completed with SEM investigations of the surface and the interior of the microspheres. The surfaces are smooth. The wall thickens of Fillite is only 5-10% of the diameter and is continuous.

The wall thicknesses of the boron-silicate-glass hollow microspheres Q-Cell 300 are between 0.4-1.5 μm (cf. Table 7.4). The bright-white coloured microspheres are added to the plastic materials to improve the mechanical properties of the final product, to increase the thermal shock resistance and the dimension stability, to increase the thermal insulation properties as well as to improve the rheology of the final products [20].

Table 7.4. Composition, properties and structure of the used non-metallic hollow microspheres Fillite PG and Q-Cel 300 [16-20]

Data from manufacturers	*Fillite PG*	*Q-Cel 300*
Composition [%]	Shell: 27-33 Al_2O_3 55-65 SiO_2 max. 4 Fe_2O_3 Gas: 70 CO_2, 30 N_2	Shell: Borosilicate glass Gas: Inert gas
Grain size [mm]	5-300	10-200
Melting point [°C]	-	1200-1350
Density [g/cm³]	0.6-0.8	0.21
Bulk density [g/cm³]	0.35-0.45	0.12
Thermal conductivity [W/m K]	0,11	-
Compression strength [MPa]	-	3.45
Price [€/kg]	0.8	8
SEM-Photos of the surface of microspheres	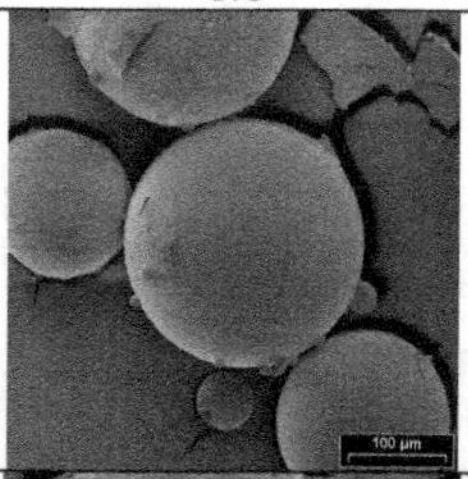	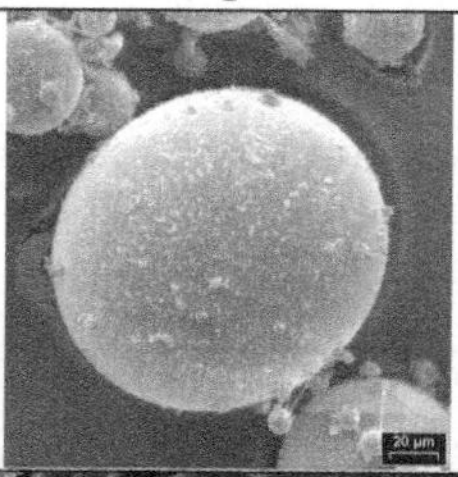
SEM-Photos of the interior of microspheres	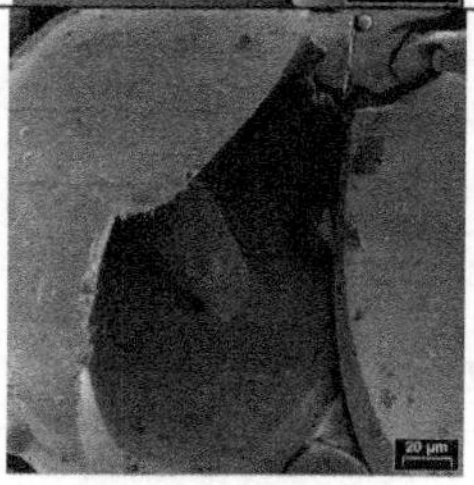	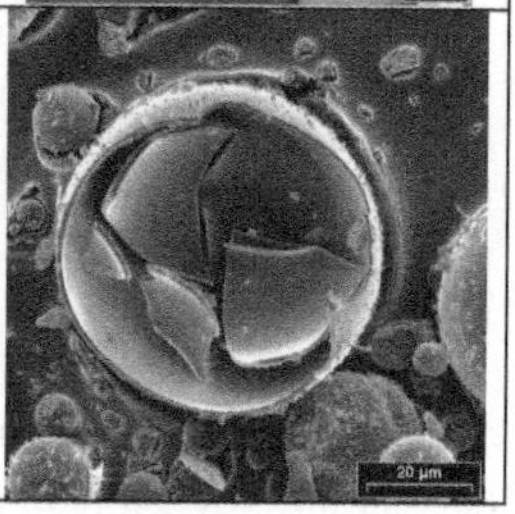

7.3 Production and Nomenclature of MHSC

Hollow-sphere-composite components are manufactured in a cold moulding process. For each forming process the mould is treated with a parting compound layer. Resin and hardener are mixed together and the filler is introduced, followed by the transfer of the mixture to the mould (cf. Fig. 7.2).

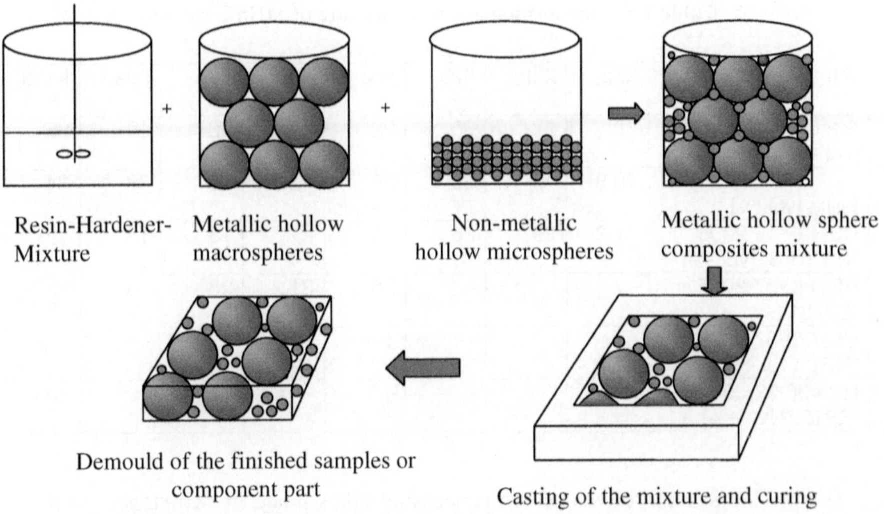

Fig. 7.2. Scheme of the important stages in the producing of MHSC

The curing process takes place at ambient temperature and does not require any additional energy consumption, drying and curing processes. Due to the low shrinkage of epoxy resins the resulting components are of very high dimensional accuracy. An additional advantage of this production process is that reinforcing elements can be simultaneously integrated into the moulding process.

The correct choice of the material content is important to achieve high specific values and to reduce the amount of resin, by increasing of packing density with fillers of HSCs. In the case of polymer concrete, the high packing density is achieved by the selection of varying sphere size. The perfect round shape in comparison to other filler types, e.g. silicate, quartzite sand and stone, provides the lowest volume of voids between the spheres. If spheres of uniform size are used they can only achieve a maximum packing density factor of 0.74 in a hexagonal package. If the spheres of a specimen are distributed according to different defined sieve sizes, the smaller spheres will occupy the spaces between the larger spheres and the resulting package density will be greater. The particular sphere types of the investigated HSC were mixed to different defined sieve sizes, thereby producing up to 0.78 packing density factor [4].

The nomenclature and composition of some selected variants of MHSC are presented in Table 7.5. The sample variants were named in the following way: II2aAR-1

- II : Number of the stage investigations
- 2: Number of variants
- a: another composition of resin and fillers
- A: capital letter from the first letter of the used adhesive (A: Araldit, E: Ebalta, P: Palatal)
- R: manufacture by using of a vibrating table
- 1: Number of the sample of the same variant

Table 7.5. Composition and nomenclature of MHSC-variants

Sample-No.	II2a_	II2b_	II2c_	II3a_	II5a_	II7a_	II9a_	II11a_
Resin-Hardener [vol.-%]	22	28	35	22	22	22	19	25
Fillite [vol.-%]	46.8	43	39	35	46.8	35	38	34
HS-420-15-2005 "E3261-3" [vol.-%]	31.2	29	26	13	-	13	13	12
HS-440-27-2005 "E2261/2"	-	-	-	20	31.2	20	20	19
HS-770-32-2005 "HF1" [vol.-%]	-	-	-	10	-	-	-	-
HS-430-32-2005 "E2152" [vol.-%]	-	-	-	-	-	10	10	10

The production and development processing takes place in two stages. In the second stage (Stage II) lighter MHS (bulk density of 0.42 g/cm³) and with smaller grain size (1.5 mm) are available from the Glatt Company. In the first stage (Stage I) the hollow spheres 05-01 HF 2 have the smallest grain sizes with 1.75 μm and are heavier than in the second stage having bulk a density of 0.8 g/cm³. All the variants, which are made with this filler type, must be repeated to determine the properties of the new similarly changed variants in the second stage.

7.4 Properties of MHSC

The component parts in machine tool manufacturing are exposed mainly to static loadings. The condition for calculating the static demanding is that the density, stiffness and strength of the materials should be known. However, these values must be determined for MHSC. Because of this reason the used testing methods for HSC are checked if they are valuable for metallic materials or MHSC. Besides, the testing methods are extended with new testing methods for the determination of the magnetic properties of MHSC.

7.4.1 Mechanical Properties of MHSC

The density characterises the specific properties of the materials and is a criterion for weight-assessments. In physics the density (ρ) of a body is the ratio of its mass (m) to its volume (V). For monolithical materials the pure density will be measured. For multi-phase materials such as fibres, pores or grained materials the compacted density will be determined. Consequently, the compacted densities were determined in the case of MHSC.

The material densities of MHSC were determined from the samples prepared for mechanical testing. The lightweight materials MHSC have low-densities between 0.78 and 0.95 g/cm³ depending on the used type of hollow-spheres (cf. Fig. 7.3).

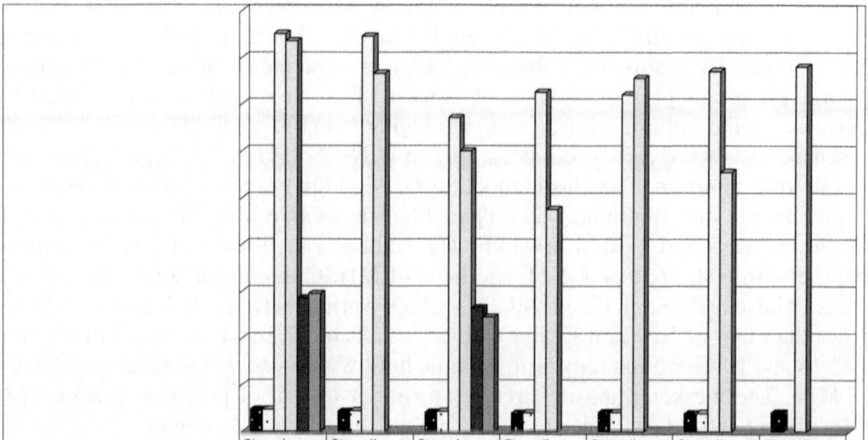

	Stage I	Stage II	Stage I	Stage II	Stage II	Stage II	Stage II
	I_3a_		I_5a_		II_7a_	II_9a_	II_11a_
Density of MHKK-A [g/cm³]	0,95	0,87	0,82	0,79	0,82	0,78	0,83
Density of MHKK-E [g/cm³]	0,95	0,89	0,87	0,81	0,83	0,8	
Tensile strength of MHKK-A [MPa]	17	16,9	13,4	14,5	14,4	15,4	15,6
Tensile strength of MHKK-E [MPa]	16,7	15,3	12	9,5	15,1	11,1	
E-Modulus of MHKK-A [GPa]	5,7		5,3				
E-Modulus of MHKK-E [GPa]	5,9		4,9				

Fig. 7.3. Density, tensile strength and E-Modulus of some selected MHSC-variants

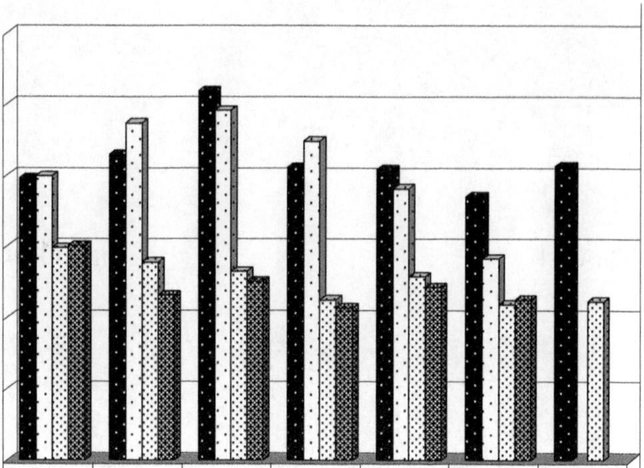

	Stage I	Stage II	Stage I	Stage II	Stage II	Stage II	Stage II
	I_3a_		I_5a_		II_7a_	II_9a_	II_11a_
Compression strength of MHKK-A [MPa]	39,6	42,9	51,8	41,1	40,85	37	41,3
Compression strength of MHKK-E [MPa]	39,9	47,3	49,1	44,8	38,1	28,3	
Bending strength of MHKK-A [MPa]	29,8	27,8	26,5	22,5	25,8	21,9	22,3
Bending strength of MHKK-E [MPa]	30,1	23,2	25,1	21,4	24,3	22,5	

Fig. 7.4. Compression and bending strength of some selected MHSC-variants

In the second stage the densities of the MHSC are lower than in the first stage because of the lighter MHS. The MHSC are lighter materials than the following components: the raw EP-resins (cf. Table 7.2), polymer concrete (densities varied between 2.1 and 2.4 g/cm³ [1]) or of HSC-14 consist of ceramic hollow spheres (density-values by similar volume fraction resin are higher than 0.9 g/cm³ [4]).

Tensile tests were carried out according to DIN EN ISO 527-4 Type 3 [21] with a test rate of 5 mm/min. The elastic modulus (E-Modulus) in tensile tests is carried out with a test rate of 1 mm/min. The prepared tensile sample was 250 mm in length, 10 mm in thickness and 25 mm in width. The additions of fillers to EP-resins improve both the tensile strength and the E-Modulus of MHSC in comparison to the EP-resin Ebalta. The tensile strength of MHSC, which varied between 9.5 and 17 MPa, is higher than that of EP-resin Ebalta with 10.8 in Table 7.2 and are better than that of HSC-14 and HSC-5 filled only with ceramic hollow spheres, which values are 8.6 and 6.7 MPa. The measured tensile strength is similar to that of polymer concrete (17.8 MPa) by only a third or a half of the densities of the polymer concrete.

The compression test samples were of 100 mm in length and 30 mm in width and thickness. During the course of compression stress-strain experiments, a test speed of 1 mm/min was used according to DIN EN ISO 604 [22] and the results are shown in Fig. 7.4.

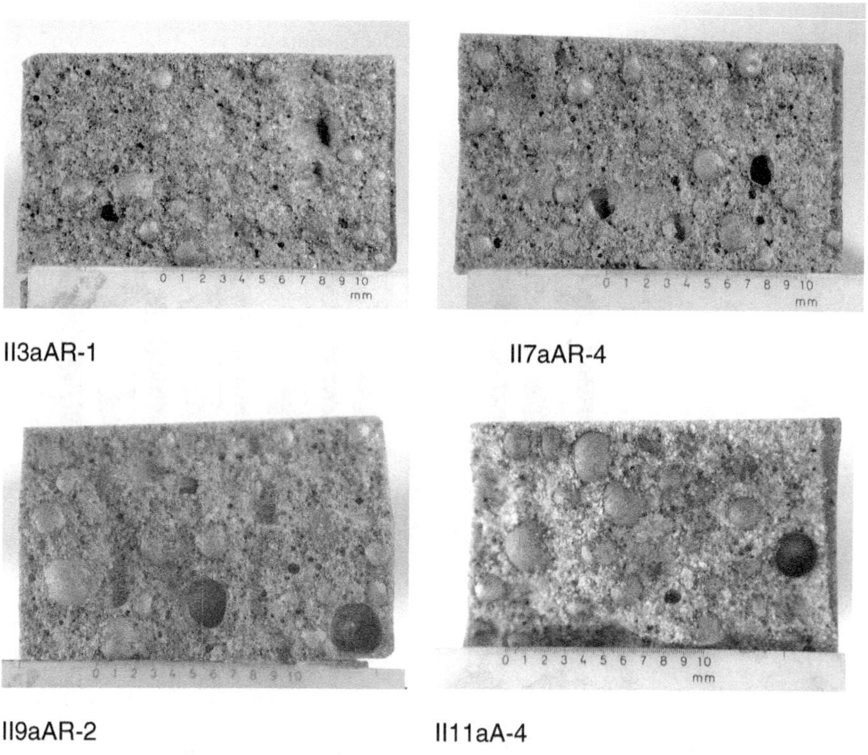

Fig. 7.5. Photos of broken surfaces of the bending samples of MHSC

The MHSC presents the highest strength under compression similar to other cellular materials, due to the crack propagation which can be rapidly stopped by impediments, such as spheres or pores. It should be noted that the adhesion bonds between fillers and binders are of paramount importance. If the stiffness of the spheres is higher than that of the resin, then cracking will start in the resin and vice-versa. The compression strengths in the second stage of the variant II3E with 47 MPa are higher than in the first stage with the heavier MHS. The achieved compression strengths by MHSC are similar to that of HSC-14 [4], but they are only the half of the compression strength of EP-resins (cf. Table 7.2) or of polymer concrete [4].

Furthermore, the diagram in Fig. 7.4 includes the measured bending strength values of MHSC. The bending tests were carried out at a speed of 4.8 mm/min [23]; the bending samples were 250 mm in length, 10 mm in thickness and 25 mm in width. The bending strength values varied between 21.4 MPa for MHSC-II5aE and 30.1 MPa for MHSC-I3aE. The bending strength of MHSC is higher than that of HSC-14 or HSC-5, which values are 20.8 and 17.7 MPa. The bending strength of MHSC differs only a little bit from the measured values of polymer concrete or epoxy resins (cf. Table 7.2) [4]. Although they did not show great differences between the bending strength of the sample variants II7aA and II3aA, the magnification photos of the broken surfaces of the bending samples clarify that the lighter MHS (HS-430-32-2005) are broken, whereas the heavier MHS (HS-770-32-2005) are pulled out of the matrix.

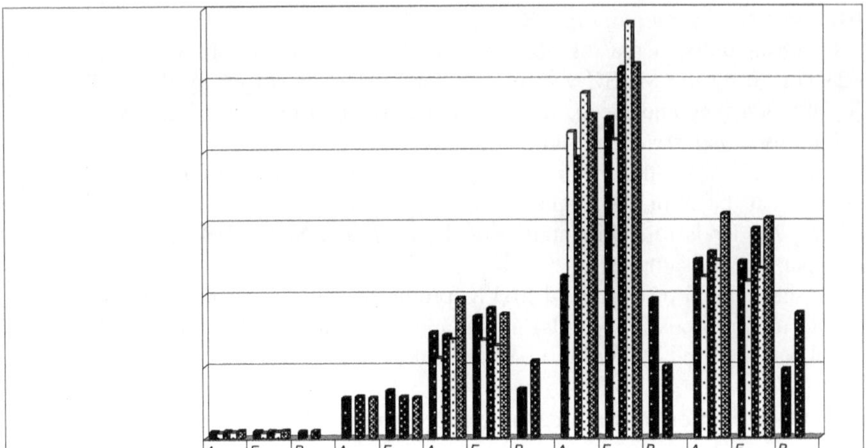

	Density [g/cm³]			E-Modulus [GPa]		Tensile strength [MPa]			Compression strength [MPa]			Bending strength [MPa]		
	A_	E_	P_	A_	E_	A_	E_	P_	A_	E_	P_	A_	E_	P_
■ I_2a_ (22 vol.-% resin)	0,92	0,95	0,92	5,6	6,6	14,7	17	6,8	22,6	44,7	19,4	24,9	24,6	9,5
□ II2a_ (22 vol.-% resin)	0,76	0,78				11,1	13,7		42,7	41,6		22,5	21,8	
■ I_2b_ (28 vol.-% resin)	0,98	0,98	0,99	5,8	5,7	14,4	18,1	10,7	39,2	51,7	10	25,9	29,2	17,4
□ II2b_ (28 vol.-% resin)	0,83	0,89				13,8	12,9		48,2	58		24,7	23,6	
▩ I_2c_ (35 vol.-% resin)	1,02	1,02		5,6	5,6	19,5	17,3		45,1	52,3		31,2	30,6	

Fig. 7.6. Influence of the percentage by volume of resins of the mechanical properties of MHSC

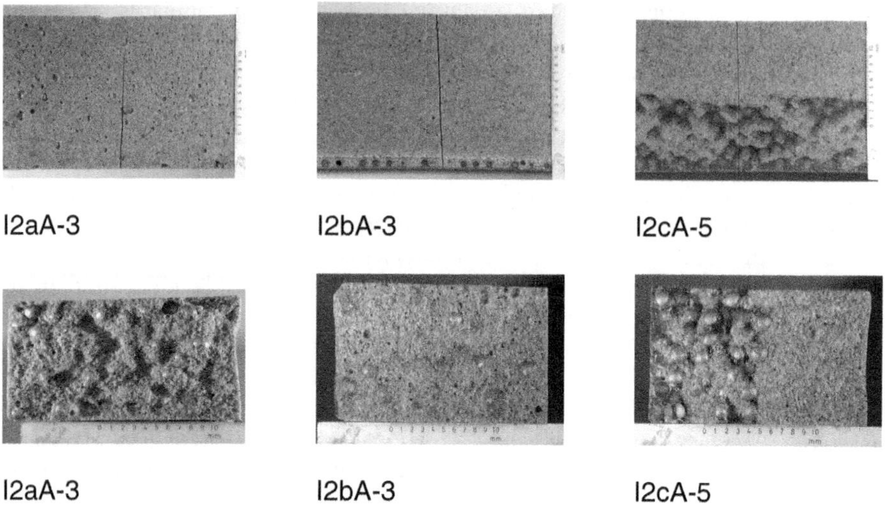

Fig. 7.7. Surfaces (above) and the surfaces of fracture (below) of the flexural samples

By increasing the volume percent part of EP-Resin Araldit higher densities and strength can be obtained by the variant II11aA in comparison to the variant II7aA.

The influence of the resin and filler concentration on the mechanical properties of MHSC can be observed in Fig. 7.6.

By substantially increasing of resin from 22 vol.-% by the MHSC-variants _2a_ up to 28 vol.-% by the MHSC-variant _2b_ and up to 35 vol.-% by the MHSC-variant _2c_ the densities and mechanical strengths are significantly increased, whereas the E-Modulus values are decreased.

Figure 7.7 shows that by increasing of the volume percent part of resin the pore fractions can be reduced. Higher volume percent (35 vol.-%) resin by the MHSC-variant _2c_ leads to sedimentation of the resin and MHS (HS-420-15-2005) in the lower parts of the samples.

The mechanical properties of MHSC produced with EP-resin Ebalta or Araldit are superior to that produced with UP-resin Palatal. Because of these results, the UP-resin Palatal was not used for further investigations.

7.4.2 Optimisation of the Mechanical Properties of MHSC

To improve the mechanical properties of MHSC in the sample variant 5aE short fibres of glass were introduced with a cutting length of 4 mm and Polyacrylonitrile (PAN) fibre with a cutting length of 2 and 4 mm [24].

In such a way the combination of this material is called MHSC-Structure (MHSC-S). The composition and names of the MHSC-S with short fibres are given in Table 7.6.

For comparison of the results of the mechanical testing of MHSC-S with MHSC, 22 percent by volume of EP-resin Ebalta are used. The fillers Q-Cells were added in the mixture of the sample variant I5aE to achieve a higher packaging density and so improved mechanical properties by this sample variant 5aE-Q-Cel. As an alternative,

Table 7.6. Composition and names of the MHSCS-samples with short fibres

Pr.-Nr.	*Harz* Vol.-%	*Q-Cell* vol.-%	*Fillite* Vol.-%	*PAN2* vol.-%	*MHK-2,69* vol.-%	*PAN4* Vol.-%	*GF/4* vol.-%
5aE-PAN/2-30	22	-	26.8	30	21.2	-	-
5aE-PAN/4-20	22	-	36.8	-	21.2	20	-
5aE-GF/4-20	22	-	36.8	-	21.2	-	20
5aE-QCel	22	20	26.8	-	31.2	-	-
5aE-PAN2-4-QCel	22	10	16.8	20	21.2	10	-

long fibres of glass or carbon were used in some MHSC-sample for flexural testing [25]. They were introduced in the mould at the opposite side of applied flexural force. With short fibre reinforced MHSC-S, the mechanical properties cannot be significantly improve. However, it is possible to identify the progress of the flexural strength of bending samples laminated with long carbon or glass fibres as shown in the Figure 7.8.

With carbon fibre reinforced flexural samples of the variant 5aE can withstand several percent strains and a flexural stress of 94 MPa can be obtained. The fracture surfaces of the flexural samples 5aE with and without fibre are presented in Fig. 7.9.

The MHS with grain size of 2.69 mm in the bending sample 5aE are not broken, as it is illustrated in the fracture surfaces of this sample. The 30 volume percent of PAN fibre in the sample 5aE-PAN2-2 has different orientation around the MHS. The only

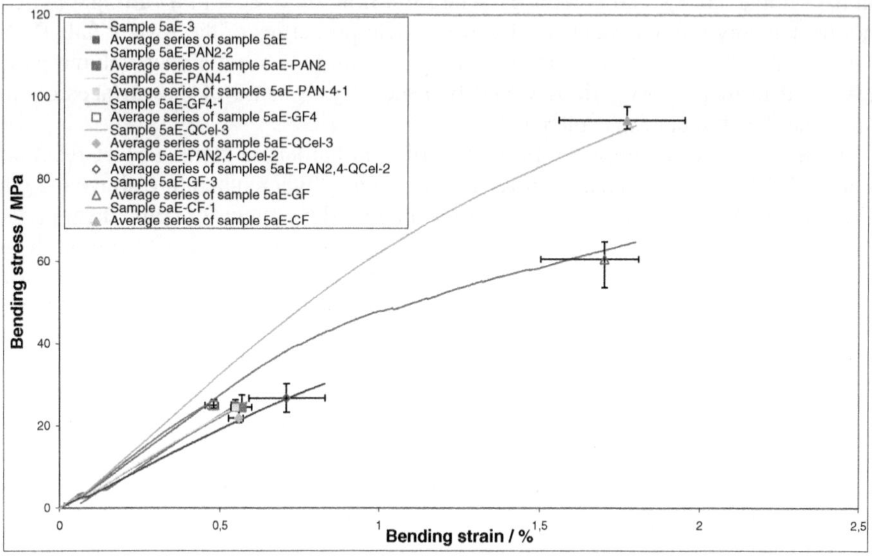

Fig. 7.8. Flexural stress-strain-curves of MHSCS

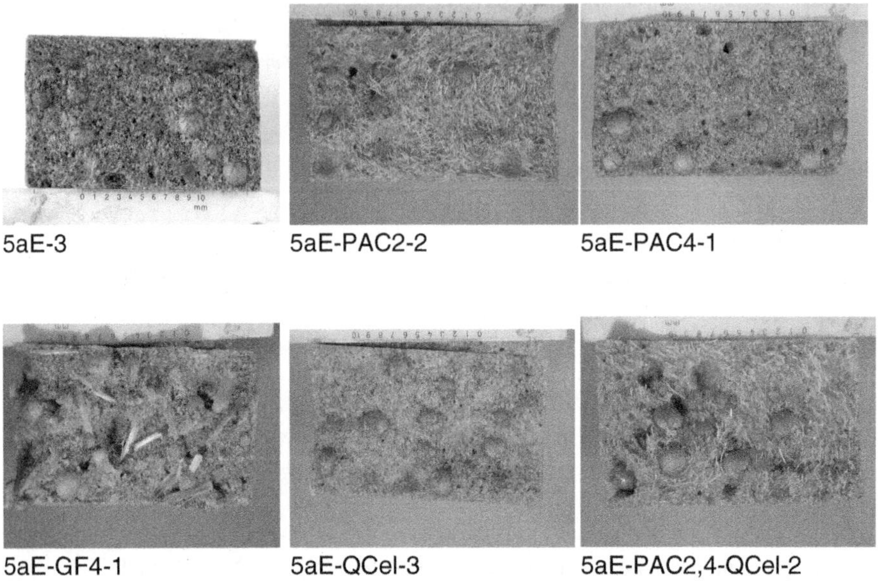

Fig. 7.9. Fracture surfaces of the bending samples of MHSC and MHSCS

20 volume percent of PAN fibre with a cutting length of 4 mm are not visible in the fracture surface of the sample 5aE-PAN4-1. In contrast to this, glass fibres having the same length and in the same concentration are anisotropic oriented in the surrounding substance. Through addition of the fine grain sized Q-Cell to the matrix the porosity in the sample 5aE-Q-Cell is reduced. The PAN fibre of 2 and 4 mm cutting length did not present any improvements of the mechanical properties of the variant 5aE-PAN2, 4-Q-Cel-2 and the fracture surface shows pores. In this case a higher volume part of resin and further investigations would be necessary, because of the surfaces of fibre are greater than that of the spheres.

Further way to optimise the properties of cold hardening resins and of resin containing MHSC are the reheating after cold hardening by a temperature above the glass transition temperature of the resin for a few hours. The samples or component parts of machine tools can be demoulded one day after manufacturing. But a core hardening sustain even a week, while the strength and the heat form stability will increase or the fracture strain will decrease. Through heat supply the "frozen" molecules and the unreactioned rest molecules will be moved and new chemical compounds will be built. It results in a higher grade of network in a shorter time. In addition to improving of hardness, stiffness or chemical stability, the improving of brittleness occurs. Because of this reason the large component parts will not be cured to avoid the high brittleness. The curing of the MHSC-samples with EP-Resin Araldit 24 hours by 60°C was not so effective as the reheating of the MHSC-samples with EP-resin Ebalta for 5 hours by 90°C. In this case the mechanical properties can be improved in comparison to the uncured variant II5aE up to 58% in compression tests.

Some samples in the second stage are manufactured in a table vibrator. The comparison of the values of the samples which are manufactured on a table vibrator with the values of the samples which are manually manufactured show, that the densities of the vibrated samples are only a little bit smaller but the mechanical properties cannot be significantly improved. It results that the MHSC must not be absolutely manufactured on a table vibrator, but they must be well pressed in the mould.

7.4.3 Magnetical Properties of MHSC

For testing of magnetical properties, samples were developed and manufactured as toroidal cores with 60 mm inner diameter, 70 mm outer diameter and 5 mm depth.

Dependence on the relative permeability (μ_r) the materials can be classified in: diamagnetic if μ_r is between 0 and 1; paramagnetic if μ_r is between 1 and 2 and ferromagnetic if μ_r is between 2 and 10^6. All the μ_r-values of MHSC in Table 7.7 are greater than 1, which means that the MHSC are ferromagnetic, but only to a low degree. The variant II7aA presents with 4.2 the highest ferromagnetic property in comparison to all other investigated MHSC-variants in Table 7.7, because of the high volume fraction and packaging density of the metallic spheres and of the higher surface of the metallic component in the matrix.

A great electrostatic charging was observed for loose MHS. The electrostatic charging is reduced by increasing of the non-metallic spheres fraction of Fillite or of EP-resin. Generally speaking, polymeric materials are diamagnetic. Because of the metallic parts included in the MHS, this property has been modified to low ferromagnetic. Probably the pores reduced a greater ferromagnetic effect of the metallic spheres in the MHSC.

Table 7.7. Magnetical properties of some selected MHSC-variants

MHKK-variant	*Coercivity* ($-_BH_C$) [A/m]	*Remanency* (B_r) [T]	*Relative Permeability* (μ_r)	*Magnetical Susceptibilty* (χ)
II2aA	63.1	0.0002	2.7	1.7
II5aA	205	0.0009	3.2	2.2
II7aA	317.4	0.0021	4.2	3.2
II11aA	146	0.0008	3.8	2.8

7.4.4 Thermal Properties of MHSC

It was mentioned in chapter 7.2.2 that the thermal behaviour refers to the shape stability and that the thermal degradation of HSC coincides with the thermal behaviour of the used resin. Thus, the same thermal behaviour can be predicted for MHSC.

For certain applications of components parts, the thermal expansion coefficient, specific heat capacity and the thermal conductivity are necessary. These values were not being measured in this project, but they can be simulated [26] or can be calculated.

The linear coefficient of thermal expansion (α) can be calculated with the following rule:

$$\overline{\alpha} = \Sigma V_i \cdot \alpha_i , \tag{7.1}$$

where V_i are the volume fractions of the raw components and α_i are the linear coefficients of thermal expansion of the raw components. Through this formula the α-values of the raw materials and of MHSC can be approximately calculated.

The α-value of $71 \cdot 10^{-6}$ K^{-1} was determined for EP-resin Ebalta [4]. This value amounts to $34.5 \cdot 10^{-6}$ K^{-1} for HSC-5 with spheres Fillite. The α-value of the raw material Fillite resulted as $24.2 \cdot 10^{-6}$ K^{-1}. The calculated α-values of MHSC are presented in the Table 7.8 in comparison to α-values of the raw materials, HSC, PC or some selected metal foams (Cymat, Alporas, ERG).

The α-values of MHSC are greater than that of HSC, because of the greater α-values of Fillite. However, they are necessary to improve the grain distribution.

The specific heat (c) represents the energy required to change a unit mass of material by one degree in temperature. In the thermal investigations of HSC, the c_p-values were measured through DSC (Netzsch Company, Germany) [4]. For EP-resin Ebalta, a c_p-value of 1.55 J/(g·K) was measured. This c_p-value corresponds to values in the literature, which are 1 J/(g·K), respectively 1.7-2.0 J/(g·K) [27, 28]. The determined c_p-value

Table 7.8. Calculated α-, c_P- und λ-values of MHSC in comparison to HSC, PC and metal foams

Materials	Linear coefficient of thermal expansion (α) [10^{-6} K^{-1}]	Specific heat (c_P) [J/(g·K)]	Thermal conductivity (λ) [W/(m·K)]
EP-Resin Araldit	72	1.55	0.15
Steel	12	0.5	45
Spheres Fillite	24.2	1	0.11
MHSC-II2aA	30.9	1.12	0.35
MHSC-II2bA	34.04	1.15	0.33
MHSC-II5A	30.9	1.12	0.35
MHSC-II7A	29.5	1.02	0.48
MHSC-II9A	28.03	1.01	0.48
MHSC-II11A	31.15	1.05	0.35
HSC-14 (Fillite+corundum)	22.3	1.33	0.33
Polymer concrete	10-20	0.9-1.3	1.3-3
Metal foam (Cymat)	19-21	0.83-0.87	0.3-10
Metal foam (Alporas)	21-23	0.83-0.87	3.5-4.5
Metal foam (ERG)	22-24	0.85-0.95	6-11

of corundum and E-spheres hollow spheres are 1 J/(g·K) [4]. This c_p-value corresponds to the values in literature for glass or ceramic [29]. The chemical composition of Fillite-spheres is similar with the chemical composition of E-spheres. Thus, a c_p-value of 1 J/(g·K) for Fillite was considered. The c_p-value of steel is less than 0.5 J/(g·K) [1].

For filled polymer systems with inorganic and powdery fillers a rule of mixtures can be written as

$$c_{p(T)} = (1-\psi)\cdot c_{pM}(T) + \psi \cdot c_{pF}(T), \tag{7.2}$$

where ψ represents the weight fraction of filler and c_{pM} and c_{pF} the specific heat of the polymer matrix and the filler, respectively [30]. The formula (7.2) was used with the *volume* fraction and a very good agreement with the measured values of HSC was obtained. The calculated c_p-values of MHSC are included in Table 7.8.

When analysing thermal processes, the thermal conductivity (λ) is the most commonly used property that helps in quantifying the transport of heat through a material. In the literature exits many expressions which help to compute the thermal conductivity of foamed or filled plastics. A rule of mixture, suggested by Knappe, commonly used to compute thermal conductivity of composite materials is expressed by the following equation:

$$\lambda_c = \frac{2\lambda_m + \lambda_f - 2\phi_f(\lambda_m - \lambda_f)}{2\lambda_m + \lambda_f + \phi_f(\lambda_m - \lambda_f)} \cdot \lambda_m, \tag{7.3}$$

where, ϕ_f is the volume fraction of filler, and λ_m, λ_f and λ_c are the thermal conductivities of the matrix, filler and composite, respectively [30].

The thermal conductivities of MHSC in Table 7.8 are calculated with Eq. (7.3). This value must be experimentally investigated by some samples of MHSC to validate the calculated results. The calculated λ-values of MHSC varied between 0.33 and 0.48 W/(m·K) and are similar to the measured HSC λ-value of 0.33 W/(m·K). The calculated λ-value of HSC is 0.312 W/(m·K). The MHSC and HSC can be considered as thermal *insulators* because of their low thermal conductivities. The polymer concretes transport faster the heat (see Table 7.8), but they are also considered as thermal insulators. Steel has a very good thermal conductivity which has a value of 45 W/(m·K). For other metal foams (Cymat, Alporas, ERG) [31-35] the specified thermal values are presented in Table 7.8, which depend on the densities of the foams. Nevertheless, they are similar to that of MHSC.

7.5 Application of MHSC

The selection of materials in lightweight construction in mechanical engineering is of paramount importance because the materials must have in addition to a small density a high stiffness. For comparing the materials the relation elastic modulus to density is used. Figure 7.10 presents the E-modulus, specific E-Modulus and the buckle goodness number of MHSC in comparison to this number of other lightweight materials.

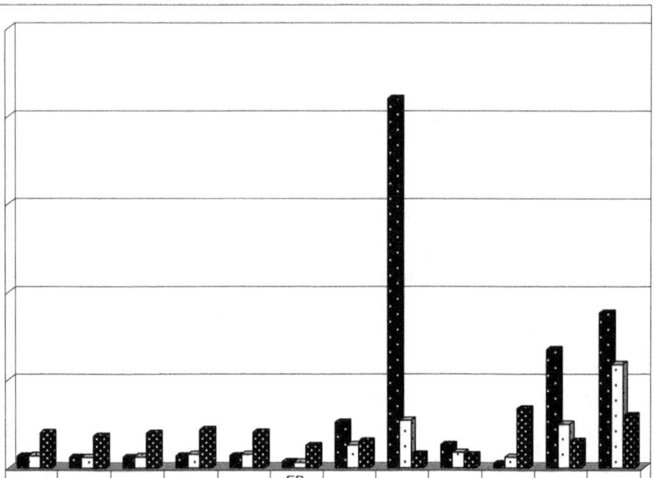

	MHSC-I2aE	MHSC-I2bE	MHSC-I3aE	MHSC-I5aE	HSC-14	EP-Resin Ebalta	PC	St	St 37-Foam	Al99,5-Foam	Al99,5	CRP
E-Modulus Gpa	6,6	5,8	5,9	6,6	6,8	3,5	25,7	210	13	2,4	67	88
Spezific E-Modulus GPa g/cm³	6,9	5,9	6,4	7,6	7,6	3	13	26,9	8,7	6	24,8	58,7
Buckle goodness MPa1/3 cm³/g	19,8	18,1	19,7	21,5	20,1	12,6	14,9	7,6	7,3	33,5	15,1	29,7

Fig. 7.10. Lightweight goodness number of MHSC in comparison to this number of other lightweight materials

The buckle goodness number characterises the behaviour or stability of the bar against buckling. All materials with buckle goodness number higher than that of aluminium are suitable for using them in constructions as lightweight materials. The buckle numbers of MHSC are greater than that of aluminium.

7.5.1 Application of MHSC as Material in Composites

The production of component parts of machine tools of MHSC takes place in analogy to the procedure of polymer concretes. The condition is that a fixed mould has to be developed and has to be manufactured. This procedure is similar to the processing of iron casting or aluminium alloy with the difference that for producing of MHSC simple form materials such as wood or artificial wood can be used, because the heating it is not so high as by casting of metals. The advantage is that the energy consumption for heating in comparison to casting alloys is cancelled. Furthermore, several connections, reinforced or functional elements can be integrated in the mould.

The stage of producing MHSC component parts will be demonstrated in the case of substitution and production of the SCARA robot arm (ADEPT Company). At the first step, it was necessary to demount the original arm from the robot. The dimensions of the original arm were obtained by using a measuring system in 3 directions from the Leitz Company. The dimensions in the region of the connection with the rest of the robot must be kept constant for the MHSC robot arm. Keeping these dimensions of

a. Upper mould

b. Lower mould filled with MHSC

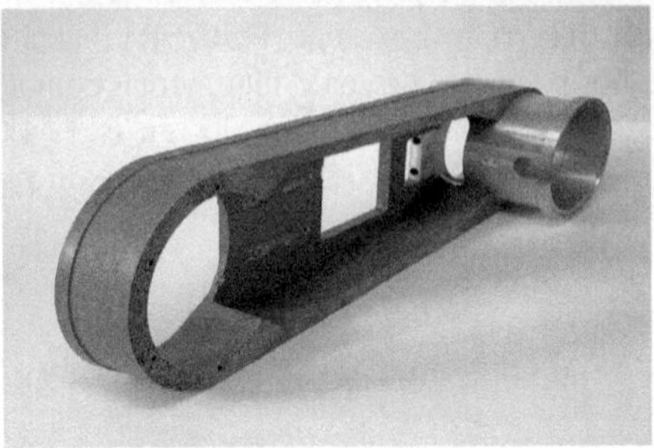

c. Upper half of robot arm

d. Robot arm of MHSC

Fig. 7.11. Mould for producing and the robot arm made of MHSC with carbon reinforcing and aluminium alloy connection elements

the arm constant, the new MHSC robot arm can be dimensioned through analytical or numerical methods. The robot arm is loaded with a bending load of 700 N. The finite element models were loaded with a flexural stress of 4000 N. The simulation indicates a wall thickness of the MHSC arm of 15 mm. The original arm has a wall thickness of 6 mm. The shape and structure of MHSC were modified in comparison to the original arm. To achieve higher stiffness and strength, the MHSC was partially reinforced with carbon bars and the connection elements were made of aluminium alloy. Based on the calculated dimensions and the imagination about the shape and structure the final casting CAD-model was created. Trough the CAD-design the mould for the casting of the robot arm was constructed of artificial wood and aluminium alloy (cf. Fig. 7.11a).

The connection elements and the carbon reinforcements were introduced and fixed in this mould (cf. Fig. 7.11b).

Furthermore, the MHSC of the variant II7aE were pressed in the prepared mould with an anti-adhesive solution. After one day the half of the robot arm was demoulded (cf. Fig. 7.11c).

The halves of the robot arm are bonded together with the same bonding medium such as the raw bonding epoxy resin Ebalta. The finished robot arm of MHSC with carbon reinforcement, shown in Fig. 7.11d, is 22 % lighter than the original arm of aluminium alloy. In addition, the stiffness-mass relation is only a half compared to the original arm.

With further reinforcements the stiffness of the MHSC robot arm can be improved. This robot arm was the first application of MHSC in combination with other materials and serves as example for further applications. Previous research has clearly demonstrated that with the reduction of the mass of the robot arms their speed can be increased and a reduction of cycle time as well as higher accuracy and energy consumption can be achieved [36-38].

Other example for applicability of MHSC as lightweight materials in combination with steel is the usage in a round rotating table of a pallet (rotor) in direct-drive design with magnetical guide and loading bearing system. Recently, direct-drive designs, with corresponding control systems, result in a substantial improvement in machine dynamics. The development of new lightweight composite construction for direct drive systems opens up new opportunities for the further reduction of masses thus allowing higher accelerations. Models in combination of polymer concrete (PC), hollow-sphere-composites (HSC) and HSC with fibre reinforcement (HSC/FR) based on law similarity [39] were developed in a previous work. The scale factor for geometrical and thermal similarity of the model in comparison to the main design is 1: 4.09 and for mass similarity 1: 68.4. Based on the developed models for comparison purposes a model of MHSC was produced and tested.

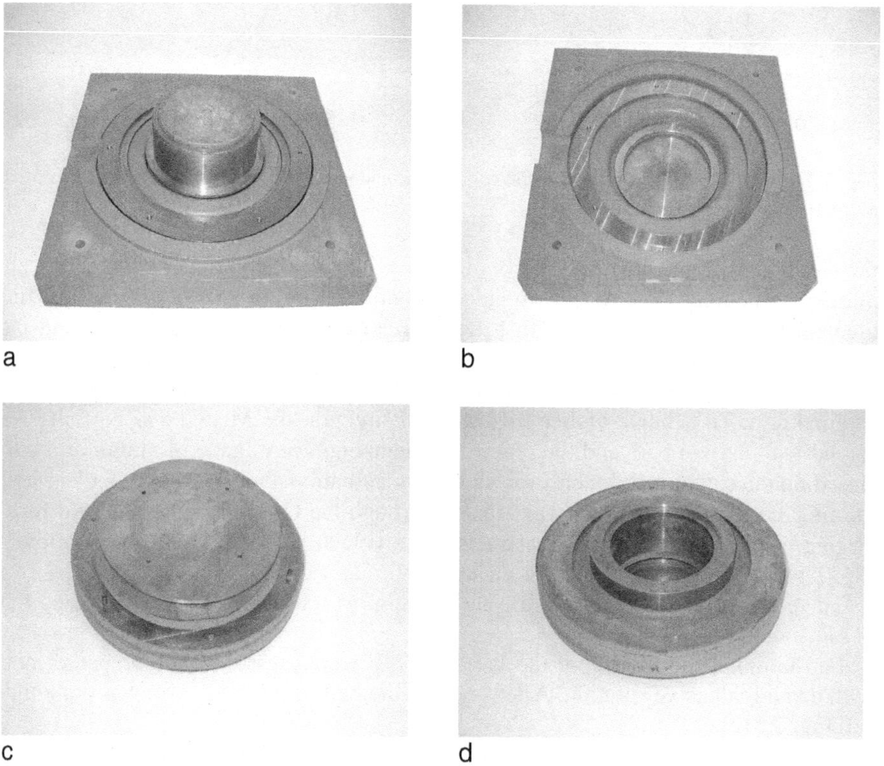

Fig. 7.12. Bottom (a) and top (b) part of the casting mould with insert components, Bottom (c) and top (d) view of MHSC-Steel-model of the round table

The magnet bearing of the round table requires the application of metal plates both radial and axial. The rotor of the torque-motor must remain of metal. The measuring rings of the positioning controller unit must be carried out of steel. The cover plate must be made of metal. For centring the cover plate a ring of steel must be availed. The rest of the volume can be substituted with light-weight materials. Figures 7.12a and 7.12b show the casting mould with inserted steel component and the core, which are screwed in the mould. A half bottom part was casted in the mould. After one day, the two half parts of the mould were screwed together and in the opening of the upper sides of the mould, the MHSC material was casted in the mould. For production of this round table model the variant II1aA with lower viscosity was selected. The bottom and upper view of the round table realised with MHSC and steel can be see in Figs. 7.12c and 7.12d.

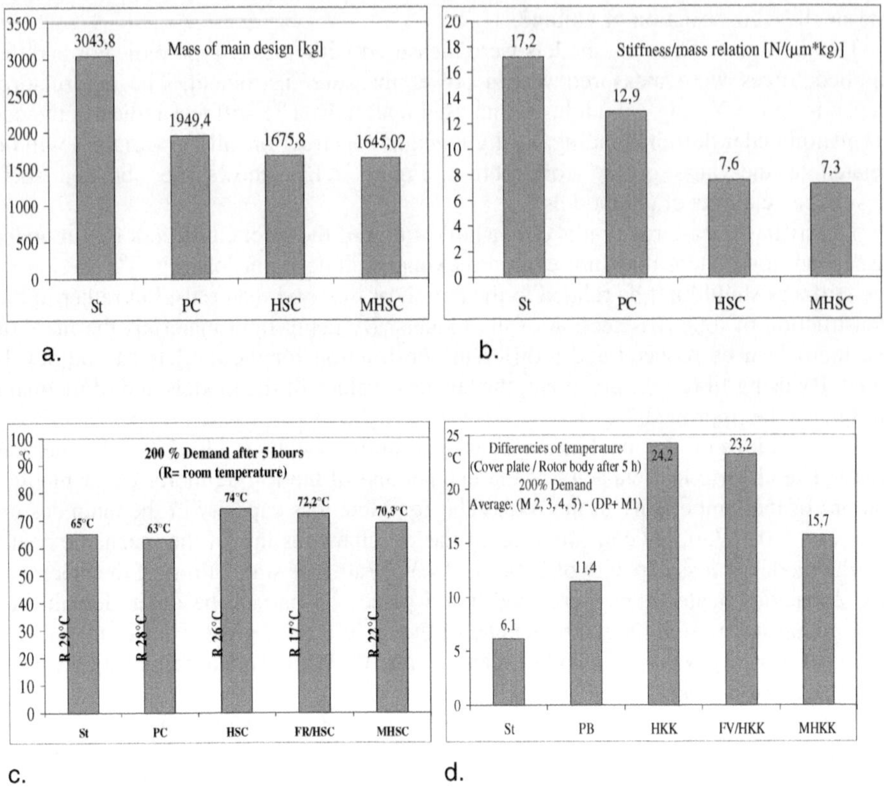

Fig. 7.13. a. Mass of main design of steel (St), polymer concrete (PC), hollow-sphere-composite (HSC) and metallic-hollow-sphere-composite (MHSC) and b. stiffness/mass relation of the models, c. achieved maximal temperature of all models by the measure place M3 and d. differences on the temperature by all the models

The model of MHSC without cover plate has a mass of 17.55 kg. Therefore, the metal insert parts have a mass of 15.6 kg. The rest in weight, i.e. less than 2 kg, represents the mass of the MHSC. Due to the scale factor of the mass similarity (1:68.4) between model and main design, the mass of the main design of steel (St), polymer concrete (PC), HSC and MHSC were obtained as shown in Fig. 7.13a.

The models were measured under static load in order to compare the static deformation and the stiffness. On the main design a load of 35000 N acts. This corresponds to a load of 8500 N in the case of the models. This static load was applied as a distributed load in the middle of the cover plate. The rigid constraint of the rotor was carried out in the place of the axial bearing. The static deformation is interesting by a central loading in the vicinity of the deficient zone of the model. In the nearness of the construction of the cross-section the greater deformations will be estimated. In these places, the deformations of the models will be measured. The construction of the cross-section was necessary in consequence of the placement of the torque-motor by the smallest construction of volume.

The deformations of the models were measured with a lever-type indicator and the applied forces were measured with a power measure instrument. The deformation depends on the Young's modulus of the used materials. The stiffness indicates the deformation under definite loading. By lightweight construction, all the variables will be related to the mass of the construction. Figure 7.13b shows the obtained stiffness/mass relations of the models.

The stiffness/mass relation is only a half of that of the steel construction. But under real conditions, there does not exist a maximal middle point loading. The deficit of the stiffness should not be related to the deficit of the used materials, but rather to the construction of the cross-section of the models. By using light materials the mass of the motor can be reduced and a different construction for the models can be developed. By using fibre reinforcement, the stiffness values of the models and of the main design can be improved.

Deformations on the machines appear in addition to static loads also under thermal loads. The electric loss capacity of the magnet and of the torque-motor occur modifications of the temperature in the rotor. The complete loss capacity of the main design arises to 2100 Watt. By consideration of the simultaneous law of the mechanic result for the models a loss capacity of 2100:68.4= 30 Watt. For simulations of the electrical loss capacities heater mats were used. On 5 places on pair in the outer sides heater mats radial were used. On the underside of the table were 2 greater heater mats used. The heat capacity was regulated between 50 and 100 Watt. The thermo elements were fixed on 5 critical places.

Figure 7.13c shows the average of 3 measurements of the achieved maximal temperature on the measuring place M3 in the outer margin with a demand of 200 % (100 Watt).

The achieved values in the exterior margin by the model of MHSC after 5 hours are greater than that by the model of steel or PC. However, they are smaller than that of the models of HSC or HSC with fibres. The same tendencies show the differences of the achieved temperatures of the model in Fig. 7.13d.

The better thermal behaviour of the model of MHSC can be probably attributed to the metal character of the hollow spheres and depend on the thermal properties (conductivity) of MHSC (cf. Table 7.8). The cellular structure of the spheres avoids a better thermal behaviour.

For improvement of the abandonment of production with regard to function, accuracy, costs etc., jigs with high stiffness and damping behaviour are required. Klaeger demonstrated that by substituting of the base materials steel for jig by PC in combination with steel insert parts results in positive effects such as increased endurance, damping behaviour and decreased deflection [3]. The realised jig of MHSC with steel in the scale factor of 2:1 can be seen in Fig. 7.14a and 7.14b.

The jig consists of 6 MHSC parts, which are bonded together with the same bonding medium such as the raw material, i.e. EP-resin Ebalta. The variant II7aE was used to produce this jig. The metal functional surface and the metal bar are put into the mould. Then, the MHSC mixture in the composite is cast with metal parts. The weight of the jig of MHSC-composite-construction is 1700 g and the weight of the similar jig made of steel reaches 3485 g.

For the extension of the application field of MHSC, a housing of an electromotor was developed together with companies of electro mechanical engineering. Figure 7.14c shows a housing of a three-phase alternating current motor in cage rotor type made of MHSC.

The base plate is made of steel. The electricmotor was tested with three different loading types. The runs of the speed-torque characteristics were satisfying and agree

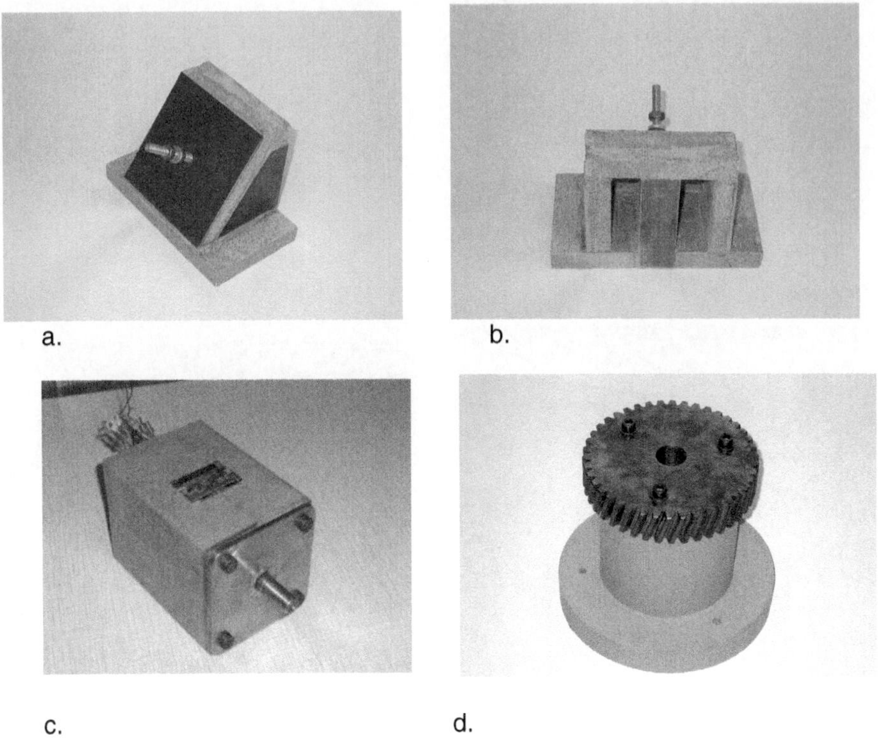

Fig. 7.14. a. and b. jig made of MHSC and steel, c. housing of electromotor made of MHSC and steel, d. gear fixture

with the characteristics of motors with same constructions made of steel. Furthermore, thermal inquires by 100 % permanent loads were carried out. The achieved temperature underlies the acceptable limit of 80°C. For applications of MHSC in electromechanical engineering further investigations are needed.

Figure 7.14d shows a further practical example of a holding fixture for gear processing.

The basal body consists of MHSC variant II9aA. The gear of the basal body was put about for gear processing.

7.5.2 Application of MHSC as Core or Filled Material

Another modality is to fill hollow construction or profiles with MHSC and thus to increase the stiffness or stability of these constructions. The form made of metal or polymeric materials serves as external surfaces. In this procedure the MHSC will be not demoulded. Figure 7.15a presents a holder made of aluminium alloy, which is filled with MHSC to improve the stiffness and stability of the hollow profiles in busses.

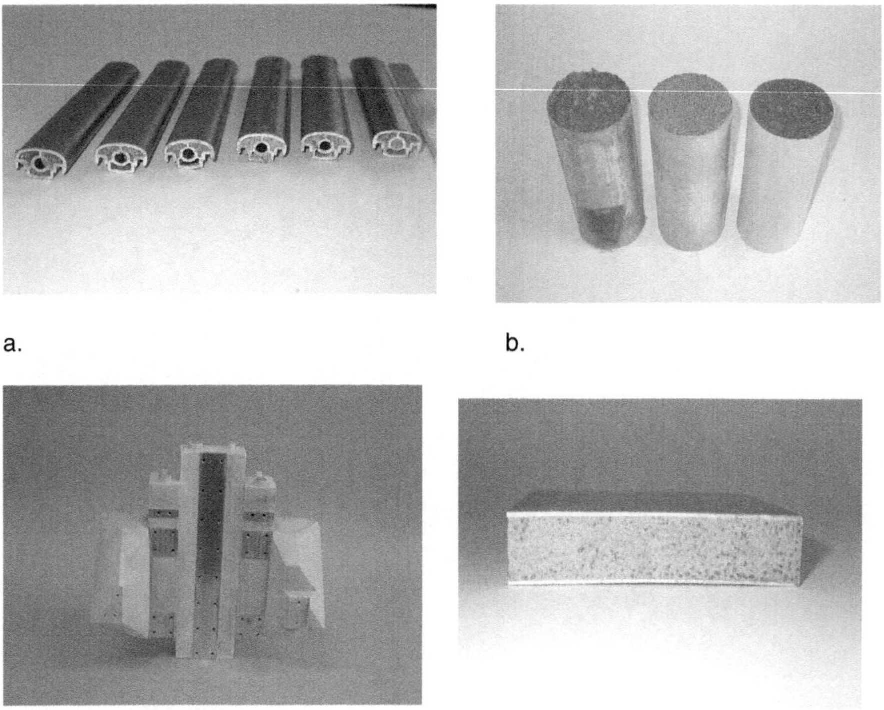

Fig. 7.15. a. Holder of aluminium alloy filled with MHSC, b. polyamide (PA) hollow cylinder filled with MHSC, c. Mirror sliding head of LASER-plant consist of shell of PA and a filler of MHSC, d. sandwich composite materials

Due to the small cavity in the interior of the profiles, the MHSC variants II2eA-II2cA with the smallest MHS of 1.5 mm and with a higher volume percent part of resin are used.

The stereo lithography is adequate to produce very complicated and filigree forms. Based on this method, hollow cylinders of polyamide (PA) with 100 mm in length and 250 mm in diameter were produced. These cylinders are filled with MHSC-II2aE and PC (cf. Fig. 7.15b).

The mechanical behaviour was investigated by the SITEC Company and demonstrated very good results. In this way, a mantel (shell) of a mirror sliding head of a LASER-plant [40] was produced. The metal functional parts are bonded on the PA-shell. Since the openings for filling are very small, the form is very difficult. As a result, the MHSC-variant II2bA with 28 volume percent part of EP-resin Araldit must be used (cf. Fig. 7.15c).

This technology produces so complicated component parts and further investigations with a shell with higher wall thickness are necessary.

Another possibility to use MHSC is as core material between two overlays of metal or fibre laminates. Figure 7.15d presents such a sandwich composite materials.

As core material MHSC can absorb the compression, due to their very good compression behaviour, while the cover plates have to react of tensile strength. Advantageous by a sandwich construction with MHSC is that the joint between cover and core can be omitted. The cover plate or fibre laminate can be integrated into the mould and the binder between plate and core of MHSC can be substituted with the bonding medium of MHSC. For determination of the characteristic values of the sandwich structure further investigations are necessary.

7.6 Conclusions

Metallic hollow spheres can be used in a polymeric material such as epoxy resin together with another filler of smaller grain size of non-metallic nature to produce the new MHSC. The way of production and the composition were derived from polymer concrete. Polymer concrete is a polymeric based material with silicate filler, which is frequently used to produce different component parts of machine tools. MHSC is predestined to be used for lightweight parts of machines, because of their lower density caused by the hollow spheres. In this research were developed 50 variants with different grain sizes, volume fractions or fillers. Some mechanical properties were principally determined. It could be shown that the filler improved the Young's modulus and reduced the strength as well as the thermal expansion coefficient in comparison to the unfilled basic resin material. For mechanical engineering application, it is important to have a higher Young's modulus to assure a higher stiffness of the component part. As a result, the machine can achieve a higher accuracy during operation. It was demonstrated that some properties such as density, tensile or bending strength of MHSC are superior to them of HSC based only on non-metallic hollow spheres. A help by selection of materials for lightweight applications are their characteristic values, such as the buckle goodness. Because this value is for MHSC higher than for aluminium alloys, it confesses that MHSC can be very advantageous for lightweight constructions. For improvement of the mechanical behaviour of MHSC, the material can be cured or reinforced with fibres. The fibre reinforcement or metal plates for sandwich core materials can be integrated into the mould. This is a further advantage of MHSC in addition to the low production cost. Because of the application of cold-hardening resins, no external heat consumption is needed.

The exemplarily realised machine parts demonstrate the way of application of MHSC respectively of MHS in a polymeric composite material. They can be used as casing or housing (such as in the case of the robot arm or of the electromotor housing), in fixture constructions (such as the gear holding fixture, jig, round table) or as core material to fill hollow profiles, structures as well as in sandwich structures. It was demonstrated that the mechanical and thermal behaviour in case of the round table of MHSC are appropriate to them of PC and the thermal behaviour of the construction of MHSC are superior to them of HSC and HSC with fibre. This can be attributed to the better thermal behaviour of MHSC based on MHS. The thermal properties are only calculated. This was the first work to use MHS as filler and the characteristic properties are not completely investigated and they must be deeper elaborated for the real application MHSC in series production. Because of the lightweight character of MHSC, the energy consumption can be reduced which can positively influence the ecology. Furthermore, MHSC can be used for traffic, production or logistic. In addition to a good thermal insulation they have very good damping properties. This can be another advantage of MHSC for application in machine tools. However, this must be still demonstrated.

References

[1] Weck, M.: Werkzeugmaschinen und Fertigungssysteme 2: Konstruktion und Berechnung. Springer, Berlin (1997)

[2] Weck, M.: Werkzeugmaschinen und Fertigungssysteme 1: Maschinenarten und Anwendungsbereiche. Springer, Berlin (1998)

[3] Wiele, H., Westkämper, E., et al.: Verbesserung des thermischen und dynamischen Verhaltens von Werkzeugmaschinenbauteilen durch Verbundkonstruktionen aus Metall und Mineralguss. Die Maschine 5(6), 55–59 (1994)

[4] Klaeger, U.: Untersuchungen zum Einsatz von Mineralguß für Werkstückspannvorrichtungen in der spanende Fertigung. Thesis work, Otto-von-Guericke-University of Magdeburg, Magdeburg (1994)

[5] Baumeister, E.: Hohlkugelkomposit - Charakterisierung thermischer und mechanischer Eigenschaften eines neuen Leichtbauwerkstoffes. Thesis work, Otto-von-Guericke-University of Magdeburg, Magdeburg, cf (2004), `http://diglib.uni-magdeburg.de/Dissertationen/2004/eribaumeister.pdf`

[6] Baumeister, E., Klaeger, S., Kaldos, A.: Characterisation and application of Hollow-sphere-composite lightweight materials. JMDA57, Proc. IMechE Part L: J. Materials: Design and Applications, IMechE 219, 207–216 (2005)

[7] Menz, P.: Maschinenbaugruppen aus Kompositmaterial, DE 1952367 (1996)

[8] Baumeister, E., Klaeger, S., Lierath, F.: Reactive resin bonded light-weigth materials for mecahnichal engineering applications. In: Gyenge, C.S. (ed.): Technical University of Cluj-Napoca, Annals of MTeM for, & Proceedings of the 6th International Symposium, Cluj-Napoca, pp. 35–38 (2003)
[9] Flemming, M., Ziegmann, G., Roth, S.: Faserverbundbauweisen. In: Fertigungsverfahren mit duroplastischer Matrix. Springer, Heidelberg (1999)
[10] N. N.: Data from Ebalta Company, 2007
[11] N. N.: Data from Bodo-Müller-Chemie about Araldit (2005)
[12] N. N.: Data from DSM Company about UP-Resin Palatal (2005)
[13] Bergmann, H.W.: Konstruktionsgrundlagen für Faserverbundbauteile. Springer, Berlin (1992)
[14] N. N.: Data from Glatt Company (2007)
[15] DIN 1045: Beton, Stahlbeton, Bemessung und Ausführung (July 1988)
[16] `http://www.fillite.com` (2008)
[17] N. N.: Data from Omya (1999-2001)
[18] N. N.: Data from Osthoff-Petrasch (1999)
[19] N. N.: Data from Gößl+Pfaff (1999)
[20] `http://omegaminerals.com` (2008)
[21] DIN EN ISO 527-4: Kunststoffe. Bestimmung der Zugeigenschaften. Teil 4: Prüfbedingungen für isotrop und anisotrop faserverkstärkte Kunststoffverbundwerkstoffe (1997)
[22] DIN EN ISO 604: Kunststoffe - Bestimmung von Druckeigenschaften (1997)
[23] DIN EN ISO 178: Bestimmung der Biegeeigenschaften – Kunststoffe (1996)
[24] N. N.: Technical data from the Schwarzwälder Textil-Werke (2006)
[25] N. N.: Faserverbundwerkstoffe. Neue Technologien. Neue Werkstoffe. Handbuch. R G GmbH, D-71111 Waldenbuch, `http://www.r-g.de`
[26] Öchsner, A., Murch, G.E., Lemos, M.J.S.: Cellular and porous materials. Thermal properties simulation and prediction. Wiley-VCH Verlag GmbH, Weinheim (2008)
[27] Gibson, L.J., Ashby, M.F.: Cellular Solids. Structure and Properties. Cambridge University Press, Cambridge (2008)
[28] VDI-Gesselscahft Verfahrenstechnik und Chemieingeniuerwesen, VDI-Wärmeatlas. Berechnungsblätter für den Wärmeübergang. Springer, Heidelberg (1997)
[29] Petzold, A.: Anorganisch-nichtmetallische Werkstoffe. Charakteristik-Eigenschaften-Anwendungsverhalten. Deutscher Verlag für Grundstoffindustrie, Leipzig/Stuttgart (1992)
[30] Osswald, T.A., Menges, G.: Material Science of Polymers for Engineers. Hanser Publishers, Munich (1996)
[31] Ashby, M.F., Evans, A., Fleck, N.A., Gibson, L.J., Hutchinson, J.W., Wadley, H.N.G.: Metal Foams: A Design Guide. Butterworth Heinemann, Boston (2000)
[32] Bahnart, J., Ashby, M., Fleck, N.: Cellular Metals and Metal Foaming Technology. MIT Verlag, Bremen (2001)
[33] Degischer, H.P., Kriszt, B.: Handbook of Cellular Metals. Production, Processing, Applications. WILEY-VCH Verlag, Weinheim (2002)
[34] Bahnart, J., Ashby, M.F., Fleck, N.A.: Metal Foams and Porous Metal Structures. In: International Conference on Metal Foams and porous Metal Structure, June 14-16. Verlag Metal Innovation Technologies, Bremen (1999)
[35] N. N.: ERG Duocel aluminium foam. ERG Corporate Literature and Reports (accessed 1998), `http://ergaerospace.com/lit.html`
[36] Gerstmann, U., Livotov, P.: Systemverhalten von Industrierobotern mit massenreduzierten Strukturen. VDI-Z 135, Nr. 4: 99--104 (1993)

[37] Hohlman, N., Livtov, P.: Wirtschaftliche Leichtbaustrukturen auf Basis von Faserverstärkten Kunststoffhalbzeugen. VDI-Berichte Nr. 1080, VDI-Gesellschaft Werkstofftechnik, Werkstofftag: Leichtbaustrukturen und leichte Bauteile, Duisburg. VDI Verlag, Düsseldorf, 323–343 (1994)
[38] Lee, D.G., Kim, K.S., Kwak, J.K.: Manufacturing of a scara type di-rect-drive robot with graphite fiber epoxy composite material. Robotica 9, 219–229 (1991)
[39] Bär, J.: Einsatz ausgewählter Polymerwerkstoffe im Werkzeugmaschinenbau. Dissertation, Otto-von-Guericke-University Magdeburg. Shaker Verlag, Aachen (2007)
[40] SITEC Industrietechnologie GmbH, Bauteil, DE 202007004683U1 (2007)

8

Dynamic Behaviour of Metallic Hollow Sphere Structures

Matej Vesenjak[1], Thomas Fiedler[2], Zoran Ren[3], and Andreas Öchsner[4]

[1] University of Maribor, Faculty of Mechanical Engineering, Smetanova 17, SI-2000 Maribor, Slovenia
m.vesenjak@uni-mb.si
[2] The University of Newcastle, School of Engineering, Callaghan, New South Wales 2308, Australia
thomas.fielder@newcastle.edu.au
[3] University of Maribor, Faculty of Mechanical Engineering, Smetanova 17, SI-2000 Maribor, Slovenia
ren@uni-mb.si
[4] Technical University of Malaysia, Faculty of Mechanical Engineering, 81310 UTM Skudai, Johor, Malaysia
andreas.oechsner@gmail.com

Abstract. The chapter focuses on the dynamic behaviour of metallic hollow sphere structures that constitute an innovative group of lightweight materials, combining high specific stiffness, good damping properties and the ability to absorb large amounts of energy at a constant low stress level. The chapter explains the methodology and results of computational experimenting to clarify and determine the individual influences on the macroscopic behaviour of MHSS, especially under dynamic loading conditions. In the beginning, the material strain rate dependency is described and formulated with several constitutive models. A very important factor at impact loading is material deformation capability and impact energy absorption, which directly influences the deceleration of impacting objects. The impact energy absorption of hollow sphere structures due to their plastic deformation under impact loading is emphasized. The second part of this chapter presents the computational results of metallic hollow sphere structures and their macroscopic behaviour under uniaxial dynamic loading conditions with additional material characterisation considering large strains. Furthermore, the influence of gas inside the metallic spheres on behaviour of metallic hollow sphere structures and their capability of impact energy absorption is addressed. Computational simulations show that it is possible to achieve different dynamic response of metallic hollow sphere structures when subjected to dynamic loading. The topology, wall thickness of spheres and strain rate sensitivity can be combined in a way that the structure response is adapted to a given engineering problem. The chapter concludes with a discussion of the advantages, disadvantages and limitations of dynamically loaded metallic hollow sphere structures and their computational models.

Keywords: Metallic hollow sphere structures, dynamic behaviour, topology, morphology, strain rate, computational simulations.

8.1 Introduction

Metallic foams have been increasingly used in various industrial engineering applications. Consequently, several manufacturing processes have been developed to increase

A. Öchsner, C. Augustin (Eds.): Multifunctional Metallic Hollow Sphere Struct., EM, pp. 137–158.
springerlink.com

their quality and efficiency [9]. However, there are still many imperfections (e.g. missing cell walls, variable intercellular wall thicknesses, cracks, irregular topology) present in the metallic foams that influence their mechanical and thermal properties. The hollow sphere structures with constant cell size and wall thickness are a promising alternative [9]. The manufacturing procedure provides for high uniformity and regularity of hollow spheres distribution without major scattering of their material properties [2]. There is a wide field of potential applications with the well known advantages, e.g. high capacity of energy absorption, excellent damping behaviour and isolation.

Most investigations on MHSS in the past were primarily concerned with determination of the quasi-static properties [8, 25]. However, some recent investigations have shown that sintered hollow spheres in Hopkinson Pressure Bar tests denoted a distinct strength enhancement under impact loading in comparison to results of quasi static loading tests. Further analyses of epoxy syntactic foams with spherical inclusions confirm the necessity of accounting for the strain rate sensitivity of porous metals under impact conditions [19, 26]. A comprehensive experimental study of the impact behaviour of metallic cellular materials is also described by Zhao et al. [34]. Neville and Rabiei found superior compressive strengths and energy absorption capabilities for syntactic steel MHSS [21]. In general, research results of these structures show that their behaviour is strongly influenced by their topology, morphology and base material properties (including strain rate and temperature dependencies). A finite element analysis of the impact behaviour of syntactic adhered MHSS with cubic symmetries was reported by Fiedler [7], this chapter describes the methodology and results of computational experimenting to clarify and determine the individual influences on the macroscopic behaviour of MHSS, especially under dynamic loading conditions.

8.1.1 Strain Rate Sensitivity

Most materials behave differently under dynamic loading conditions because strain rate influences their mechanical properties. The effect of strong strain-rate dependence has already been observed by Ohrndorf et al. [23], by studying mechanical properties of closed-cell structures at strain rates up to 2 s^{-1}. The cellular materials have a significantly different stress-strain relationship at higher strain-rates where the characteristic plateau stress is not so prominent. Increased strain rate does not affect their Young's modulus but it increases their yield and plateau stress [10].

Most frequently cellular structures that are used in impact applications experience strain rates typically between 1 s^{-1} and 10^5 s^{-1} [10].

The extent of the strain-rate effect depends on the material and the temperature [15]. For materials with high strain-rate dependency the strain localization occurs very slowly. At high strain rates dynamic (inertial) effects increase the material strength. Gibson and Ashby [10] described three features of dynamic crushing influence: localisation (concentration of deformation; geometric softening), micro-inertia (motion of cell walls) and densification (steep increase of stress). The impact forces transmitted through the cellular structure can be evaluated with the impact sensitivity factor, representing the ratio between the kinetic energy of impacting body and the energy necessary to crush one layer of cells. The material rate-dependence is usually taken into account with straightforward incorporation into the constitutive material model [33].

For many materials the strain rate sensitivity at a constant temperature and certain strain can be described with the following power-law expression

$$\sigma = C \cdot \dot{\varepsilon}^{m}, \tag{8.1}$$

where the exponent m represents the strain rate sensitivity [15]. The relative levels of stress at two strain rates, evaluated at the same strain, are given by

$$\frac{\sigma_2}{\sigma_1} = \left(\frac{\dot{\varepsilon}_2}{\dot{\varepsilon}_1}\right)^{m}. \tag{8.2}$$

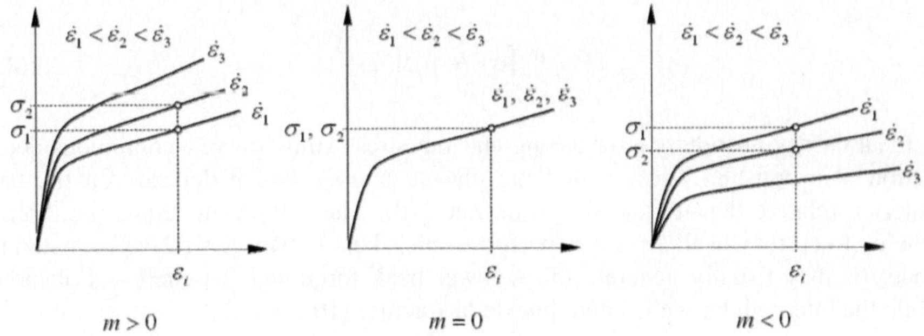

Fig. 8.1. Strain rate sensitivity of the material

The strength of the material under dynamic loading (Fig. 8.1) increases, if $m > 0$, and decreases if $m < 0$ while it remains the same, if $m = 0$. The values for the strain rate sensitivity parameter are listed in [15].

The most commonly used constitutive model in dynamic simulations accounting for strain rate sensitivity is the Cowper-Symonds model, which is based on scaling the quasi-static material stress – strain response

$$\sigma = \sigma_{stat} \cdot \left[1 + \left(\frac{\dot{\varepsilon}}{C}\right)^{\frac{1}{p}}\right], \tag{8.3}$$

where σ corresponds to the stress at strain rate $\dot{\varepsilon}$ and σ_{stat} is the quasi-static response of the material. Parameters C and p define the strain rate sensitivity [1, 3, 12, 13, 16]. An advanced constitutive model accounting for strain rate sensitivity is the Johnson-Cook model, which also considers the temperature influence [12, 13, 16, 27]. The comparison of the strain rate sensitivity for the mentioned constitutive models can be expressed as

$$\frac{\sigma_2}{\sigma_1} = \underbrace{\left(\frac{\dot{\varepsilon}_2}{\dot{\varepsilon}_1}\right)^{m}}_{\text{power law}} \equiv \underbrace{1 + \left(\frac{\dot{\varepsilon}}{C}\right)^{\frac{1}{p}}}_{\substack{\text{Copwer-Symonds} \\ \sigma_1 = \sigma_{stat}}} \equiv \underbrace{1 + c \cdot \ln\left(\frac{\dot{\varepsilon}_2}{\dot{\varepsilon}_1}\right)}_{\text{Johnson-Cook}}. \tag{8.4}$$

From Eq. (8.4) it can be concluded that the strain rate sensitivity is described in power law and Johnson-Cook model with a single parameter. In contrast, the Cowper-Symonds model uses two parameters, which allows for better calibration of the model. However, it has to be taken into account that the response of the analyzed model can be simulated with the highest accuracy, if the stress-strain curves for different strain rates obtained experimentally are directly used in a constitutive model.

8.1.2 Impact Energy Absorption

The cellular materials are able to absorb a significant amount of impact energy through their elastic and plastic deformation during the loading process, which is represented by the area under the strain-stress curve (Fig. 8.2)

$$U = \int_V \left(\int_\varepsilon \sigma \cdot d\varepsilon \right) \cdot dV \, . \tag{8.5}$$

Ideal energy absorbers have a long and flat stress-strain curve, (continuous deformation at a constant stress called the plateau stress), which depends on the base material, relative density and the strain rate [10]. The cellular materials (i.e. hollow sphere structures) in this respect perform much better than general solid materials, since: (i) they usually generate much lower peak force and (ii) energy is absorbed while the intercellular walls bend, buckle or fracture [10].

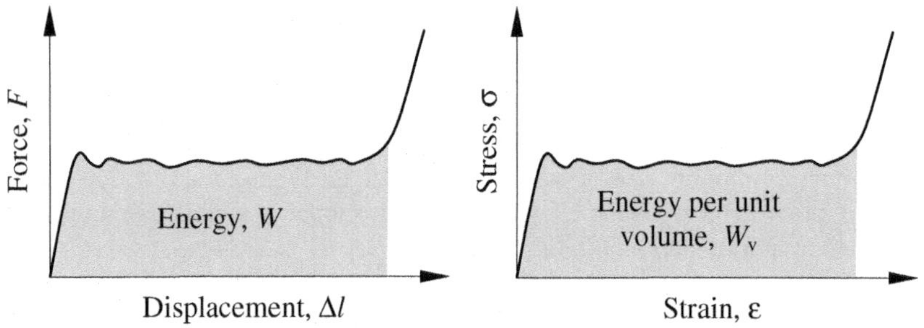

Fig. 8.2. Force-displacement (left) and stress-strain (right) curve of a general cellular material

With a proper choice of cellular material design parameters, its response can be easily adjusted to certain application demands. Elastic buckling, plastic yielding and crushing of cell walls during deformation contribute to nearly constant load resistance until the cellular material is almost completely densified. Optimum selection usually requires Rusch curves, J-factor or C-factor methods to construct energy absorption diagrams which condense the data for energy absorption as a function of peak stress, strain rate, relative density and type of the cellular material [10].

Although the strain rate and temperature effects strongly influence each other and consequently the mechanical response of the cellular structure, only the strain rate effects and sensitivity will be considered and discussed in this chapter [10].

8.1.3 Internal Pore Pressure Influence

Usually the cellular structures contain gases or liquids, which are captured in the cells at closed-cell structures and can flow through the structure at open-cell structures [10]. When the closed-cell cellular structure is deformed, the gas in cells compresses or expands. The higher the pore pressure the higher the influence on the macroscopic mechanical behaviour of the cellular structure is, at least until the gas diffusion equalises the pore pressure with the atmospheric pressure [28-30]. The pore pressure increase is also highly sensitive to the temperature change, which can additionally contribute to the pore pressure increase [27, 29, 30, 32].

8.2 Computational Models of MHSS

8.2.1 CAD and FE Model

Theoretically, the finite element model of a random MHSS structure can be generated. To obtain meaningful computational results for the overall response of a structure subjected to dynamic loading, one needs to use a large enough representative volume to account for structure randomness and to avoid the influence of free surfaces, e.g. boundary effects. The required refinement of the mesh in conjunction with its dimensions yields rather large computational models with a significant number of unknowns. The problem size usually exceeds the capacity of the available computer hardware.

In general, cellular materials exhibit a random arrangement of spheres (topology). However, the MHSS have a certain degree of regular topology, as shown in Fig. 8.3. Analysis of the pseudo-random MHSS topology shows that a cubic symmetric arrangement of the hollow spheres can be presumed in order to decrease the model size.

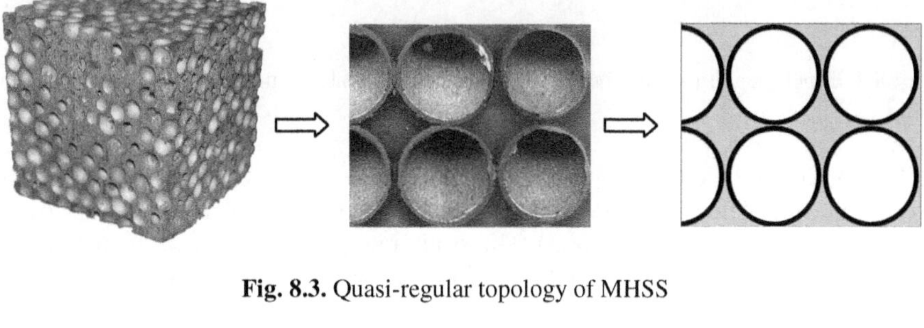

Fig. 8.3. Quasi-regular topology of MHSS

Three different cubic symmetries of hollow sphere structures are considered: (a) simple cubic, (b) body centred cubic and (c) face centred cubic arrangement (Fig. 4.4, chapter 4).

The characteristic dimensions of the MHSS were determined from measurements of experimental samples. For the chosen MHSS with syntactic and partial morphology, the outer radius *R* of the spheres is equal to 1.0 mm and the minimum distance between two neighbouring spheres is a_{min} = 0.36 mm. The thickness t of the metallic shells varied between 0.035, 0.05 and 0.075 mm.

8.2.2 Material Properties

Constitutive behaviour was described with a linear-elastic / ideal-plastic material model with the von Mises yield condition. The sintered steel sphere was characterised by Young's modulus of E = 115000 N/mm², Poisson's ratio ν = 0.3, initial yield strength σ_y = 255 N/mm² and density ρ_{St} = 6.95 g/dm^3 [4].

Experimentally determined quasi-static material properties of the L1100 epoxy resin (Fig. 6.4) were: Young's modulus E = 2460 N/mm², the 0.2% offset yield strength $k_{0.2}$ = 113 N/mm², Poisson's ratio ν = 0.36 and the density ρ_{St} = 1.13 g/dm^3 [17, 31].

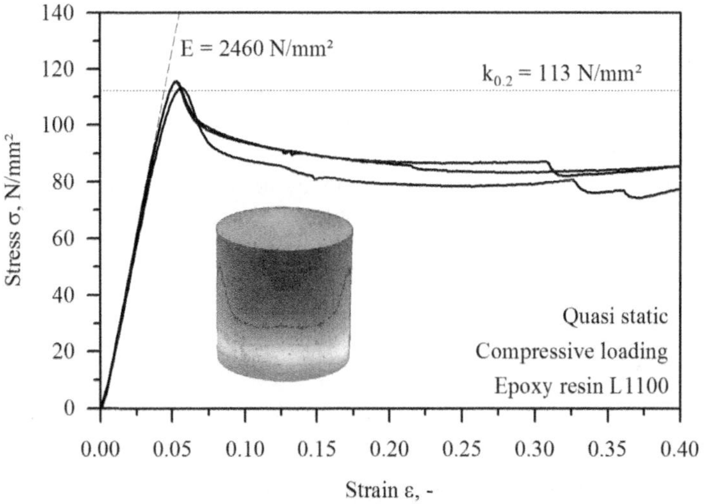

Fig. 8.4. Experimental measurements of the quasi static material properties of the L1100 epoxy resin

To confine the model complexity, creep and relaxation behaviour of the adhesive were neglected. The strain rate effects were considered by implementing the Cowper-Symonds constitutive relation (8.3). The strain rate parameters used for steel and adhesive were C = 40.4 s^{-1}, p = 5 and C = 1050 s^{-1}, p = 3.7, respectively [3, 16, 27].

8.2.3 Initial and Boundary Conditions

Assuming cubic symmetry of the geometry and the applied loads, only one eighth of a unit cell needed to be modelled, thereby significantly reducing the number of unknowns and the required computing time. Reflective symmetry boundary conditions (Fig. 4.6, chapter 4) were defined along the symmetry planes to simulate the behaviour of the whole unit cell [5]. The definition of the suitable boundary conditions allows for reduction of the required geometry within the finite element meshes. The periodic boundary conditions (repetitive boundary conditions, Fig. 4.6, chapter 4) presume that the considered unit cell is positioned in the middle of a regular hollow sphere structure, where boundary effects are negligible [36]. Additionally, it was assumed that the co-operative deformation of multiple cells and associated longer range buckling modes do not influence the buckling modes of a single unit cell [7, 22, 27, 31]. Additionally, the sphere surface elements were defined as one contact group to account for self-contact at very large deformations [12, 14]. The coefficient of friction was set to 0.1. The displacement controlled loading, corresponding to desired strain rate (up to $\dot{\varepsilon} = 1000\ \mathrm{s}^{-1}$), was applied to the upper surface of the unit cell model. Figure 4.6 (chapter 4) shows the finite element discretisation and boundary conditions of syntactic MHSS in a section normal to the *x*-axis.

Hexahedron fully integrated finite elements with quadratic shape functions were used in the discretisation of the models (Fig. 8.5 and 8.6).

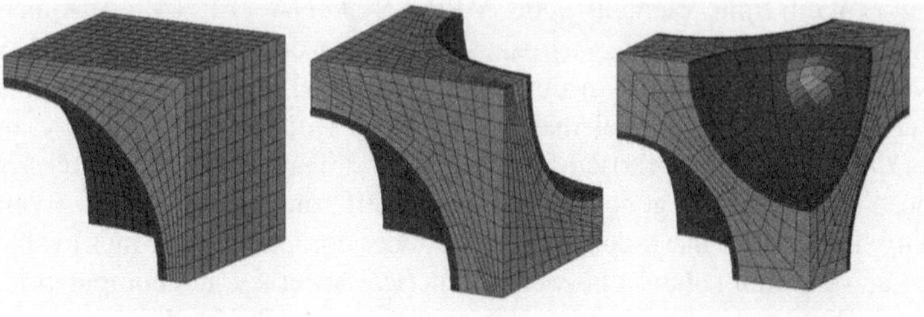

a) b) c)

Fig. 8.5. Finite element computational models of MHSS with syntactic morphology: a) simple cubic; b) body centred cubic; c) face centred cubic

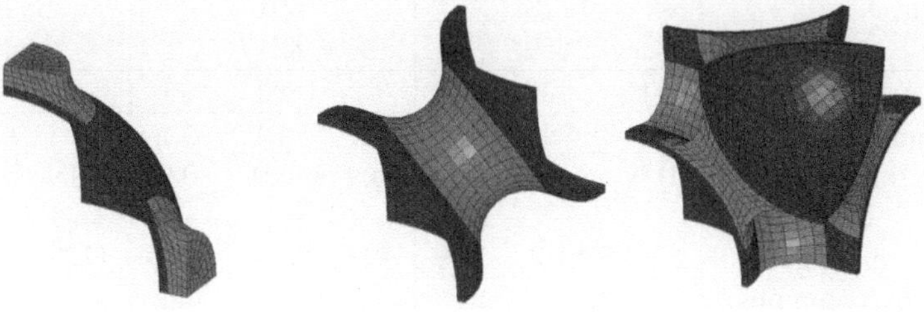

a) b) c)

Fig. 8.6. Finite element computational models of MHSS with partial morphology: a) simple cubic; b) body centred cubic; c) face centred cubic

8.2.4 Computational Simulations

Solution techniques for solving a dynamic problem in time domain can be divided into two groups: (i) implicit methods (most accurate but computationally intensive) and (ii) explicit methods (less accurate but computationally very efficient and naturally suitable for parallelisation). However, the explicit methods are only conditionally stable (time step must be smaller than the critical time step), while the implicit methods exhibit unconditional stability (time step can be arbitrary) [16, 24, 27]. For solving highly nonlinear dynamic problems under impact loading conditions or at high strain rates, the explicit methods usually perform better, since they provide the best compromise between the computational time and the result accuracy for this type of problems [27]. In explicit dynamic finite element analyses it is often the case that finite elements with reduced integration are being used to decrease the computational effort. However, since in these elements the Gaussian integration point is positioned on the symmetry axis of the element, hourglassing energy may appear in some cases. This necessitates control of the hourglass energy throughout the simulation [13, 16, 35].

Due to stress propagation through discretised structure numerical filters are commonly used during post-processing to improve results of the dynamic structural response. That way the undesired frequencies, which blur the measured signal, can be eliminated. Many up-to-date engineering codes have already integrated most common filter algorithms (e.g. SAE, Butterworth) [20, 27].

The MHSS behaviour under uniaxial dynamic loading conditions was analysed by using the explicit finite element code MPP LS-DYNA [12, 14]. Adequate element types, mesh densities, and time step sizes have been determined by initial parametric analyses to establish sufficiently accurate computational results insensitive to space and time discretisation. The finite element model with simple cubic topology and a shell thickness t = 0.05 mm was chosen for this purpose. Three model discretisations of different mesh densities were generated using two different MPP LS-DYNA versions and re-meshing. Increase of mesh density yields longer computational times (Table 8.1) and therefore, an optimum solution between numerical accuracy and computation time had to be found. The results of these analyses are summarised in Fig. 8.7.

Table 8.1. Comparison of the mesh density and time step sensitivity

	Mesh I			Mesh II			Mesh III	
Elements	1728			6912			55296	
Nodes	2169			8217			60401	
Solver parameters	#1	#2	#3	#1	#2	#3	#1	#2
Init. time step [µs]	4.1	4.1	4.1	4.1	4.1	4.1	2.1	4.1
Comp. time frame [ms]	0.9	0.9	0.9	0.9	0.9	0.9	0.9	0.9
Comp. time [min]	52.3	51.2	64.9	79.0	71.4	88.0	1122.3	958.3
Normalised computational time	1.00	0.98	1.24	1.51	1.36	1.68	21.44	18.31

Remarks: Hardware: PC-cluster of 8 units with Intel Pentium IV 3200 MHz processors and 1 GB RAM each (#1 – MPP LS-DYNA v971; #2 – MPP LS-DYNA v971 accounting for re-meshing; #3 – MPP LS-DYNA v970).

It is obvious that the results from mesh II and III coincide. The accuracy of the results using mesh I is decreasing after the densification strain. From Table 8.1 it can be also noticed that the version of the used computational code (#1, #3) does not influence accuracy but improves computational time. This can also be observed by re-meshing (#2), since re-meshing allows for larger possible time steps. The re-meshing algorithm repositions the element nodes to avoid element distortion and consequently avoids high deformation levels and results in larger distances between element nodes.

According to these results the mesh density and time step from the mesh II and solver parameter #2 have been chosen for further parametric computational simulations.

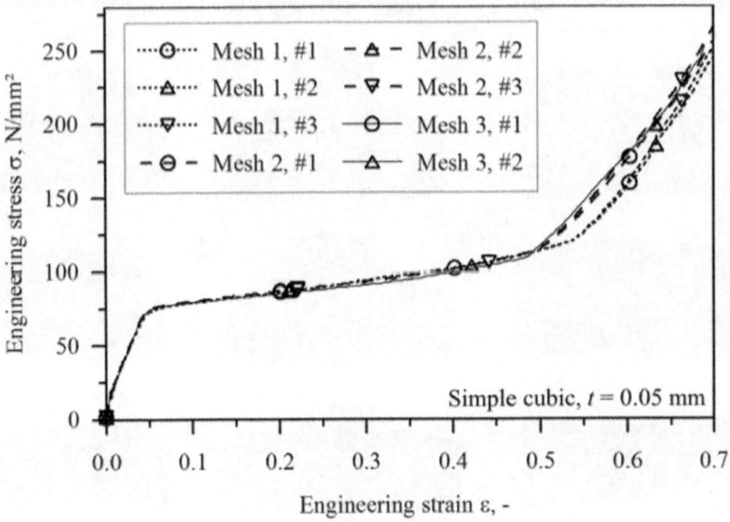

Fig. 8.7. Mesh density and time step sensitivity analysis

8.3 Dynamic Behaviour of MHSS – Computational Results

The results of the parametric finite element computational simulations are shown in stress – strain diagrams. The macroscopic engineering stress of MHSS unit cells in the loading direction (*y*-axis) is derived as a sum of reaction forces (R_y) divided by the initial unit cell cross-section ($A_0 = a \cdot a$):

$$\sigma_y = \frac{R_y}{A_0}. \tag{8.6}$$

The macroscopic engineering strain is calculated as displacement in the loading direction (*y*-axis) divided by the initial unit cell length:

$$\varepsilon_y = \frac{\Delta a_y}{a}. \tag{8.7}$$

8.3.1 Behaviour of MHSS with Syntactic and Partial Morphology

The computationally determined behaviour of syntactic and partial hollow sphere structures with simple cubic and face centred cubic arrangements under dynamic compressive loading is presented in Fig. 8.8, where the lighter regions denote increasing plastic deformation.

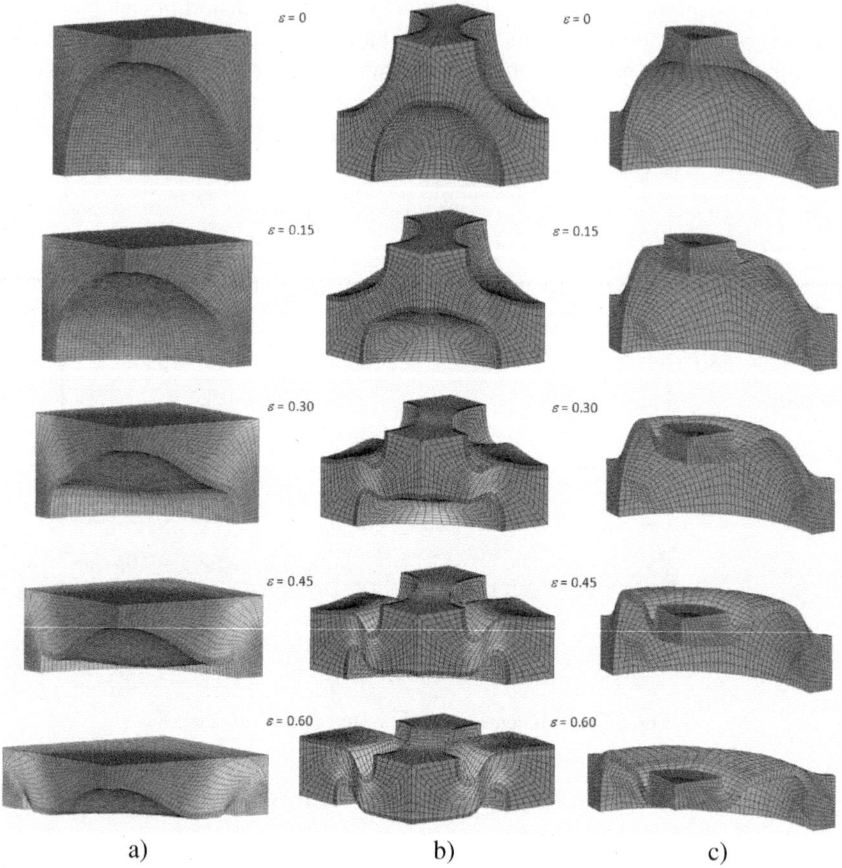

Fig. 8.8. Deformation of MHSS under compressive dynamic loading a) syntactic simple cubic; b) syntactic face centred cubic; c) partial simple cubic

The deformation of the structures is visualised (scale factor: 1:1) for five different engineering strains $\varepsilon_y = \Delta a_y/a = 0, 0.15, 0.30, 0.45$ and 0.6. Limited local buckling of the spherical shells can be observed at low strains ($\varepsilon < 0.15$) for both simple cubic structures. In the case of face centred cubic structure, a distinct localised shear strain deformation within the adhesive matrix can be observed. This deformation is concentrated in the centre of the outer surfaces of the visualised geometries, whereas the remaining matrix experiences much smaller deformations. The observations of these computational results show that inner wall surfaces of the face centred spheres contact at lower strains in comparison to the simple cubic arrangement. After the initial self-contact, the plastic deformation is mostly concentrated in the contact area. With increasing strain ($\varepsilon > 0.30$) the deformation of the partial MHSS is clearly restricted inside the metallic sphere wall at the connection to the adhesive material in the upper part of the structure, which results in local buckling. This phenomenon has already been observed by Karagiozova et al. for single hollow spheres [18].

In the case of syntactic simple cubic MHSS the strain distribution is less localised and the sphere wall bends until it contacts the opposite sphere surface, which initiates the densification of the structure. Densification of simple cubic topology can be observed for syntactic morphology at a strain of $\varepsilon_D \approx 0.50$, whereas the partial MHSS still buckle at this strains and densification starts at larger strains ($\varepsilon_D \approx 0.68$).

8.3.2 Influence of the Sphere Thickness and Topology

The behaviour of the simple cubic MHSS for three different considered sphere wall thicknesses t is shown in Fig. 8.9.

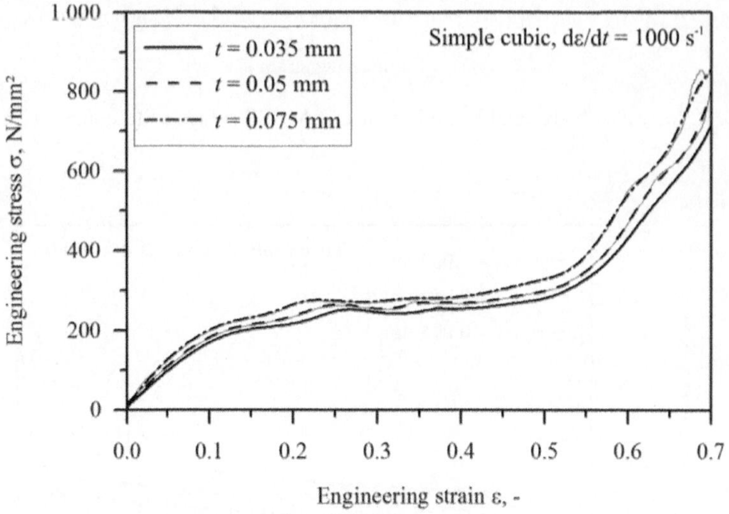

Fig. 8.9. Response of simple cubic syntactic MHSS subjected to compressive dynamic loading

The initial linear-elastic behaviour is followed by the transition zone, then by the stress plateau, where the stress oscillations due to sphere walls buckling can be observed. At the critical strain $\varepsilon_D \approx 0.53$ the inner surfaces of the spherical shell touch, which is followed by the densification process and rapid stress increase. It is obvious that increase of sphere wall thickness results in higher structural stiffness and thus higher macroscopic stress levels. The computational results are in good correlation with the results gained from similar computational simulations under quasi-static compressive loading, which have been validated and confirmed by an experimental testing programme [8].

The computational results for body centred cubic and face centred cubic topologies with syntactic morphology and three different shell thicknesses t are shown in Fig. 8.10 and 8.11.

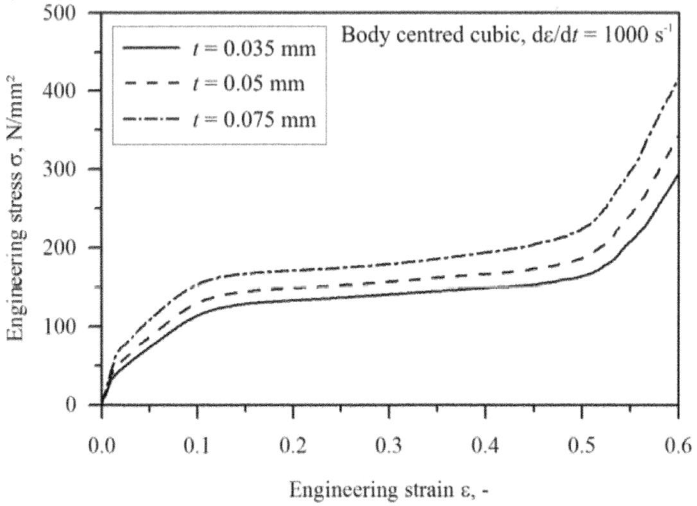

Fig. 8.10. Response of body centred cubic syntactic MHSS subjected to compressive dynamic loading

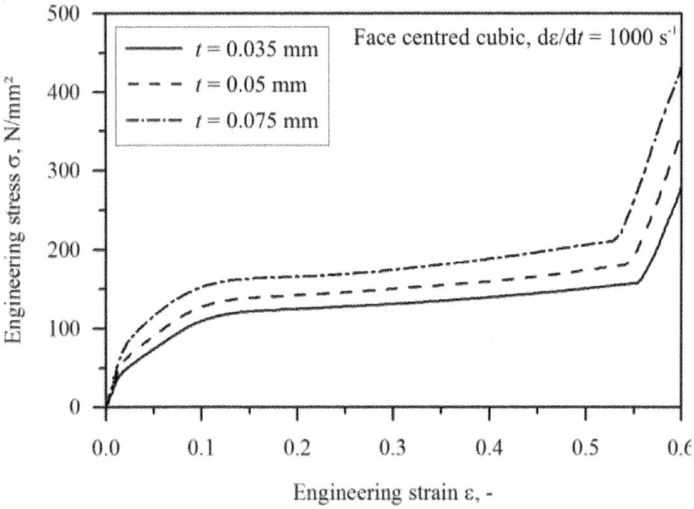

Fig. 8.11. Response of face centred cubic syntactic MHSS subjected to compressive dynamic loading

As observed for the simple cubic topology, increase of sphere wall thickness results in higher structural stiffness and thus higher macroscopic stress levels. In the case of the face centred cubic arrangement, the stress level abruptly increases at the beginning of the densification, which is attributed to the large initial contact of the sphere surfaces. This phenomenon is more pronounced in comparison to the body centred cubic and simple cubic arrangements.

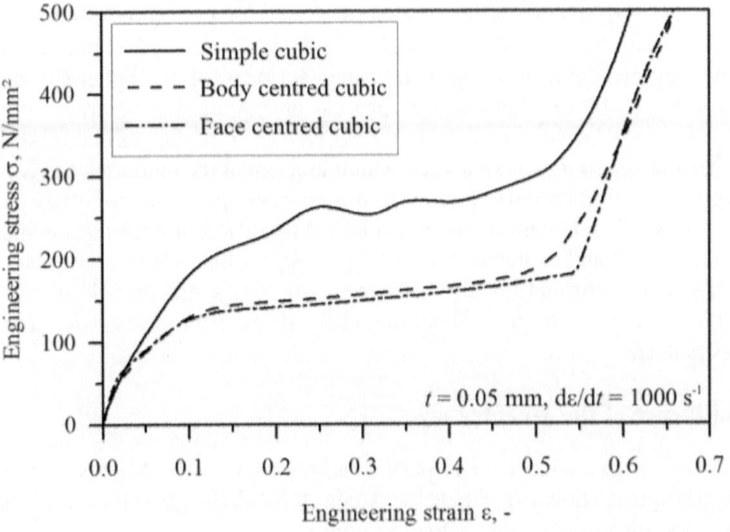

Fig. 8.12. Macroscopic stress-strain relationship for different syntactic MHSS topologies

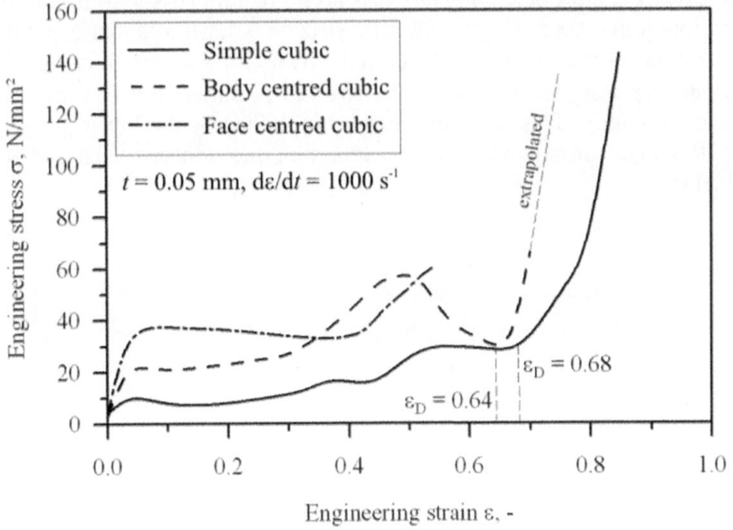

Fig. 8.13. Macroscopic stress-strain relationship for partial MHSS

Figure 8.12 permits a direct comparison of the stress-strain relations of syntactic MHSS for the three different topologies with the constant sphere wall thickness $t = 0.05$ mm. As expected, the simple cubic structure exhibits the highest stiffness due to the lowest porosity of this topology. In comparison, the porosity of the body centred cubic MHSS and face centred cubic MHSS is 1.31 and 1.43 times higher, respectively, which consequently results in smaller structural stiffness during the deformation process.

A different dependence on the topology is found for partial morphology, as shown in Fig. 8.13. At the critical strain $\varepsilon_D \approx 0.64$ (body centred cubic topology) and $\varepsilon_D \approx 0.68$ (simple cubic topology) the inner surfaces of the spherical shell touch, which is followed by the densification process and rapid stress increase. Simulations show that increase of sphere wall thickness results in higher structure stiffness and thus higher macroscopic stress levels, which consequently increases the capability of the energy absorption by deformation. In this case, simple cubic topology exhibits the lowest stresses. This deviation can be explained by different deformation mechanisms observed in partial and syntactic MHSS (Fig. 8.8). Partial MHSS show a strong localisation of plastic deformation (i.e. buckling), whereas syntactic MHSS deform more uniformly. The partial structures buckling also causes strong stress oscillations within the stress plateau.

8.3.3 Influence of the Morphology

The influence of the syntactic and partial morphology of the MHSS on their macroscopic behaviour is shown in Fig. 8.14. Again, three different sphere wall thicknesses are considered using the simple cubic topology.

The results are in agreement with similar computational simulations under quasi-static compressive loading [7, 8]. From the Fig 8.14 it can be observed that the syntactic MHSS exhibit much higher stress levels in comparison to the structure with partial morphology. Accordingly, syntactic structures have a higher capability of energy absorption by deformation than partial MHSS. In addition, densification of the material and exponential increase of the stresses is initiated at lower strains ε_D for syntactic morphology. This behaviour can be explained by the lower porosity of syntactic MHSS and consequently lower support of the adhesive material in case of partial MHSS.

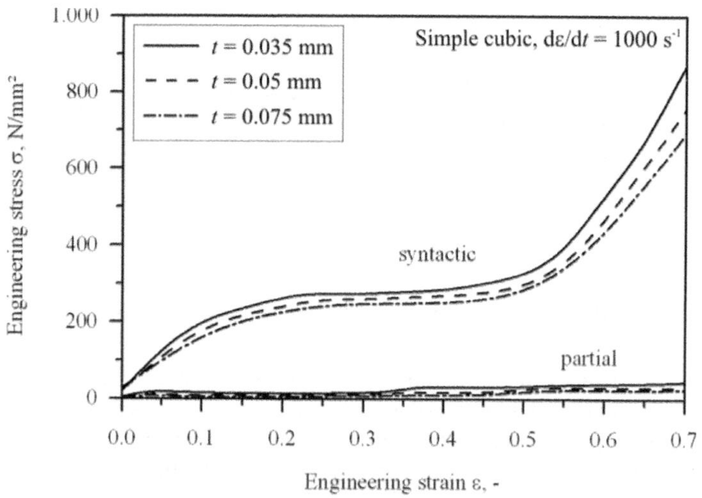

Fig. 8.14. Response of simple cubic MHSS with syntactic and partial morphology subjected to compressive dynamic loading

8.3.4 Influence of the Strain Rate Sensitivity

In this section the influence of the strain rate sensitivity on the macroscopic behaviour of partial and syntactic MHSS is discussed. The behaviour of simple cubic MHSS with syntactic morphology under different strain rates is shown in Fig. 8.15.

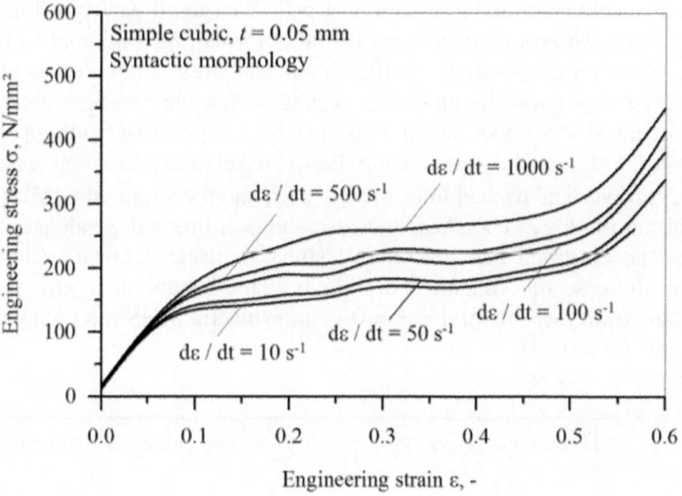

Fig. 8.15. Influence of the strain rate on macroscopic stress strain response of simple cubic MHSS with syntactic morphology

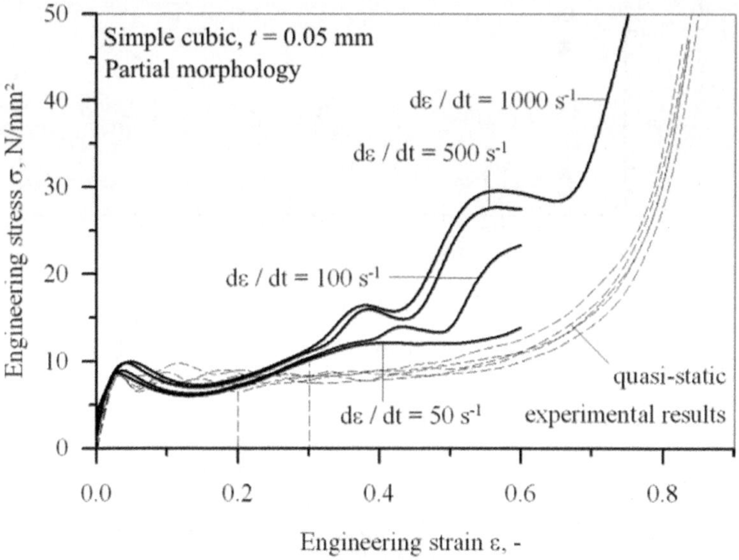

Fig. 8.16. Influence of the strain rate on macroscopic behaviour of partial MHSS

It is evident that higher strain rates result in higher structural stiffness and corresponding stress levels. The strain rate influence on the compressive behaviour of partial MHSS is shown in Fig. 8.16. Again, it is obvious that with higher strain rates the structural stiffness and corresponding stress levels increase which consequently results in higher capability of energy absorption by deformation. This effect becomes more pronounced at higher strains. Additionally, computational results show that the deformation mechanism at large strains ($\varepsilon > 0.4$) is significantly influenced by the strain rate. Large strain-rate dependent lateral deformations (in x and z direction) can be observed, which cause strong oscillations in the stress-strain curves at high strain rates. At low strain rates, the numerical results correspond to experimental measurements of partial MHSS under uniaxial quasi-static compressive loading [7].

Figure 8.17 shows the relationship between relative stress increase at strains $\varepsilon = 0.2, 0.3$ (cf. vertical dashed lines in Fig. 8.16) and the strain rate $d\varepsilon/dt$ for syntactic and partial morphology. The stress increase shows a linear dependence on the strain rate. It can be observed that the relative stress increase is considerably more pronounced in the case of syntactic morphology. This behaviour is also reflected in a much higher strain rate sensitivity $m = 0.11$ of syntactic MHSS in comparison to partial structures ($m = 0.04$).

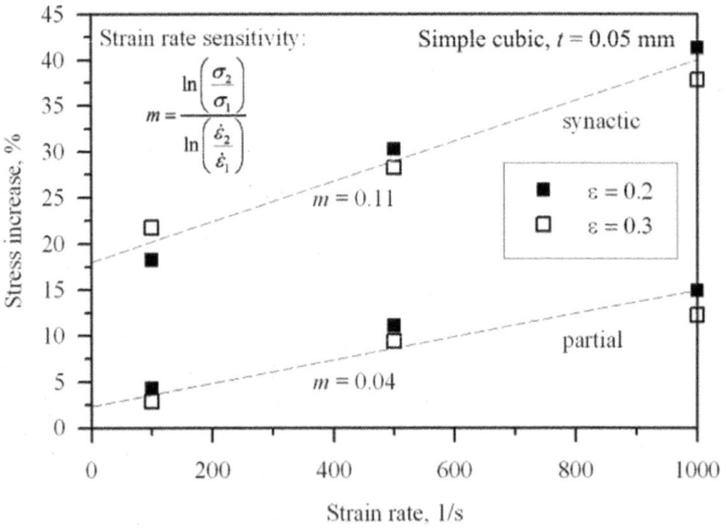

Fig. 8.17. Strain rate sensitivity of the syntactic and partial MHSS

8.3.5 Influence of the Internal Pore Pressure

As already mentioned, the internal pore pressure due to the gas trapped in the hollow spheres, might influence the global behaviour of MHSS.

However, as shown in Fig. 8.18, the influence of the internal pore pressure is negligible through the deformation of the MHSS. Furthermore, it must be considered that the cell walls of the hollow spheres exhibit a certain micro porosity, therefore a high initial pore pressure cannot be expected.

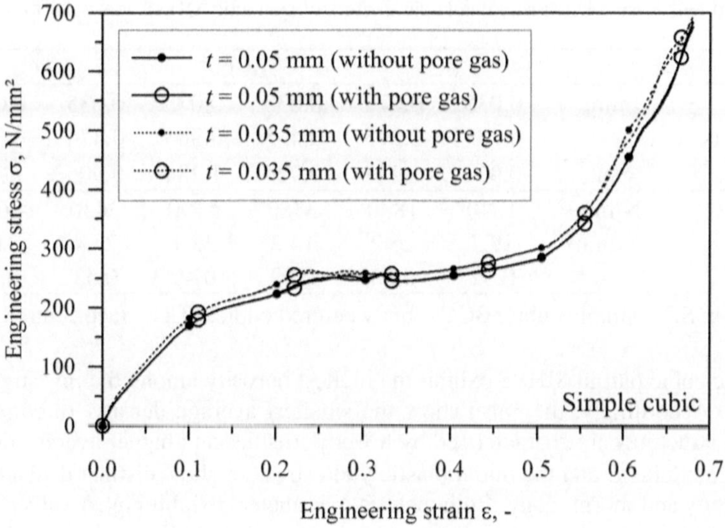

Fig. 8.18. Influence of the internal pore pressure

8.4 Discussion

The results of the computational simulations were used to derive the macroscopic material parameters of simple cubic, body centred cubic and face centred cubic topologies of metallic hollow sphere structure. The porosity, the average density ($\overline{\rho}$), Young's modulus (E), the initial plastic yield stress (σ_y) and the densification strain (ε_D) are summarised in Tables 8.2 and 8.3. The average density of MHSS is calculated according to $\overline{\rho} = (V_S \cdot \rho_{St} + V_M \cdot \rho_{Ep}) / V_U$ where V_S is the volume of the metallic hollow spherical shell, V_M the matrix volume and V_U the unit cell volume.

Table 8.2. Macroscopic materials parameters of MHSS at $\dot{\varepsilon} = 1000\ s^{-1}$

		Partial MHSS			Syntactic MHSS		
		SC	BCC	FCC	SC	BCC	FCC
t	mm	0.05	0.05	0.05	0.05	0.05	0.05
Porosity	-	0.907	0.863	0.811	0.320	0.415	0.452
$\overline{\rho}$	g/cm³	0.414	0.557	0.651	1.078	1.063	1.057
E	N/mm²	590	840	1300	1540	3770	3630
σ_y	N/mm²	5.6	9.1	14.2	21.1	21.2	26.2
ε_D	-	0.68	0.64	-	0.53	0.49	0.54

Remark: SC – simple cubic, BCC – body centred cubic, FCC – face centred cubic.

Table 8.3. Macroscopic mateials parameters of syntactic MHSS at $\dot{\varepsilon} = 1000\ s^{-1}$

		SC		BCC		FCC	
t	mm	0.035	0.075	0.035	0.075	0.035	0.075
Porosity	-	0.335	0.295	0.435	0.383	0.474	0.417
$\bar{\rho}$	g/cm^3	0.971	1.247	0.924	1.284	0.906	1.298
E	N/mm²	1340	1830	3380	5190	3430	4660
σ_y	N/mm²	19.7	26.2	14.3	32.1	23.4	31.1
ε_D	-	0.52	0.53	0.48	0.49	0.53	0.55

Remark: SC – simple cubic, BCC – body centred cubic, FCC – face centred cubic.

Simple cubic partial MHSS exhibit the highest porosity among the investigated geometries. Accordingly, they also show the smallest average density. In comparison, syntactic structures are characterised by lower porosities and higher average densities. Young's modulus E and the initial plastic yield stress σ_y show distinct dependence on the topology and morphology. Both material parameters exhibit higher values for syntactic MHSS. In addition, the highest values are found for FCC topology, followed by BCC and SC arrangement of spheres. Finally, due to their higher porosity, partial MHSS exhibit significantly higher densification strains. By increasing the metal sphere thickness, the macroscopic Young's modulus, yield stress, plateau stress and plateau modulus increase. From these results it can be observed that there is a very low influence of the sphere thickness on the densification strain ($\Delta\varepsilon_D = \pm\ 0.05$), which assures large plateau regions.

Figure 8.19 shows Young's modulus plotted versus the average density. For comparison, the values of the commercial cellular metallic foams Alporas® and M-Pore®

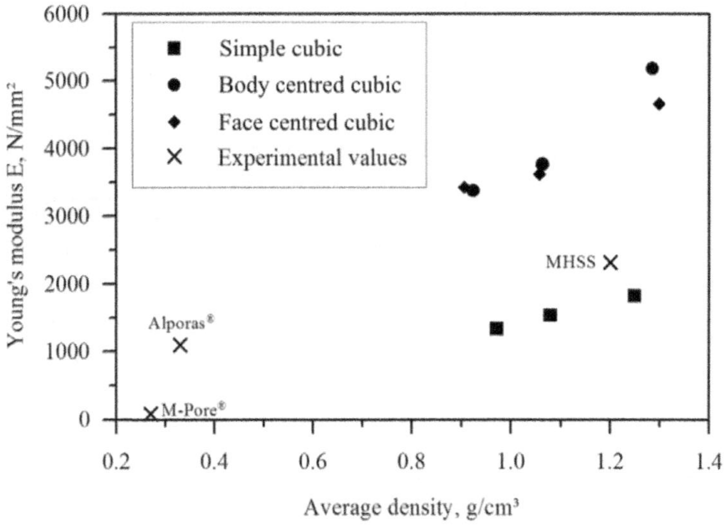

Fig. 8.19. Young's modulus E plotted versus the average density; Alporas [6], M-Pore [11]

are also shown. Furthermore, the experimentally determined Young's modulus of MHSS with random arrangement of spheres is shown, as well. The computational results indicate a distinct dependence of Young's modulus on the MHSS topology. The body centred and face centred cubic arrangements of spheres yield distinctly higher values in comparison to the simple cubic MHSS. The experimentally obtained value is between these limits and closer to the simple cubic results.

The MHSS show good specific values of the yield stress σ_y (Fig. 8.20) in comparison to the commercial aluminium foams. However, the experimental data was obtained by quasi-static testing and therefore cannot be directly compared to the computational results of dynamically loaded MHSS (strain rate $\dot{\varepsilon} = 1000\ s^{-1}$). The computational results show a distinct dependence of the specific yield stress on topology. At low densities, the face centred cubic arrangement of spheres yields the highest yield stress for a specific density. At higher densities, the body centred cubic MHSS show superior performance with a higher rate of yield stress increase.

Additionally, for all three topologies, almost linear dependence of the plateau stress σ_{pl} on the average density was observed (Table 8.4).

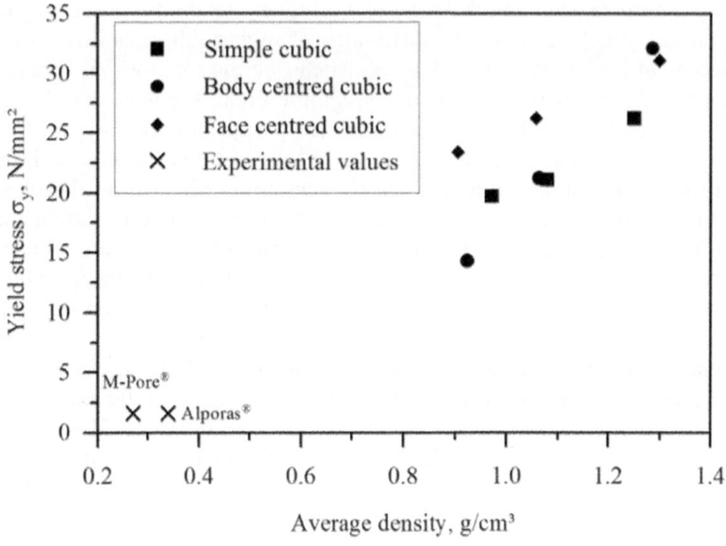

Fig. 8.20. Yield stress σ_y plotted versus the average density; Alporas [6], M-Pore [11]

Table 8.4. Macroscopic materials parameters of syntactic MHSS at $\dot{\varepsilon} = 1000\ s^{-1}$

		Simple cubic			Body centred cubic			Face centred cubic		
t	mm	0.035	0.05	0.075	0.035	0.05	0.075	0.035	0.05	).075
$\bar{\rho}$	g/cm³	0.971	1.078	1.247	0.924	1.063	1.284	0.906	1.057	l.298
σ_{pl}	N/mm²	205	218	237	127	142	165	118	135	l60

8.5 Conclusions

In this chapter the dynamic behaviour of metallic hollow sphere structures has been studied with use of computational simulations. The computational models of metallic hollow sphere structures fully and partially embedded within an adhesive matrix with different sphere distribution and wall thicknesses subjected to uniaxial dynamic compressive loading conditions were built for this purpose. The computational results provide better insight into behaviour and mechanical properties of dynamically loaded MHSS.

The macroscopic stress-strain behaviour of MHSS exhibits typical porous material characteristics. By means of computational simulations it was shown that the response of MHSS subjected to dynamic loading depends on the strain rate, topology, sphere wall thickness and morphology. A strong influence of MHSS morphology on its properties was observed. Syntactic structures are characterised by lower porosity and therefore exhibit distinctly higher stiffness and stresses in comparison to partial MHSS. Partial hollow sphere structures show a pronounced localisation of plastic deformation (i.e. buckling) and deform at relatively low stresses. Due to their high porosity, partial MHSS have larger densification strains in comparison to syntactic structures. Therefore, syntactic MHSS have a higher capability of energy absorption, whereas partial structures have the advantage of higher deformability. Additionally, the properties of hollow sphere structures depend on the topology of spheres. Due to different plastic deformation mechanisms, simple cubic topology yields the highest stresses for syntactic and the lowest stresses for partial morphology. Furthermore, it was observed that the increase of the sphere wall thickness is reflected in higher stress levels. However, the sphere thickness has a very low influence on the densification strain, which assures large stress plateau regions. Finally, it was shown that higher strain rates lead to increase of stresses and consequently higher capability of energy absorption. This effect is more pronounced in the case of syntactic morphology. For low strain rates good agreement with experimental data obtained in compressive quasi-static testing was found.

Computational simulations show that it is possible to achieve different dynamic response of metallic hollow sphere structures when subjected to dynamic loading. The topology, wall thickness of spheres and strain rate sensitivity can be combined in a way that the structure response is adapted to a given engineering problem.

Acknowledgments

The Slovenia-Portugal cooperation in science and technology, as well the financial support of the Slovenian Research Agency are gratefully acknowledged.

References

1. Altenhof, W., Harte, A.M., Turchi, R.: Experimental and numerical compressive testing of aluminum foam filled mild steel tubular hat sections. In: Proceedings 7th International LS-DYNA Users Conference (2002)
2. Ashby, M.F., Evans, A., Fleck, N.A., et al.: Metal foams: a design guide. Elsevier Science, Burlington (2000)

3. Bodner, S.R., Symonds, P.S.: Experimental and theoretical investigation of the plastic deformation of cantilever beams subjected to impulsive loading. J. Appl. Mech. 29, 719–728 (1962)
4. Cias, A., Mitchell, S.C., Watts, A., et al.: Microstructure and mechanical properties of sintered (2-4)Mn-(06-08)C steels. Powder Metallurgy 42, 227–233 (1999)
5. Cook, R.D., Malkus, D.S., Plesha, M.E.: Concepts and applications of finite element analysis. John Wiley & Sons, New York (1989)
6. Enterprise Gleich GmbH: Data Sheet. Kaltenkirchen (2006)
7. Fiedler, T.: Numerical and Experimental Investigation of Hollow Sphere Structures in Sandwich Panels, Ph.D. thesis, University of Aveiro, Aveiro (2008)
8. Fiedler, T., Öchsner, A., Grácio, J.: Numerical Investigations on the Mechanical Properties of Adhesively Bonded Hollow Sphere Structures. J. Compos. Mater. (accepted for publication) (2007)
9. Friedl, O., Motz, C., Peterlik, H., et al.: Experimental Investigation of Mechanical Properties of Metallic Hollow Sphere Structures. Metallurgical and Materials Transactions B 39, 135–146 (2008)
10. Gibson, L.J., Ashby, M.F.: Cellular solids: structure and properties. Cambridge University Press, Cambridge (1997)
11. Gierlich, D., Khn, C., Hackeschmidt, K., et al.: Mechanical properties of open-pored aluminium foams (Mechanische Kenndaten offenporiger Aluminiumschäume). Konstruktion 9, 69–74 (2004)
12. Hallquist, J.: LS-DYNA Keyword User's Manual. Livermore Software Technology Corporation (2007)
13. Hallquist, J.O.: LS-DYNA theoretical manual. Livermore Software Technology Corporation. Livermore, California (1998)
14. Hallquist, J.O.: LS-DYNA keyword user's manual. Livermore Software Technology Corporation. Livermore, California (2003)
15. Hosford, W.F.: Mechanical Behaviour of Materials. Cambridge University Press, New York (2005)
16. Jacob, P., Goulding, L.: An Explicit Finite Element Primer. NAFEMS, Glasgow (2002)
17. Jeandrau, J.P.: Analysis and design data for adhesively bonded joints. Int. J. Adhes Adhes 11, 71–79 (1991)
18. Karagiozova, D., Yu, T.X., Gao, Z.Y.: Modelling of MHS cellular solid in large strains. Int. J. Mech. Sci. 48, 1273–1286 (2006)
19. Kim, H.S., Khamis, M.A.: Fracture and impact behaviours of hollow micro-sphere/epoxy resin composites. Compos: Part A 32, 1311–1317 (2001)
20. Litwin, L.: FIR and IIR digital filters. IEEE potentials magazine 19, 28–31 (2000)
21. Neville, B.P., Rabiei, A.: Composite metal foams processed through powder metallurgy. Mater. Design 29, 388–396 (2008)
22. Öchnser, A.: Experimentelle und numerische Untersuchung des elasto-plastischen Verhaltens zellularer Modellwerkstoffe, Ph.D. thesis. University Erlangen, Nuremberg (2003)
23. Ohrndorf, A., Schmidt, P., Krupp, U., et al.: Mechanische Untersuchungen eines geschlossenporigen Aluminiumschaums. In: Proceedings Werkstoffprüfung (2000)
24. Ren, Z.: Progressive fracturing under dynamic loading conditions, Ph.D. thesis. Swansea University, Swansea (1993)
25. Sanders, W.S., Gibson, L.J.: Mechanics of hollow sphere foams. Mater Sci. Eng. A 347, 70–85 (2003)
26. Song, B., Chen, W., Weerasooriya, T.: Quasi-Static and Dynamic Compressive Behaviors of a S-2 Glass/SC15 Composite. J. Compos. Mater 37, 1723–1743 (2003)

27. Vesenjak, M.: Computational modelling of cellular structure under impact conditions, Ph.D. thesis. Faculty of Mechanical Engineering, Maribor (2006)
28. Vesenjak, M., Öchsner, A., Ren, Z.: Behaviour of closed-cell foams under impact. In: Proceedings 2 Workshop on Advanced Computational Engineering Mechanics, pp. 195–224 (2005)
29. Vesenjak, M., Öchsner, A., Ren, Z.: Thermal Post-Impact Behavior of Closed-Cell Cellular Structures with Fillers – Part I. Mater. Sci. Forum 553, 196–201 (2007)
30. Vesenjak, M., Žunič, Z., Öchsner, A., et al.: Thermal Post-Impact Behavior of Closed-Cell Cellular Structures with Fillers – Part II. Mater. Sci. Forum 553, 202–207 (2007)
31. Vesenjak, M., Fiedler, T., Ren, Z., et al.: Behaviour of Syntactic and Partial Hollow Sphere Structures under Dynamic Loading. Adv. Eng. Mater. 10, 185–191 (2008)
32. Vesenjak, M., Žunič, Z., Öchsner, A., et al.: Heat conduction in closed-cell cellular metals. Mater. Sci. Forum 36, 608–612 (2005)
33. Xue, Z., Vaziri, A., Hutchinson, J.W.: Materials with Application to Sandwich Plate Cores. CMES - Comp. Model Eng. 10, 79–95 (2005)
34. Zhao, H., Elnasri, I., Abdennadher, S.: An experimental study on the behaviour under impact loading of metallic cellular materials. Int. J. Mech. Sci. 47, 757–774 (2005)
35. Zienkiewicz, G.C., Taylor, R.L.: Finite element method. The basis, vol. 1. Butterworth - Heinemann, Oxford (2000)
36. Zienkiwicz, O.C., Scot, F.C.: On the principle of repeatability and its application to the analysis of turbine and pump impellers. Int. J. Num. Methods Engng. 4, 445–452 (1972)

9

Fatigue of Metal Hollow Spheres Structures

Eric Maire[1], Oliver Caty[1], and Régis Bouchet[2]

[1] Université de Lyon, INSA-Lyon, MATEIS CNRS UMR5510 F-69621 Villeurbanne, France
eric.maire@insa-lyon.fr
[2] DMMP, ONERA Chatillon, 92322 CHATILLON, France
regis.bouchet@onera.com

Abstract. This chapter presents the study of the fatigue behaviour of metal hollow spheres structures (MHSS). It starts with a brief litterature review on cellular metals and specifically MHSS. Then is describes the fatigue study carried out on several of these materials. This study has involved the determination of the macroscopic fatigue behaviour of several MHSS. The local damage mechanisms have been studied qualitatively and quantitatively using Xray computed tomography. Finally a model has been developed to calculate the local stresses from a finite element calculation performed on the tomographic image.

Keywords: X-Ray, Tomography, Fatigue, Finite Element, fatigue tests, macroscopic fatigue behaviour, S-N curves, damage mechanisms, XRCT setup, densification, modelling, shell meshing.

9.1 Introduction

As explained in earlier chapters of the present book, metal hollow spheres structures (MHSS) are assemblies of hollow spheres stuck or soldered together to form a cellular structure with multi functional properties. The resulting material has good acoustic performances and can be designed for structural application and for high temperature use. The hollow spheres used can exhibit a mono disperse size and thickness, the resulting material can then be rather homogeneous. This type of material allows a remarkable flexibility in terms of microstructural design. The thickness and diameter of the spheres, the nature of the metal used (nickel, steel), the pattern for the piling up (random, centered face cubic, simple cubic) can be varied. This allows to tailor the microstructure in order to reach the target properties imposed by the design on the material. Before being able to tailor properties and because these materials are completely new, it is necessary to know the relation between the varying parameters and the resulting properties. In the case of the present chapter the fatigue properties are more specifically studied.

Fatigue can be observed when a material is subjected to a cyclic loading. In most of the potential applications of MHSS, this type of loading is certainly present. Fatigue induces micro-damage which nucleate, grow and then propagate until they coalesce in the form of a large crack that leads to the complete collapse of the structure. The danger of this damage mode is that it remains silent for a long period of the life of the component. The reason for this is that it develops at load levels which, when applied in one single cycle, are far from leading to a macroscopic collapse, and even

A. Öchsner, C. Augustin (Eds.): Multifunctional Metallic Hollow Sphere Struc., EM, pp. 159–182.
springerlink.com

far from leading to detectable plasticity. Fatigue is then a key problem in engineering design. For bulk metals, a lot of problems related to the fatigue behaviour (short cracks, crack closure, etc.) are still unsolved although they have been studied for many decades. In the field of cellular materials the situation is even worse because of the youth of these materials, that explains the relatively small number of published studies in this field.

The purpose of the present chapter is to analyze and model the effect of the microstructure of MHSS on their fatigue properties. Standard compression/compression and tension/tension fatigue tests were performed in order to measure the macroscopic behavior of different materials made of stainless steel (314 and 316) with different structural parameters. Ex situ fatigue tests coupled with X-ray computed tomography (XRCT) were also performed in order to investigate the microscopic damage mechanisms. A modeling of these properties based on the calculation of the stress concentration inside the complex microstructure was finally developed. This finite element (FE) model reproduces the exact microstructure as seen from the tomographic data.

9.2 Literature Review

Only a few papers have been published on the standard fatigue properties of cellular metals and even less on MHSS in particular. Ashby et al. in the design guide of metal foams [3] have summarized the results of different studies dealing with the fatigue behaviour of different metal foams [16, 21, 25] but not on MHSS. In tension or shear, all the authors observed a single macroscopic crack developing at the weakest section and progressing rather fast. The measured fatigue limit, in tension and even more in shear, is quite low. In the compression loading mode, for a sufficient stress level and after an initiation period, large plastic strains gradually develop leading to the progressive collapsing of the cellular material.

Motz et al. have recently published a paper on fatigue crack propagation in MHSS [22]. These authors mostly present results on compact tension specimens in tension–tension or in tension–compression loadings but they also estimate the endurance LIMIT in compression on standard samples of two kinds of MHSS (cubic samples made of 316 stainless steel with a density of 300 g/l, diameters of 2 and 4 mm). They measured a ratio of the fatigue strength divided by the yield stress in the range of 0.5 to 0.65 for MHSS in compression and in the range of 0.2 to 0.25 in tension.

Table 9.1. Parameters of the samples tested in the study

Acronym	Material	d (g/l)	D (mm)	t (μm)	p_w	E (MPa)	σ_y (MPa)
316D2.8t90	316	800	2.8	90	Porous	1000	4
314D2.8t90	314	800	2.8	90	Dense	2000	6.8
314D2.8t65	314	600	2.8	65	Dense	1600	5.4
314D2.8t45	314	400	2.8	45	Dense	1000	3.2
314D1.7t45	314	800	1.7	45	Dense	2000	8

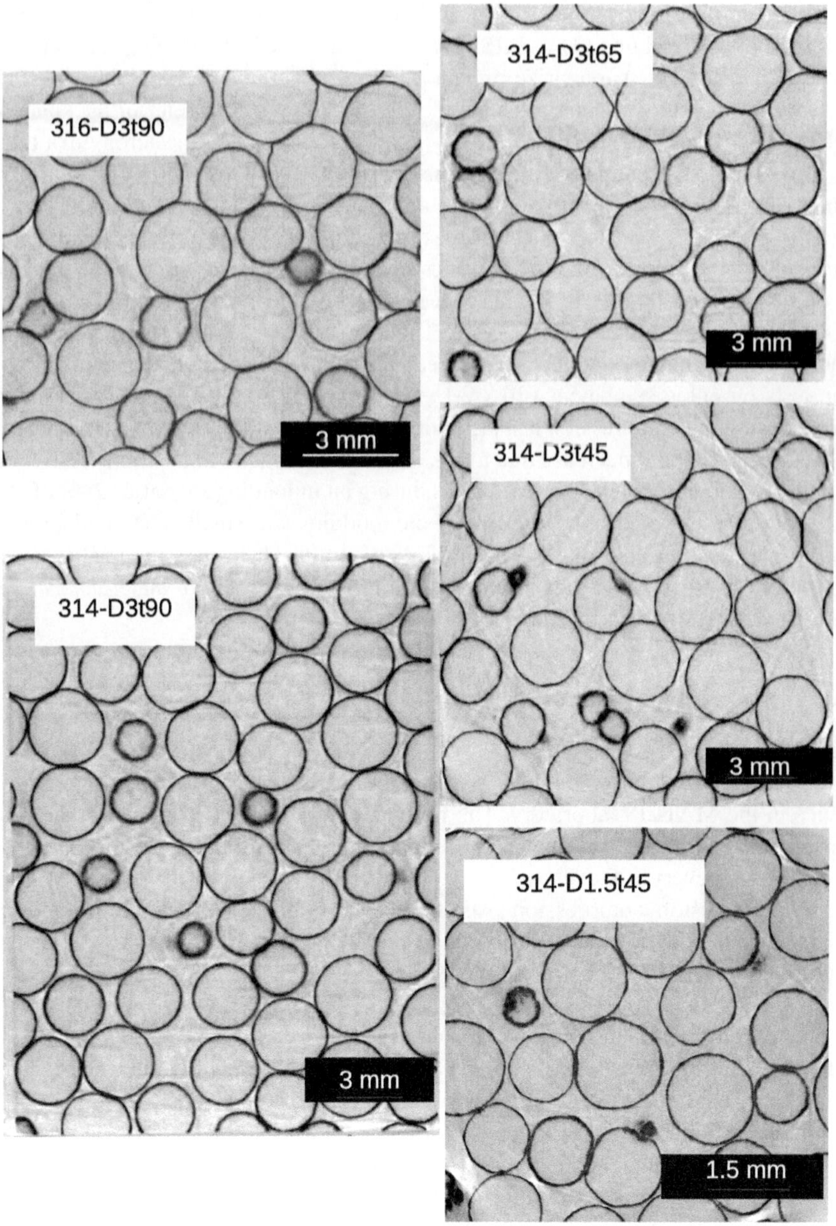

Fig. 9.1. Reconstructed slices of the five kinds of samples studied

9.2.1 Materials

Our investigations were carried out on five different kinds of MHSS. These were kindly provided by the PLANSEE Austrian company. Two kinds of constitutive materials were available, the first one (316 - AISI 1.4404 respec. EN X2CrNiMo18-8) with porous walls and the second one (314 - AISI 1.4841 respec. EN X15CrNiSi25-20) with more dense walls. Table 9.1 summarizes the morphological parameters of all the MHSS used in terms of constitutive material, density of the structure d, diameter of the spheres D, wall thickness t and wall porosity p_w. The table also summarizes the acronym of the material used in the subsequent sections. The construction of these acronyms summarises somehow the morphological parameters and then helps to understand quickly which material is concerned. Most of the values presented were given by the manufacturers and confirmed by our morphological investigation performed by processing and analysis of 3D datasets obtained by means of X ray tomography (see [7] for the measurement procedures and manufacturing process). The macroscopic Young's modulus E and the macroscopic yield stress σ_y were measured by Fallet et al. [11] with standard compression tests on cubic specimens (30 x 30 x 30 mm^3). The unloading Young's modulus varied during the compression test [13] with an increase at the beginning followed by a plateau and by a final decrease due to the effect of damage at high strain. The Young's modulus was then chosen to be measured during an unloading step after 2 % of strain, i.e. in the plateau where the evolution of the modulus was small. The yield stress was also defined as the value of the stress for 2 % of strain. These measured values are also reported in Table 9.1. A reconstructed slice of each of these five materials as seen in XRCT (see section 9.3.1) is shown in Fig. 9.1.

9.2.2 Macroscopic Fatigue Properties

9.2.2.1 Fatigue Tests

All the fatigue tests presented in the paper were carried out on a hydraulic MTS fatigue device in the MATEIS laboratory. The frequency was set to 10 Hz and the tests were carried out with a load type control. The maximum and minimum positions and loads were recorded every 100 cycles.

For compression–compression tests (fatigue load ratio $R = 10$), cubic samples (side 15 mm) machined by electrical discharge machining were compressed between two plates (see figure 9.2). Specimens with twice larger dimensions were also prepared to investigate the effect of the sample size on the fatigue properties.

The tension specimens were dog bone shaped in order to concentrate the stress and localize the cracks. The useful section was about 15 x 15 mm^2. For the tension–tension tests ($R = 0.1$), the samples were gripped with Wood's metal. This solution permits an automatic alignment and an easy removal of the samples. The Wood's metal was melted in small pots machined in the grips. While the metal was liquid (at a temperature higher than 70°C), 2 cm of the extremities of the sample were dipped into the molten alloy. The whole set was then cooled down until the metal solidified. During this step the load was controlled and set to zero. The dimensions of the samples and the design of the grips are shown in figure 9.2.

9.2.2.2 Macroscopic Fatigue Behaviour

The purpose of this section was to built the $S-N$ curves of the studied MHSS. This curve is an important information for the design of the material. In order to achieve this,

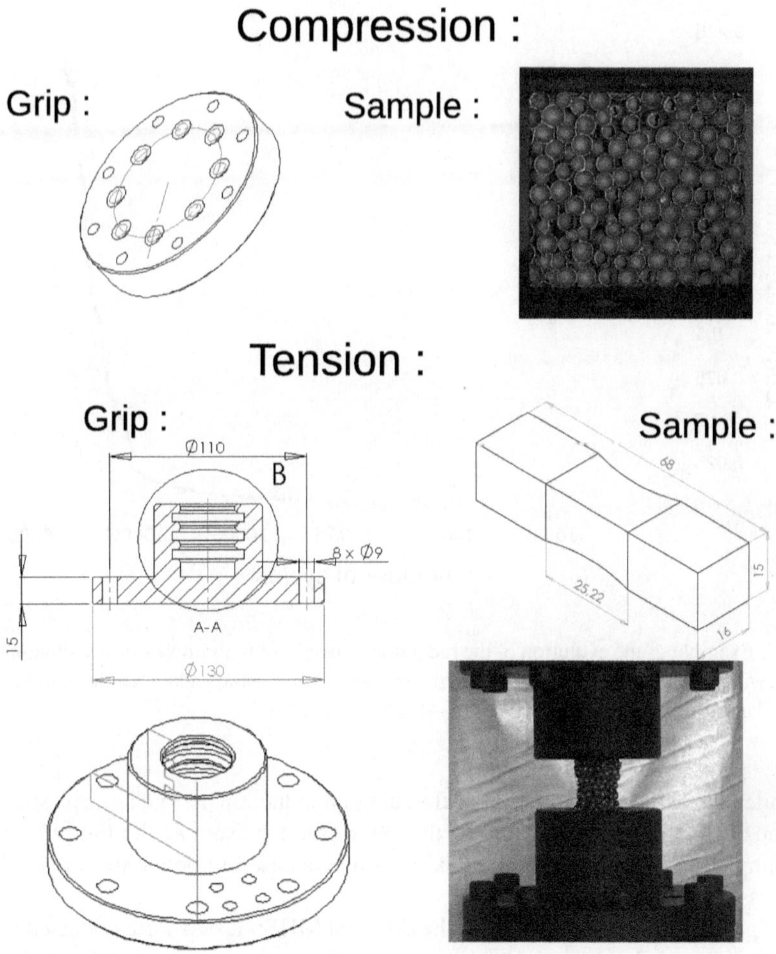

Fig. 9.2. Geometry of the samples used for the compression–compression and tension–tension fatigue experiments. The grips are also sketched. Notice the pots to grip the tensile samples using Wood's metal.

samples were tested at a given stress level until failure (value recommended by [3]). The number of cycles to failure N_f was then plotted on a (N_f–σ) log–linear axes chart. In tension, N_f was defined as the number of cycles to break the sample. In compression, N_f was determined from the ε–N curve. ε was defined according to the expression:

$$\varepsilon = \frac{h(\varepsilon) - h_0}{h_0} \tag{9.1}$$

where $h(\varepsilon)$ is the current height and h_0 the initial height of the sample.

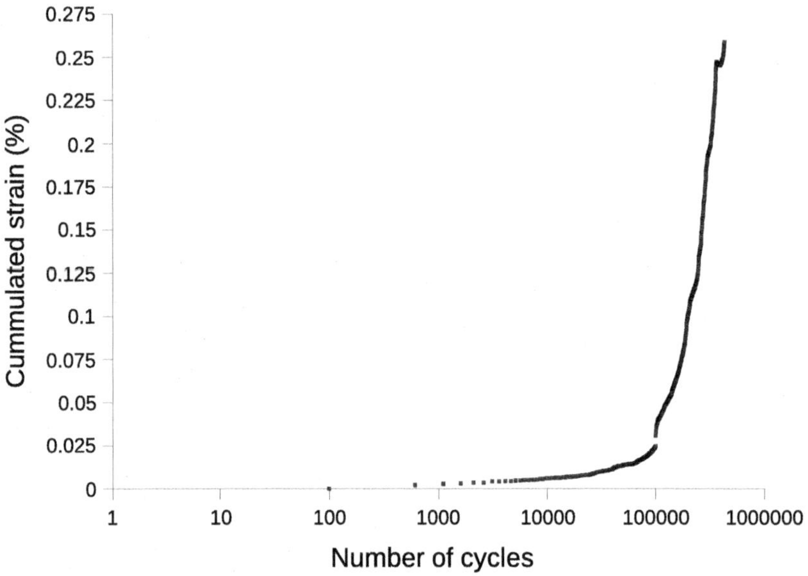

Fig. 9.3. Example of the evolution of the reduction in height of the sample during a compression–compression fatigue test, showing the sharp knee used to define the endurance limit in the present study. The example shown concerns the 314D2.8t90 material.

While increasing the number of cycles, after an induction period, large plastic strains develop and a sharp knee appears on the curve (see for example the Fig. 9.3 showing an example of ε–N curve). This inflexion point was used to define the value of N_f in compression.

In Fig. 9.4 all the $S-N$ curves of the different MHSS tested were collected to compare the fatigue behaviour and understand the effect of the parameters on fatigue life. The scatter width of the number of cycles to rupture for a given stress is about one decade for all the MHSS tested. This value is close to those measured for dense materials [5] and for other metal foams [3]. The table 9.2 summarizes the values of the fatigue limit and of the stress sensitivity measured from these $S-N$ curves.

In compression, the fatigue limit $\sigma_{\text{fatigue}}^{\text{compression}}$ of the MHSS measured is approximately equal to the half of their yield strength σ_y

$$\frac{\sigma_{\text{fatigue}}^{\text{compression}}}{\sigma_y} \approx 0.5 \tag{9.2}$$

Table 9.2. Fatigue limit and stress sensitivity measured from the $S-N$ curves shown in figure 9.4

Material	Fatigue limit (MPa)	Stress sensitivity (MPa/cycles)
316D2.8t90	3	0.4
314D2.8t90	3.3	0.4
314D2.8t65	2.7	0.4
314D2.8t45	1.9	0.4
314D1.7t45	3.5	0.54
314D1.7t45 (Tension)	1.75	0.54

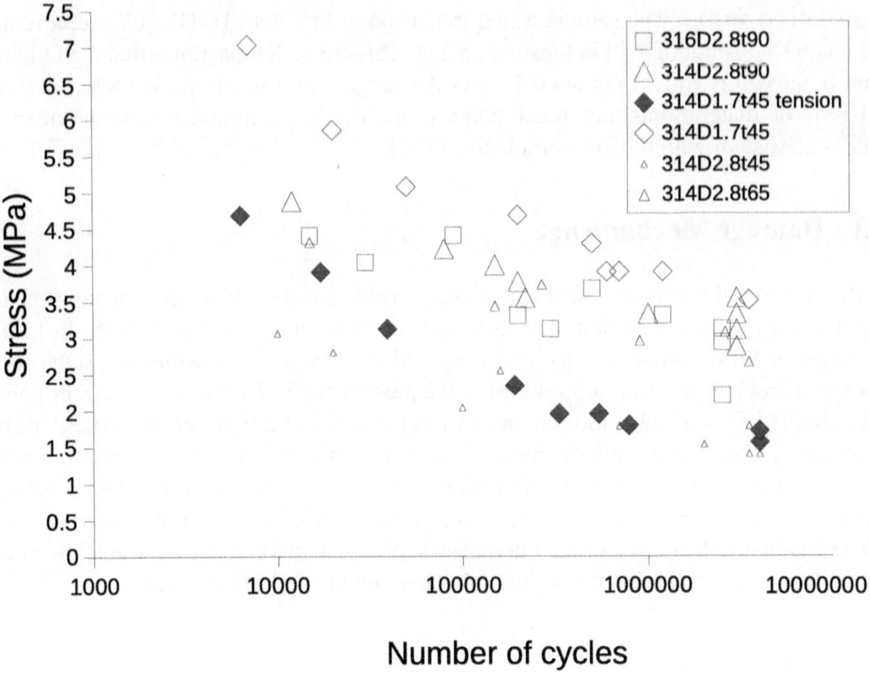

Fig. 9.4. $S-N$ curves measured for all the studied materials

9.2.2.3 Comments on the Effect of the Parameters

SAMPLE SIZE: It has been checked that all fatigue lives measured with the larger samples (30 mm side) are in the uncertainty range of the fatigue lives measured with smaller samples (15 mm side). One can conclude that the small samples used for the main part of the present work are big enough to be representative in fatigue. This is consistent to what was found in [11].

MATERIAL: Samples 316D2.8t90 and 314D2.8t90 MHSS have close structural parameters (diameter, thickness and density, see Table 9.1) but they are made of two different constitutive materials. These two steels in their bulk form have similar static and cyclic properties. The corresponding MHSS shows a similar fatigue behaviour (the difference in endurance limit is only 10 %).

GEOMETRY: Four samples made of the same constitutive material (314) and loading conditions (compression) but with different morphologies were tested: two samples have the same density but different diameters and three have the same diameter but different densities (one can notice that thickness is obviously correlated to both density and diameter). Keeping diameter constant but varying densities (314D2.8t90, 314D2.8t65 and 314D2.8t45 samples) shows that density has a first order effect on the fatigue strength but not on the stress sensitivity (i.e. the slopes of the $S-N$ curves are similar for these MHSS). Then keeping density constant but varying diameters (314D1.7t45 and 314D2.8t90 samples) shows that diameter acts on the stress sensitivity but not so much on the fatigue strength.

LOADING MODE: The sample tested in tension in Fig. 9.4 (314D1.7t45) represents the unique $S-N$ curve built in tension/tension. This curve demonstrates that the fatigue limit in tension is equal to about 0.5 times the fatigue limit in compression for a same MHSS. The material has then much better properties in compression than in tension. This is consistent with the measurements of [22].

9.3 Damage Mechanisms

In this section, the damage mechanisms responsible for the global deformation measured in the preceding section, are inspected for some of the studied MHSS. In [22], by means of SEM, fatigue cracks were observed to propagate in tension. The same sort of experiment has not been reproduced in the present study. The fracture surfaces have been simply inspected post mortem in order to try to detect fatigue streaks. These trials to observe features at the microstructural scale were not successful. The present study has then rather focussed on the modification of the meso–structure of the MHSS during fatigue, as observed by X-Ray computed tomography (XRCT). This rather new method allows to obtain 3D images of the internal structure of a material non destructively. It is particularly well adapted to the study of the architecture of cellular solids.

9.3.1 XRCT Setup

The standard laboratory tomograph available in our group is a commercial product sold by the `Phoenix X-ray` company. It contains two principal components:

- The X-ray source is an open transmission nanofocus X-ray tube operated between 40 and 160 kV. The source/detector distance is fixed to 80 cm but the resolution (defined in the present chapter as the voxel size in the reconstructed image) can be varied by changing the position of the rotating object between the source and the detector. The thin transmission target bombed by focused electrons is made of tungsten or molybdenium. The size of the focus can be easily varied from 1 to 6 μm.
- The X-ray detector is a Paxscan^{TM} amorphous silicon flat panel initially developed for medical application. It is composed of about 1900 rows and 1500 lines of sensitive pixels, the size of each is 127 x 127 μm^2.

The source/detector distance is fixed to 80 cm. For the present study, the samples were scanned with a voxel size of 15 μm. The tomograph was operated at 140 keV and 120 μA with a filtering of 0.2 mm of Cu.

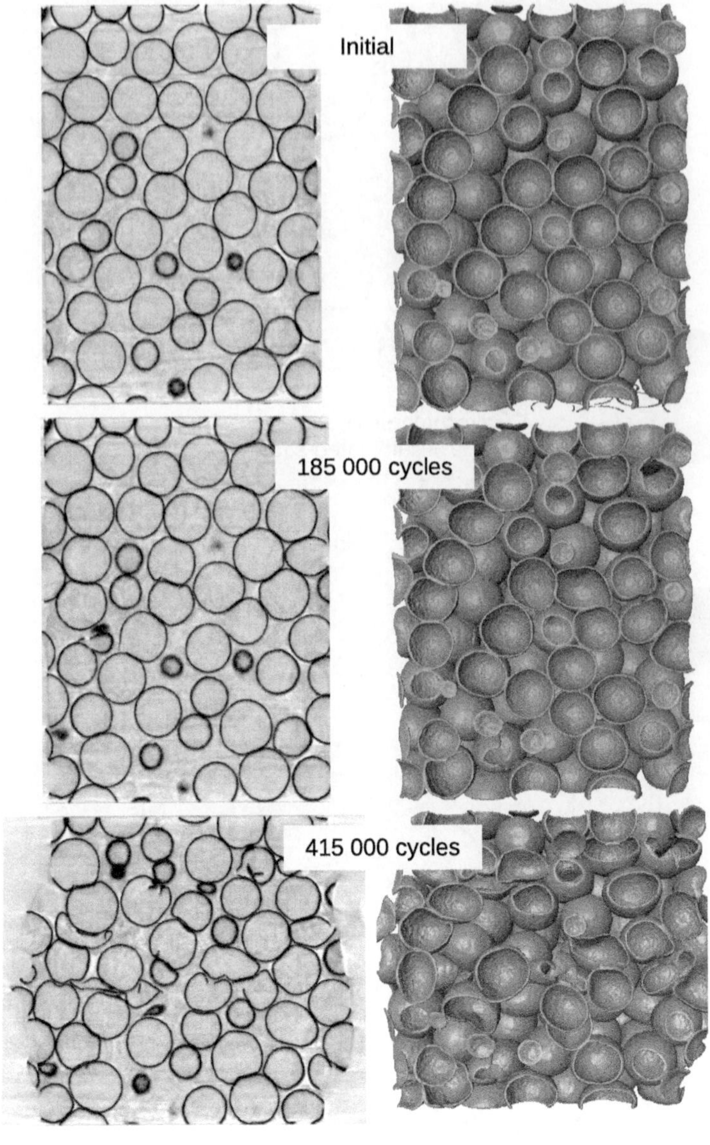

Fig. 9.5. Evolution of an extracted slice (left) and a 3D view (right) during an ex situ fatigue test. Testing conditions: σ^{max} = 4.5 MPa in compression (i.e. 1.5 times the fatigue limit).

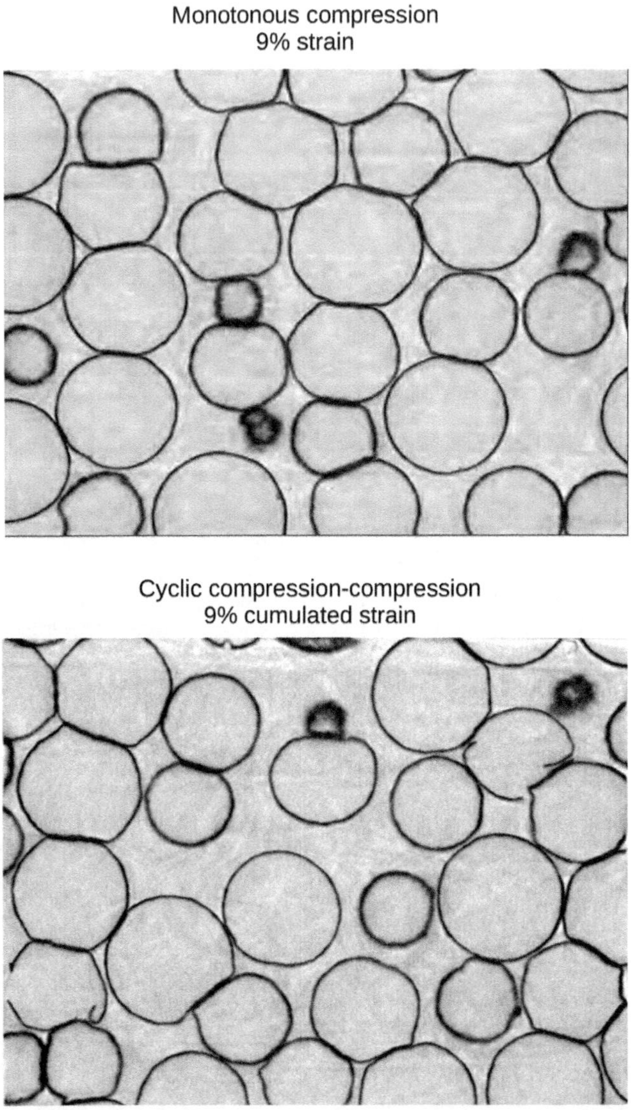

Fig. 9.6. Comparison of the microstructure of a sample deformed in monotonous compression (top) and in a the compression–compression fatigue mode (bottom) for a similar total deformation of about 9%. Microcracks are visible in the case of a cyclic deformation only.

9.3.2 Qualitative Damage Mechanisms

9.3.2.1 Compression

In Fig. 9.5, the results of an ex situ fatigue test are presented: the left pictures show a 3D visualization of an outside border of the MHSS and the ones on the right show reconstructed tomographic slices extracted from the interior of the sample. The cut was

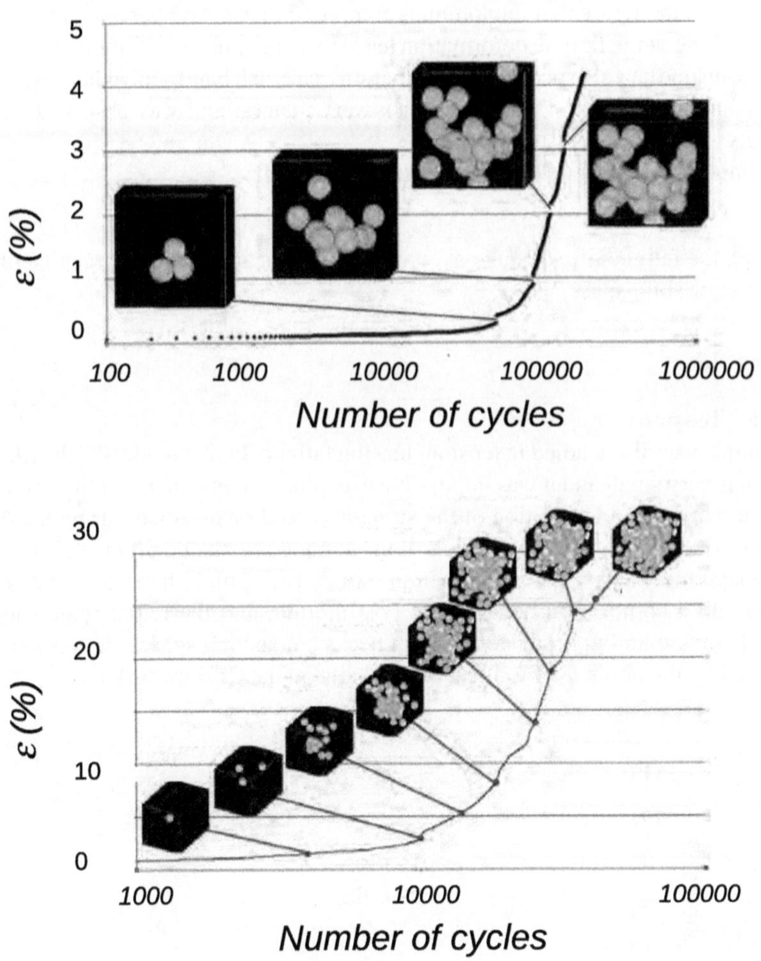

Fig. 9.7. Spatial distribution of the broken spheres superposed to the ε–N curve for sample 314D2.8t65 (top) and 314D2.8t90 (bottom)

chosen to show and follow all the different kinds of fatigue damage observed in the cellular material during loading cycles. The 3D visualization was taken on the outside border because fatigue phenomena are well known to appear at the surface of a specimen. It permits to reveal the collapsing of the entire specimen during the fatigue test.

One can notice that the material present initial defects which subsequently deform and break. Wall buckling and wall bending can also be observed. The micro-cracks can appear in several areas in the MHSS corresponding to the highly loaded regions. Crack development is not the unique phenomenon, but it is really specific to the cyclic loading studied here. Fig. 9.6 shows for example the comparison of a similar material compressed up to 10 % of strain in one cycle (Fig. a) and in fatigue (Fig. b). It is obvious from the figure that monotonous compression is accompanied with only minor microcracking while fatigue deformation leads to a high number of microcracks. After a 20 % strain, one can also observe the formation of a crush band somewhat perpendicular to the loading direction. This behaviour is very general and was observed for all the materials tested in the present study.

An important finding of the present study is then that

> during the fatigue life, damage in compression appears by micro-cracking of different initiation sites.

9.3.2.2 Tension

One sample was also studied in tension–tension fatigue. In this mode, the ductility of the material is very small and it was impossible to capture the last stages of the deformation. As a consequence, no evolution of the structure could be detected during the different scans performed along the fatigue life of the sample and the qualitative observation of the damage mechanism have not been possible. The sample has been analysed after fracture and a comparison between the post mortem and the initial stage shows that the final crack mainly propagates through necks connecting spheres in tension. This is evidenced by the absence of long pieces of walls on the fracture surface.

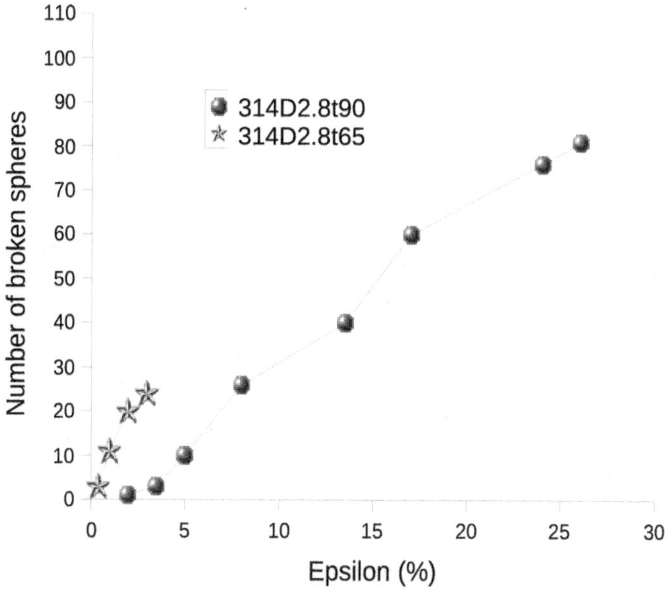

Fig. 9.8. Evolution of the cummulated number of broken spheres measured by manual marking of the XRCT images with the number of cycles N for samples 314D2.8t90 (circles) and 314D2.8t65 (stars)

9.3.3 Quantification of the Damage Mechanisms

In the case of two similar materials with different densities (314D2.8t90 and 314D2.8t65), for different deformation states in terms of number of cycles, the cracked spheres (i.e. all the spheres submitted to damage initiation by cracking) were manually marked in a duplicated image of their initial state. This allowed us to analyse the spatial distribution of the damage initiation. This distribution is shown in Fig. 9.7 in superposition with the ε–N curve of the studied samples. This figure shows that damage initiation is quite homogeneous in space i.e. damage initiates at random in several locations of the loaded sample. This is consistent with the qualitative images shown in section 9.3 where damage initiates all along the height of the sample. Figure 9.8 shows that the cummulated number of broken spheres is quasi linear with the number of cycles.

9.3.4 Quantification of the Densification

For the same materials as in the preceding section, the evolution of the density of the samples calculated in slices perpendicular to the compression axis has been measured by image processing of the 3D images obtained by XRCT. The density profiles along h both in the initial and final stages for the two materials are shown in figure 9.9. From one number of cycle to the next, it has been checked that the total solid fraction in the tomographic block measured after thresholding of the images was constant. For the

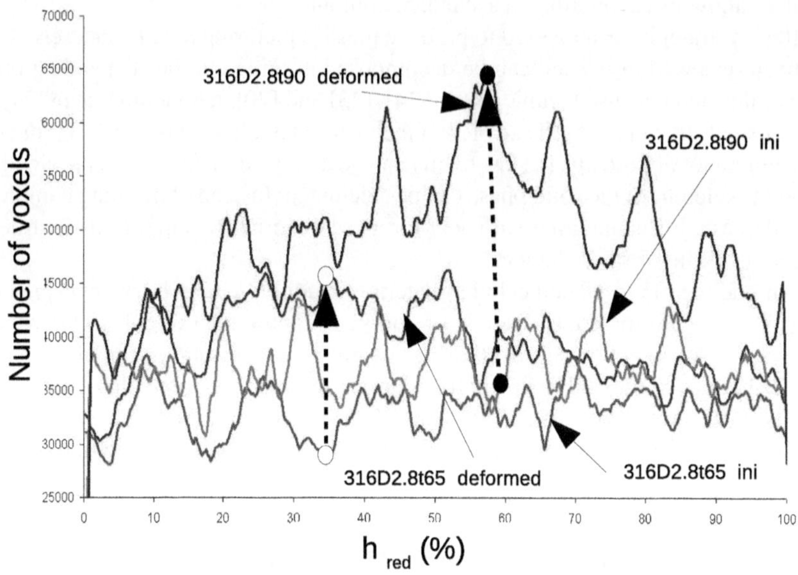

Fig. 9.9. Density profile along the reduced height h_{red} calculated for the sample 314D2.8t90 and 314D2.8t65 in tomographic slices perpendicular to the compression axis. For each sample, the initial and final deformation states are shown. The dotted arrows show that low density regions densify in each case.

sake of simplifying the comparisons, the x axis of the curve is not exactly the height h of the sample, but a reduced height h_{red} has rather been recalculated such as:

$$h_{red} = \frac{h}{H(N)} x100 \tag{9.3}$$

$H(N)$ being the height of the deformed sample at the considered number of cycle N. The advantage of the use of h_{red} is that for each deformed state, it varies between 0 and 100 %. The figure shows (see the dotted arrows) that densification occurs in bands perpendicular to the compression axis during fatigue. This densification is localised in regions where the density was initially small.

9.4 A First Step Towards Modelling: Determining the Local Stresses

From the observation of the damage mechanisms presented in section 9.3, it appears reasonable to try to model the fatigue properties of these materials by calculating the local value of the stresses. The knowledge of the stress distribution in these structures can be subsequently used to calculate fatigue criteria for damage initiation. Propagation is clearly out of the scope of the present modelling section. In order to accurately calculate the local stresses, we presented in [8] a way to use directly the exact 3D pictures of the materials obtained by XRCT. Using this method, the images can be directly used to produce Finite Element (FE) meshes of the materials, the size of which being small enough to allow its calculation on a standard computer.

Different strategies can be used to produce meshes picturing the 3D data sets. These have been reviewed in [6] and can be decomposed in three families depending on the nature of the elements used: cubic like in [14], [23] and [20], tetrahedra like in [24] and beams or shells like in [12], [9] and [10]. If the material is a low density cellular material i.e. contains a lot of porosity, the last family can be a very efficient method provided that a graph (a skeleton) of the solid phase can be calculated. Instead of meshing completely the solid phase, it is sometimes sufficient for these materials to simplify their structure by means of beam or shell elements.

The modeling of closed cell cellular materials using FE cannot be made using beam elements. For these materials one can of course use tetrahedra or cubic elements but when they have a very low density, they can contain walls that are extremely thin compared to the size of the cells. This is for example the case for some closed cell foams ([20] and [19]), syntactic foams ([24] and [2]) or for metal hollow spheres structures (MHSS) having a large diameter and a small thickness like the ones used in the present study.

9.4.1 Preamble: A Density Limit for the Meshing of Tomographic Images Using Continuum Elements

One of the first and efficient methods to obtain FE meshes reproducing as exactly as possible a cellular microstructure as seen in X-Ray tomography is the one described for instance in Youssef et al. [24]. This method uses continuum elements. But, when the number of pixels to describe the cell wall thickness in 3D images is too small, the continuum elements should reasonably no longer be used because they can cause numerical problems [20]. If the number of elements across the thickness is too small, one solution is to increase the resolution of the picture but then, the size of the studied block must be decreased and this can induce representative size problems. One can rather simply calculate a lower limit for the relative density of a cellular material d_{LL} tractable using continuum elements. Let us assume that the cells are cubic like described by Ashby et al. ([15] and [3]) for closed cell cellular solids. D is the width of the cell and t the thickness of the cell walls. The foam is digitized using a detector with lateral width W. If one wants to be representative of the overall material in an FE calculation, the sample should contain about N cells along this lateral size W. For a proper meshing using continuum element, it is desirable that there is at least v voxels across the thickness t. The lower limit of the relative density of such a foam that will allow a safe meshing using continuum elements can be estimated as:

$$d_{LL} = 1 - \left(\frac{\frac{W}{N} - v}{\frac{W}{N}} \right)^3 \tag{9.4}$$

If $W = 1000$, $N = 10$ and $v = 5$ pixels one can calculate that $d_{LL} = 0.14$. Using this rule, one can conclude that continuum FE models shouldn't be applied for closed cell foams with a relative density lower than about 0.14. If the relative density is lower, it is safer to use shell elements as will be presented in the rest of this study. This simple estimation assumes that the walls have all the same thickness. If t is scattered, the mesh should describe with a sufficient precision the smallest regions. This would have the effect to reduce the value of the lower bound d_{LL}.

9.4.2 Shell FE Model with Thickness Measurement

9.4.2.1 Shell Meshing

The method presented in this [8] is based on the skeletonization of the solid phase by thinning [18]. The final goal is to determine, by image processing, the median plane of each wall. The process described in Fig. 9.10, works with grey level images of closed cells foams. The figure shows the results on an extracted slice in 2D but the process is fully 3D and works with 3D data sets as input. The grey level image must first be binarized by thresholding. Each cell of the gaseous phase has then to be labeled. Each label can then be dilated isotropically through the solid phase until it encounters the neighboring dilating label. At the end of this dilation procedure, the solid phase has been eliminated and the frontier between two neighboring labels lies exactly in the middle of the previously existing wall between these two cells. In the example shown in the figure, one of the walls is lost during the process because the air inside the considered sphere is in contact with the air outside the spheres. This illustrates that to work properly, it is necessary that the spheres are perfectly closed. The frontier between labels can be meshed by a marching cube algorithm using surface (shell) triangular elements and can then be simplified to reduce the number of triangles while preserving a good description of the surface [26]. The linear triangular elements produced are then firstly imported into the FE code used (Abaqus^{TM} in the present study [1]). They are then translated

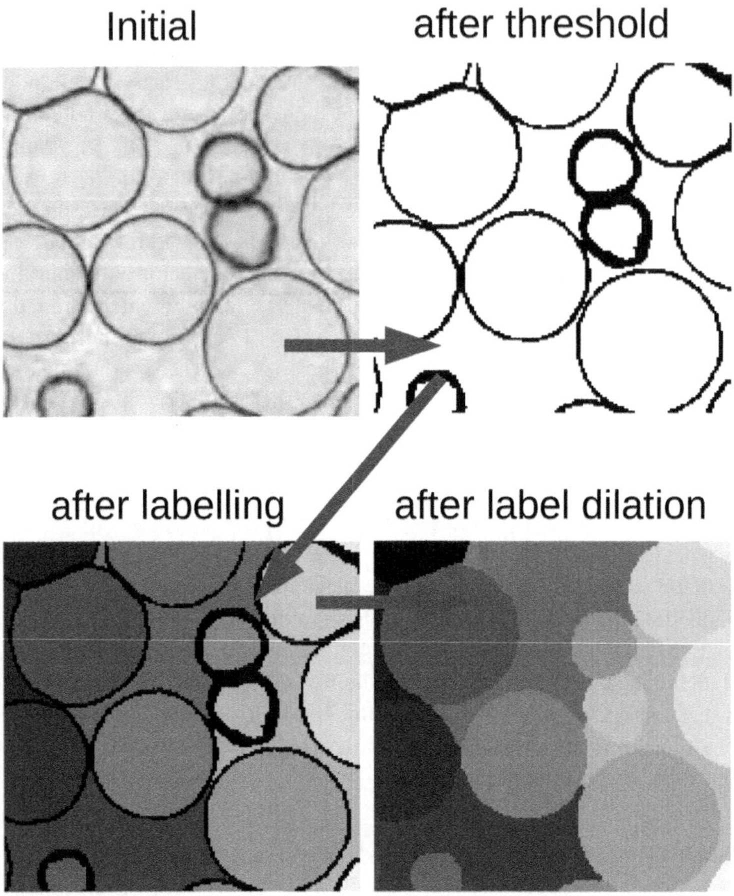

Fig. 9.10. Sequence of image processing used to skeletonize the solid phase of the cellular material. Binarisation of a slice of the initial tomographic image to separate the air (white) and the solid phase (black). Labelization of the air phase inside each cell: a different color is attributed to each cluster of 3D connecting voxels. Isotropic dilatation of labelized phases through the solid phase until it encounters a neighboring dilating label.

into quadratic elements using an available subset of the FE software (translation from S3 to STRI65 in Abaqus^{TM}).

9.4.2.2 Thickness Measurement

From the preceding step, an easy implementation is that each element can be attributed a constant thickness t in agreement with the global relative density of the material. But it would be more realistic to measure locally the actual thickness of the wall in the tomographic image and attribute it to the corresponding finite shell element [4]. The measure of the thickness was done in the present work using a mathematical morphology procedure called "3D Granulometry" (described in [19]) involving erosion/dilation

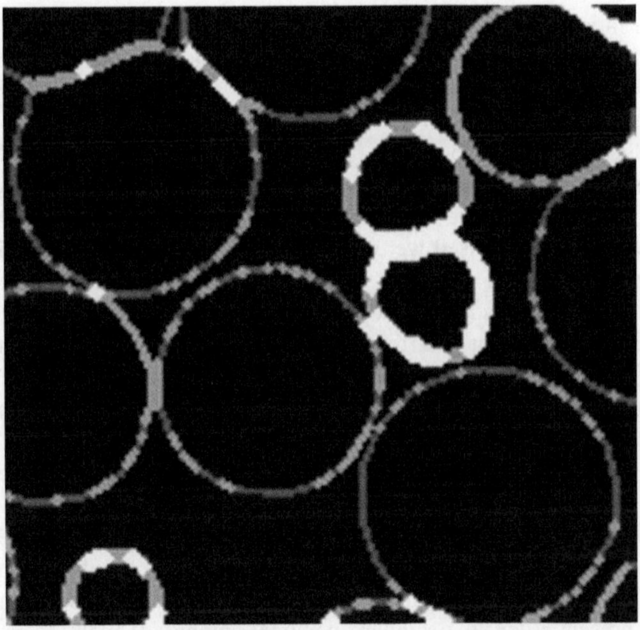

Fig. 9.11. Map of the wall thickness measured by an erosion/dilation 3D image treatment algorithm. Same slice as in figure 9.10. Higher thickness regions have a lighter colour.

steps of the binary image. The structural element used for the operation was a sphere in our case (erosion was governed by the euclidean distance to the considered voxel). The treatment provides a map of the wall thickness over the entire specimen like shown in Fig. 9.11 presenting the slice already shown in Fig. 9.10. In this figure, each voxel is colored according to the thickness measured at its location by the granulometry algorithm. The second step calculates the coordinates of the barycentre of each element in the mesh and attributes the measured thickness read in parallel on the map. The relative density of the mesh must sometimes be slightly corrected by a coefficient applied uniformly to the thickness of all the elements in order to be in exact agreement with the overall solid fraction in the material.

9.4.3 FE Model

Finally a global calculation can be performed. In this example, a very large tomographic bloc has been calculated — dimensions of the sample are 575 x 615 x 650 voxels with a resolution of 26 μm —. The behavior of the bulk material is modeled by an elasto-plastic law. Local properties of cellular materials are generally different from those of the bulk material. Nanoindentation measurements have been performed to try to measure the properties of the steel locally. The Young's modulus of the solid phase E measured is in good agreement with what can be expected for a steel (E_{steel} = 200 GPa). However, this measurement does not take into account the effect of the porosities. We chose to use adapted coefficients for the wall porosity volume fraction: E = 150 GPa and Poisson ratio $\nu = 0.3$. E is estimated by multiplying the steel modulus with the volume fraction of metal (about 75 %). This simple rule of mixture is actually an upper limit of the value of the modulus for this porous steel. Other models, permitting to calculate the Young's modulus from the level of porosity would provide lower Young's modulus values. The plastic behavior of the solid phase is discretized and modeled by a proof stress vs. plastic strain table chosen to reach a good fit for the global behavior of the foam while keeping the value in the range commonly given for porous sintered stainless steel. The values used were [(140 MPa , 0); (153.4 MPa, 0.035); (187 MPa, 0.1); (210 MPa, 0.2)] [17].

Different kinds of boundary conditions can be applied to the external faces of the mesh. The experimental test described in the present paper was a uniaxial monotonous compression test where one face was clamped and the other moved. The friction coefficient seems to be experimentally rather high as no lateral displacement of the walls in contact with the plates could be observed in the tomographic experiment. To reproduce these boundary conditions in the simulation, one face of the sample was then completely constrained while, on the opposite face, a uniaxial displacement (-1.476 mm; 10 % of height reduction) was applied along the compression direction (all other displacements being constrained). Figure 9.12 shows a deformed contour plot of the von Mises stress in the structure calculated using the procedure described above. The computer used all through the present study is a standard personal computer (Intel^R Xeon^{TM}, CPU 3.2 GHz, RAM 2GB). It is interesting to note that the new meshing procedure presented here allowed us to calculate with this modest computer the total volume of the sample

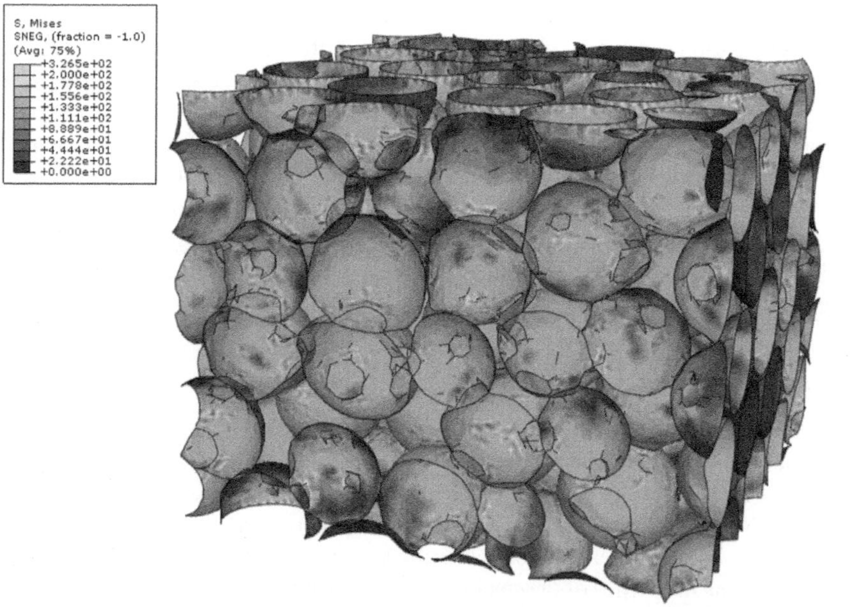

Fig. 9.12. Global calculation of a stainless steel metal hollow sphere structure sample. 3D visualization of the deformed contour plot of the von Mises stress.

tested experimentally. As a consequence, the effects of the free surfaces of the sample are completely accounted for. The resulting calculation can be post processed to determine the weak points of the microstructure under a given (an arbitrary) load. Such a realistic mesh can also be used to compute the global behavior of the studied material. It is for instance a versatile tool to simulate the behavior under multi-axial loading which is, in some cases, difficult to obtain experimentally.

9.4.4 Model Validation

The purpose in this section is to validate the whole shell FE model described. This validation is done at two scales, i.e. global and local. For the global scale, a calculation convergence study is presented. The model is developed to simulate the experimental compression test performed on the actual cellular material. Comparisons between experimental measurements and the results of the calculation are made at the two scales to check the accuracy of the model. The presented shell model is also compared with a tetrahedral calculation built with the method presented by Youssef et al. [24].

9.4.4.1 Global Validation

The procedure developed to mesh the median plane produces first a very dense triangular mesh, because the size of the triangles is directly related to the voxel size in the tomographic images. The number of triangles can be optimized using a simplification algorithm already implemented in the commercial software Amira^{TM} [26]. Decreasing the number of triangles permits to win CPU time but creates a lack of geometrical description. Four calculations were performed on a same structure using the same boundary conditions and the same material properties. To analyze the convergence, the number of triangles was decreased and the macroscopic response of each mesh was compared with the finest mesh possible to calculate. The stress Σ was calculated using the sum F of the reaction forces in the compression direction of each node of the mobile face divided by the total section S of the sample (S = 16 x 17 mm^2), ($\Sigma = F/S$). The strain ε is obtained by dividing the displacement of the mobile face Δh by the initial height of the sample h_0 ($\varepsilon = \Delta h/h_0$). With such curves one can measure the Young's modulus E and the yield stress σ_y of the modeled cellular material (σ_y is defined as the plateau stress for 2 % of total strain). These values allow to calculate the normalized difference with the same value measured using the finest mesh for the Young's modulus E^{finest} and the yield stress $\sigma_y{}^{\text{finest}}$. These relative differences are reported in Table 9.3. The table shows that the calculation converges to a constant behavior when the number of triangles is above 80000 elements. The yield strength converges more slowly. The number of cells in this numerical experiment is approximately 350. The convergence is then reached for an approximate number of 230 elements per sphere. The model using 30 398 quadratic elements with its 1.5 hours of CPU time, its 0.7 % error for the Young's modulus and 6.2% error for the yield stress appears to be an efficient tradeoff between model resolution and time consumption.

Continuum meshes produced with the method of Youssef et al. [24] and exhibiting 297 182 quadratic and linear tetrahedral elements have been compared with shell meshes with 76 445 quadratics and linear elements to highlight the influence of the type

Table 9.3. Effect of the mesh size on the calculated macroscopic Young's modulus and plateau stress

Number of elements	$\frac{E-E^{\text{finest}}}{E^{\text{finest}}}$ (%)	$\frac{\sigma_y-\sigma_y^{\text{finest}}}{\sigma_y^{\text{finest}}}$ (%)	CPU time
30 398	0.7	6.2	1 h 30
57 236	0.4	2.7	2 h 50
86 515	0.25	1.1	4 h 15
123 848	0	0	7 h 50

of elements used on the global behaviour calculated. The comparison has been carried out using the same procedure as the one used in table 9.3. The calculation case with quadratic shell elements has been used as a reference for comparing the performance of the other types of elements in terms of E^{finest} and σ_y^{finest}. The results on these four kinds of meshes are presented in Table 9.4. The model with linear tetrahedrons is much stiffer than the other models. Linear tetrahedral elements are known to be not perfectly adapted for the simulation of bending loadings because of locking problems. However all the three other computations seem to converge to the same values and give very close and global results. The advantages of the shell meshes compared to the tetrahedral ones are faster convergence, capability to calculate larger samples and lower time consumption.

9.4.4.2 Comparison between the Macroscopic Behavior Measured and Calculated

In Fig.9.13, stress/strain curves measured and calculated are reported on a same graph for sample 314D2.8t90. In the plastic strain range studied, the yield stress and the hardening of the studied MHSS is well reproduced by the shell FE calculation. The little differences are explained by the high fraction of porosity in the sphere wall and some approximations in the geometrical description of the structure with the shell elements. The bulk material Young's modulus was fixed at E = 200 x 0.75 = 150 GPa to account for the partially porous nature of the walls. The relative density of the mesh is adjusted to the relative density measured in the tomographic images by a coefficient uniformly affected to the thickness of all shell elements. The FEM computation seems to be con-

Table 9.4. Comparison on the macroscopic Young's modulus and the macroscopic plateau stress calculated with continuum meshes [24] and shell meshes proposed in the present work

Number of elements	Type of elements	$\frac{E-E^{\text{finest}}}{E^{\text{finest}}}$ (%)	$\frac{\sigma_y-\sigma_y^{\text{finest}}}{\sigma_y^{\text{finest}}}$ (%)	CPU time
297 182	Linear tetrahedrons	37	36.2	48 min
297 182	Quadratic tetrahedrons	10.3	8.8	9 h 26
76 445	Linear triangles	3.2	4.4	12 min
76 445	Quadratic triangles	0	0	1 h 19

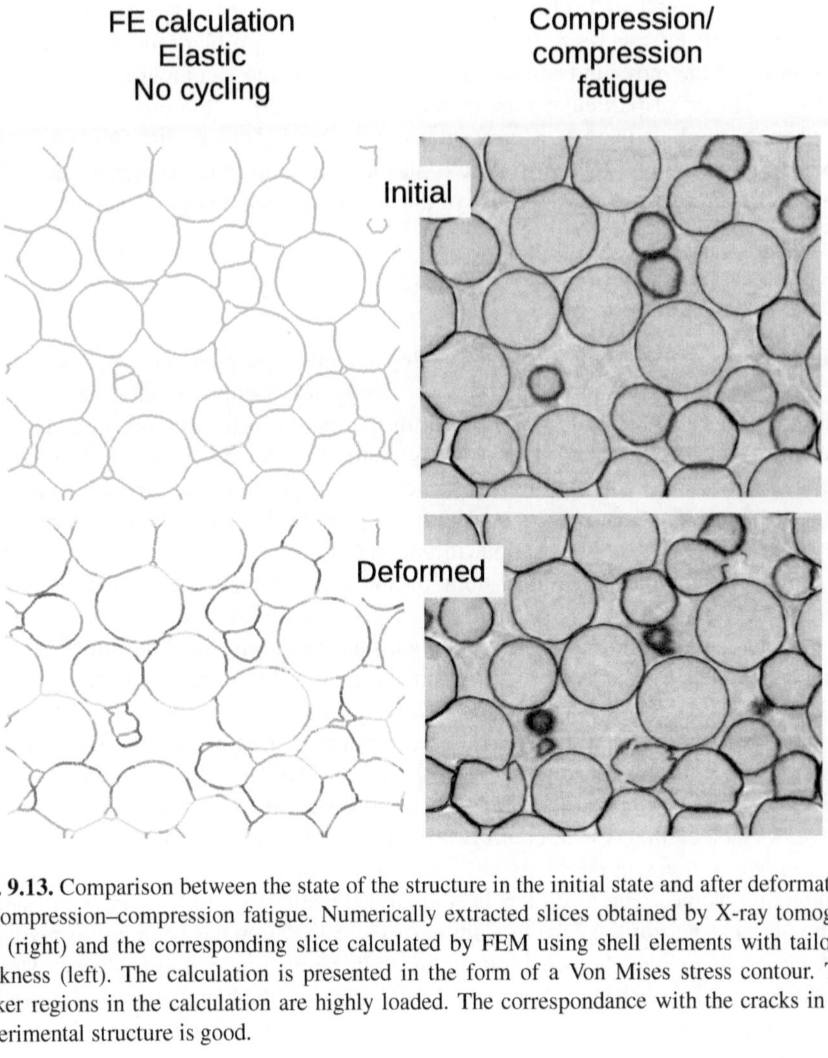

Fig. 9.13. Comparison between the state of the structure in the initial state and after deformation in compression–compression fatigue. Numerically extracted slices obtained by X-ray tomography (right) and the corresponding slice calculated by FEM using shell elements with tailored thickness (left). The calculation is presented in the form of a Von Mises stress contour. The darker regions in the calculation are highly loaded. The correspondance with the cracks in the experimental structure is good.

verged. Experimentally, it is always rather difficult to perform a good measurement of the deformation in the elastic regime. In the present case, this measurement was performed very carefully by image correlation done on an exterior face of the sample. Despite all this precautions, a gap between the Young's modulus measured and simulated remains.

The thickness attribution algorithm works with the barycentre of each element and with the map of the wall thickness. From the coordinates of the barycentre, the thickness is read on the map and affected to the element. A problem can occur in the Y types areas. Three shell elements describe the three branches of the Y, and each element is attributed a measured thickness. In some cases, this value of the thickness measured is too high. For two of the branches, the thickness is overestimated because attributed to two different elements. This leads to a slight overestimation of the sphere wall thickness of the walls of the spheres very close to the neck. This problem increases artificially the density of the mesh and can locally increase the stiffness compared with the actual material. This is of particular importance to explain part of the differences observed between the predicted and measured values of the Young's modulus. The experimental modulus is probably affected by micro-plastic deformation occurring in these Y regions and poorly reproduced because of the lower representativity of the mesh in these regions.

9.4.5 Local Validation

The preceding section shows that the developed method permits to predict more or less the global behaviour of a cellular structure. This is not essential in our opinion because other (and more simple) models do exist in the literature for this purpose. The main interest of the shell model is that it is a versatile tool for modeling local strains and stresses. This kind of simulation permits to reach the local fields for each gauss point of the structure. This is our primary goal in the present chapter as the local stresses can then in turn be used to apply a fatigue initiation criterion to the structure. A comparison with the experimental fatigue test done on the structure is carried out in the present section.

A local validation of the shell model regarding the fatigue local deformation modes can be obtained by the comparison with the actual deformation kinematics of the material observed by X-Ray micro tomography. Figure 9.13 shows the comparison of the experimentally deformed state of a particular slice of sample 314D2.8t90 in compression–compression fatigue as observed in XRCT. The figure also shows the same slice extracted from the calculation performed using the procedure described in the work (shell elements with controlled thickness). The figure shows a contour plot of the von Mises stress in the elements. The elastic displacements have been magnified to be more visible). The mesh has been obtained using the XRCT scan of the sample at the initial state. For the sake of simplicity, the calculation has been performed using just one loading cycle, the behaviour of the walls being assumed to be elastic. Despite the differences between the experimental conditions and the hypothesis used to perform the calculation, one can note two interesting things validating the proposed method at the local scale:

- the local deformation patterns predicted by the model and observed in the deformed slice are very close (the slight differences are due to the fact that it is always a bit difficult to find the same slice in the initial and in the deformed 3D image);
- the cracks are observed to appear experimentally in regions where the stresses are high in the elastic calculation (darker regions in the contour plot).

9.5 Conclusion

The fatigue behaviour of several MHSS has been studied experimentally and numerically in the present chapter. It has been measured macroscopically that the endurance limit in compression is equal to about the half of the yield strength of the materials.

The fatigue limit in tension is approximately twice smaller than the measured limit in compression. The density (conversely the wall thickness) has a first order effect on the fatigue resistance, the denser (the thicker) the walls, the higher the fatigue resistance. The spheres diameter has a direct impact on the stress sensitivity of the MHSS (the higher the diameter, the higher the slope of the $S-N$ curve).

At the microstructural scale, the compression–compression fatigue deformation mode induces microcracks in the structure whereas monotonous loading at the same cumulated strain is achieved without any detectable cracking. The nucleation of these cracks is surprisingly spatially distributed very homogeneously in the observed samples. The number of cracks varies linearly with the number of fatigue cycles. On the contrary, densification bands are observed to appear showing that the deformation presents a certain degree of localisation. These bands are observed to be perpendicular to the compression axis, in regions where the density was initially small.

In tension, the fatigue fracture is extremely rapid, precluding an easy observation of the crack propagation. Post mortem observation shows that the crack propagates close to the necks.

The local structure being a very important parameter for the fatigue damage resistance, a procedure has been implemented to calculate the local stress distribution. This procedure is based on a shell FE calculation reproducing the structure of the materials as observed in XRCT. The thickness of each created shell element is controlled by the value of the thickness measured in the tomographic image. The model has been validated at the global scale using a monotonous compression test, but more importantly at the local scale.

Future works will focus on the post processing of the obtained models using fatigue initiation criteria.

Acknowledgements. The two French research organisations, CNRS and ONERA, are acknowledged for funding this study and the company PLANSEE for kindly providing the materials. Discussions with Yves Bréchet and all partners of the MAPO project in general were also very helpful.

References

1. ABAQUS, Finite element program theory manual (1998)
2. Adrien, J., Maire, E., Gimenez, N., Sauvont-Moynot, V.: Acta Mater 55, 1667–1679 (2007)
3. Ashby, M., Evans, A., Fleck, N., Gibson, L., Hutchinson, J., Wadley, H.: Metal foams: a design guide. Butterworth-Heinemann, Boston (2000)
4. Autuori, B., Bruyere-Granier, K., Morestin, F., Brunet, M., Verriest, J.P.: IEEE T Bio-Med Eng. 53(7), 1225–1232 (2006)
5. Bathias, C., Baïlon, J.P.: La fatigue des matériaux et des structures. Hermès Editions, Paris (1997)
6. Buffière, J.Y., Cloetens, P., Ludwig, W., Maire, E., Salvo, L.: MRS Bull. 33(6), 611–619 (2008)
7. Caty, O., Maire, E., Bouchet, R.: Adv. Eng. Mater. 10(3), 179–184 (2007)
8. Caty, O., Maire, E., Bouchet, R.: Acta Mater. 56(19), 5524–5534 (2008)

9. Dillard, T., N'Guyen, F., Forest, S., Bienvenu, Y., Bartout, J.D., Salvo, L., Dendievel, R., Maire, E., Cloetens, P., Lantuejoul, C.: Cellular Metals. In: Banhart, J., Fleck, N., Mortensen, A. (eds.) Metfoam 2003, pp. 301–306 (2003)
10. Elliott, J.A., Windle, A.H., Hobdell, J.R., Eeckhaut, G., Oldman, R.J., Ludwig, W., Boller, E., Cloetens, P., Baruchel, J.: J. Mater. Sci. 37, 1547–1555 (2002)
11. Fallet, A., Lhuissier, P., Salvo, L., Bréchet, Y.: Adv. Eng. Mater. 10(9), 858–862 (2008)
12. Fazekas, A., Dendivel, R., Salvo, L., Bréchet, Y.: Mech. Sci. 44(10), 2047–2066 (2002)
13. Friedl, O., Motz, C., Peterlik, H., Puchegger, S., Reger, N., Pippan, R.: Metall. Mater. Trans. B 39B, 135–146 (2008)
14. Garboczi, E.J., Day, A.R.: J. Mech. Phys. Solids 43, 1349–1362 (1995)
15. Gibson, L.J., Ashby, M.F.: Cellular Solids. Cambridge Solid State Science Series, 52–92 (2001)
16. Harte, A.M., Fleck, N., Ashby, M.: Acta Mater. 47(8), 2511–2524 (1999)
17. Lee, W.S., Lin, C.F., Liu, T.J.: Mater. Charact. 58, 363–370 (2007)
18. Liang, Z., Ionnidis, M.A., Chatzis, I.: J. Colloid Interf. Sci. 221, 13–24 (2000)
19. Maire, E., Colombo, P., Adrien, J., Babout, L., Biasetto, L.: J. Eur. Ceram. Soc. 27, 1973–1981 (2007)
20. Maire, E., Fazekas, A., Salvo, L., Dendievel, R., Youssef, S., Cloetens, P., Letang, J.M.: Compos. Sci. Technol. 63, 2431–2443 (2003)
21. McCullough, K.Y.G., Fleck, N., Ashby, M.: Fatigue Fract. Eng. M 23, 199–208 (2000)
22. Motz, C., Friedl, O., Pippan, R.: Int. J. Fatigue 27, 1571–1581 (2005)
23. Ulrich, D., Van Rietbergen, B., Weinans, H., Ruegsegger, P.: J. Biomech. 31, 1187–1192 (1998)
24. Youssef, S., Maire, E., Gaertner, R.: Acta Mater. 53, 719–730 (2005)
25. Zettl, B., Mayer, H., Stanzl-Tschegg, S.E., Degisher, H.P.: Mat. Sci. Eng. A-Struct. 292, 1–7 (2000)
26. `http://www.tgs.com/support/amiradoc/Documentation/amira41UsersGuide.pdf`

10 Design of Functional Properties for Composite Spherical Granulates: Functional Design of Spherical Granulate Composites

Constant van Lookeren Campagne

L2Consultancy B.V., Nrd Houdringelaan 20,
NL-3722BR, Bilthoven, The Netherlands
c.vanlookerencampagne@inter.nl.net

Abstract. This chapter shows how to design a spherical formulated granule or hollow sphere in attempt to optimize performance in gas separation or catalytic processes. The focus in this chapter is on the mathematical model and the design methods based on Glatt process equipment.

Keywords: Zeolites; Design of customized zeolites; Design of catalysts; Spherical granules and granulates.

10.1 Introduction

Within the last 10 years significant progress in granulation technology made it possible to produce sophisticated ceramic spherical granules composed of a porous shell covering an inert core. The core can be solid or of a different porous material. A special type of granules can also be produced with a hollow core.

These types of granules are of interest for various applications, notably:

- Adsorbents in gas separation processes, e.g. for use in zero emission power plants,
- Catalysts in chemical reactions,
- Applications for controlled dosage of therapeutic ingredients,
- Reservoir of fragrances for room perfuming.

In gas separation processes and chemical reactions with catalysts, the activity often takes place mainly or exclusively in the outer layer of the conventional used granules. It is therefore an advantage to save expensive material and substitute the core with an inactive cheap material.

For controlled dosage systems of therapeutic ingredients it is important to have inert porous shells, which allow controlled diffusion of active components out of the core.

For room perfuming the same principles are applied, however here the active component is a fragrance with a low vapour pressure.

A. Öchsner, C. Augustin (Eds.): Multifunctional Metallic Hollow Sphere Struct., EM, pp. 183–193.
springerlink.com

Glatt GmbH is leading in granulation technologies. Machines built by Glatt allow the modification of the following properties of porous shells as a function of radius:

- The porosity,
- The composition (active matter),
- The specific surface area.

Glatt GmbH developed a continuous fluid bed granulation process. An important feature of the process is that undersized granules are given back into the machine until they have grown at target size when they will be removed. This novel type of granulation is carried out in a so called ProCell fluid bed unit.

Consequently the ProCell equipment is particularly well suited for continuous production of spherical particles within a narrow particle size distribution in the range of 100 μm to 5 cm. For use in applications mentioned above this is very important because the well defined product allows the best reproducibility in application. In parallel new nozzles have been developed, what made it possible to spray now in total 3 to 6 components at the same time. These components can react with each other as soon as they get into the granulation zone of the ProCell. New types of filters allow longer intervals of continuous processing between interrupts for cleaning.

The control of the granulation process has been improved by using new inline control equipment that can continuously monitor the distribution of particle size.

Glatt also developed a spinning disk granulator. The spinning disk equipment is very suitable for layering. This equipment is special because the granules and coating powders are brought into intense contact with each other and the process air. By modification of the disk speed the shell density can be influenced in a broad range. It becomes possible to build up compounded granules with high porosity inside but showing very low abrasion as result of a compact outer layer.

In this chapter we will provide a mathematical model to design products for such applications based on the well known non-stationary diffusion transport model for porous spheres. The model approach will be used for the application of adsorbents for gas separation since this is the most complex and challenging application. For gas adsorption zeolites are used and therefore a chapter on these zeolite materials is incorporated as well. The application is of high interest at the present time. In attempt to reduce green house gases, one approach is to install zero emission power plants. One of the used technologies is the Oxyfuel process where zeolites can be used to produce oxygen. Therefore having explained the model, some attention will be paid to the application of zero or reduced emission power plants.

10.2 Zeolites

Commonly used molecular sieves have pore sizes between two and fifteen Ångström. Within this pore size range many products in chemical industry can be adsorbed inbetween the structure (Fig. 10.1). Typical examples are the following Zeolites. Often the trade name is the poresize in Ångström:

- Zeolite 3A: pore size 3Å, Adsorbs e.g. NH_3, H_2O, (not C_2H_6), good for drying polar liquids.
- Zeolite 4A: pore size 4Å, Adsorbs e.g. H_2O, CO_2, SO_2, H_2S, C_2H_4, C_2H_6, C_3H_6, EtOH. Will not adsorb C_3H_8 and higher hydrocarbons. Good for and commonly used for drying nonpolar liquids and gases.
- Zeolite 5A: pore size 5Å, Adsorbs normal (linear) hydrocarbons to n-C_4H_{10}, alcohols to C_4H_9OH, mercaptans to C_4H_9SH. Will not adsorb isocompounds or rings greater than C_4.
- Zeolite 10X: pore size 8 Å, Adsorbs branched hydrocarbons and aromatics. Useful for drying gases.
- Zeolite 13X: pore size 10 Å, Adsorbs di-n-butylamine (not tri-n-butylamine). Useful for drying Hexamethylphosphoramide.

Fig. 10.1. Typical Zeolite Structure [1]

The characterisation is that all members of the family contain at least one silicon atom per aluminium atom, for instance:

$$\text{Na-A } Na_2O \cdot Al_2O_3 \cdot 2SiO_2 \cdot 4.5\ H_2O$$

Some very common zeolites are [1]:

- LiLSX (Lithium with low Silica content): commonly used for high efficient nitrogen sorption, separation of oxygen and nitrogen,
- 5A: mostly used for separation of oxygen and nitrogen, sorption of nitrogen, 5A: works with much higher kinetic then LiLSX but has a lower capacity.

10.3 Non-stationary Diffusion Model

The molar flow of a gas component A having a concentration ψ_A (kmol/m^3), can be defined by Fick's law as

$$j_{A,x} = -D_A \cdot d\psi_A/dx, \qquad (10.1)$$

where $j_{A,x}$ (kmol/m^2) is the molar flow of a gas per square meter of surface area, and D_A (m^2/s) is the diffusion coefficient [2].

For non-stationary situations this molar flow changes as a function of time and space. For instance, if a spherical porous sorbent particle is exposed to an instantaneous increase in concentration of a gas component A from ψ_{A0} to ψ_{A1} at the boundary of the spherical granule the weight of the sorbent particle will start to rise due to adsorption on the sorbent particle. From this weight increase we can calculate the amount of gas A adsorbed as a function of time per cubic meter sorbent. The volume of n sorbent particles will be $n \cdot 4/3 \cdot \pi \cdot R^3$. There is no production of the gas component A in this closed space. Therefore the amount of gas taken away is equal to the molar flow adsorbed over a period of time into and on the porous particle. The change as function of time of the molar gas diffusion flow can be defined as [2]

$$\partial^2 \psi_A / \partial x^2 = \partial \psi_A / \partial Fo, \tag{10.2}$$

where Fo is the mass transfer Fourier number, which is defined as $D \cdot t/R^2$. In these cases the Fourier number is often called dimensionless time. This makes sense since introducing the dimensions of this Fourier number the conclusion is that it has no dimension (which means: m^2/s · s/m^2 = 1).

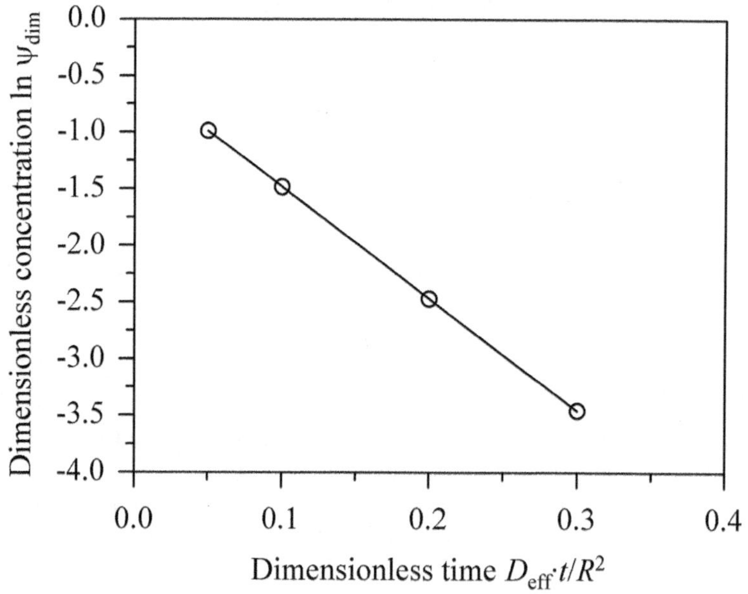

Fig. 10.2. Dimensionless concentration versus dimensionless time

The aim of this chapter is to provide a design model for spherical granules. Therefore it is better to define the differential equation (10.2) in radial coordinates, which after transformation leads to the following equation for *homogeneous* porous spheres:

On the left hand side we integrate over the radius of the particle, on the right hand side, we integrate over the time. Index m is mass:

$$\frac{3}{R^3}\int_0^R (\psi_{Am}-\psi_{A1})/(\psi_{A0}-\psi_{A1}) = 6/\pi^2\sum_{n=1}^{\infty}(\exp(-n^2\pi^2 Fo)/n^2). \quad (10.3)$$

For large values of the Fourier number, i.e. $Fo \geq 0.25$, this leads to the equation:

$$\lim_{Fo\to\infty}(\psi_{Am}-\psi_{A1})/(\psi_{A0}-\psi_{A1}) = 0.6086\cdot\exp(-9.8696\cdot Fo). \quad (10.4)$$

Figure 10.2 provides a graph of the natural logarithm of the dimensionless concentration ψ_{dim} versus the dimensionless time. It is defined as $(\psi_{Am}-\psi_{A1})/(\psi_{A0}-\psi_{A1})$.

It should be noted that the model is not valid for small Fourier numbers and is therefore truncated before it crosses the vertical axis. It is important to remind, that there are some important conditions under which this model is valid:

- There is only mass transport as a result of diffusion and no convection mass transfer.
- The Fourier number is ≥ 0.25 preferably ≥ 0.5.

10.4 Practical Model Applications: Designing the Adsorbent

10.4.1 Application: Measuring the Improved Mass Transfer Coefficient

The above derived model can be applied to design a sorbent for the separation of gas components; for instance oxygen from air.

Equation (10.4) can be written under consideration of the Fourier number as follows:

$$\lim_{Fo\to\infty}(\psi_{Am}-\psi_{A1})/(\psi_{A0}-\psi_{A1}) = 0.6086\cdot\exp(-9.8696\cdot D\cdot t/R^2). \quad (10.5)$$

It follows straight from Eq. (10.5), that if the logarithm of the increase in relative weight of an adsorbent is plotted as function of time the result will be a graph with a slope having a value equal to – D_{eff}/R^2, where R is the radius and D_{eff} is the so-called effective diffusion coefficient. This value of the diffusion coefficient is called D_{eff} since it represents the value for the time period covering the increase in weight that is effectively measured. The value of D_{eff}/R^2 is directly proportional to the mass transfer coefficient. If the value of D_{eff}/R^2 is increased by 100% the adsorption process will run twice as fast. Thanks to Eq. (10.5), the comparative performance increase in the adsorbent can be measured quite accurately. Let us assume that a number of homogeneous porous sorbent particles are exposed to an instantaneous increase in concentration of a gas component A and that the weight increase of the sorbent particles can be measured. In Tab. 10.1 below the increase in weight is given as a function of time after a jump increase of the concentration of a component A at the outer boundary layer

Table 10.1. The increase in adsorbed mass as a function of time during a simulated adsorption experiment to test the performance of a porous sorbent

Time t (s)	Weight (g)	$\psi_{A,dim}$
0	0.815	
1	1.39	0.61
1.5	1.52	0.52
2	1.63	0.45
2.5	1.72	0.39
3	1.80	0.33
3.5	1.98	0.25
4.0	2.02	0.18
6.0	2.14	0.01
600	2.29	0.00

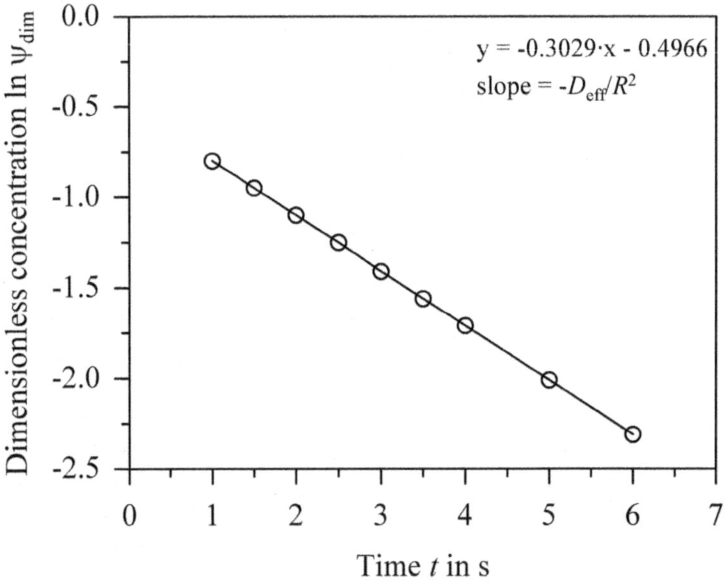

Fig. 10.3. Determination of the effective Diffusion rate (D_{eff}/R^2) of a sorbent

of the spherical sorbent particle. One second after the increase in the component A concentration, the weight of the particles is f.i. $1.39{\cdot}10^{-3}$ kg. After 600 seconds the weight of the particles is f.i. $2.29{\cdot}10^{-3}$ kg. That means for instance that at $t = 2$ seconds the value of $(\psi_{Am} - \psi_{A1})/(\psi_{A0} - \psi_{A1}) = (1.63 - 2.29)/(0.815 - 2.29) = 0.45$.

In the plot below the simulated values of Tab. 10.1 are given and a trend line is drawn indicating the slope of the graph and thus the value of D_{eff}/R^2. It should be noted that the values are simulated values. In practical situations the variation coefficient depends on the reproducibility and accuracy of the equipment used for the analysis.

From the above chart it is possible to see whether the sorbent adsorbs at a fast rate and is very effective or not. The steeper the downward slope the more effective the sorbent. The chart is made out of Table 10.1 but corrected using empirical data.

10.4.2 Application: Increasing the Mass Transfer (D_{eff}/R^2)

If the objective is to increase the mass transfer of the sorbent particle, it follows from Eq. (10.5), that if we reduce the radius of the spherical sorbent particle by a factor $\sqrt{2}$ we will reach the same amount of component adsorbed in half the time. This infers that the mass transfer coefficient has doubled in value. However this comes at a cost, since a column filled with these adsorbent having a significantly reduced size will also have a considerably higher resistance and pressure drop for the gas flow.

The effect of the higher pressure drop can be negated by providing a hollow porous sphere or a sphere with an inert core. The length of the diffusion path will also be reduced and therefore the mass transfer will be increased. If the differential Eq. (10.3) is integrated over the radius value from R_1 to R_2 the result can be used to calculate the effect on the mass transfer rate, knowing that R_1 is the radius of the inert spherical core and R_2 is the outer radius of the spherical particle. However since we have taken away part of the sorbent particle the capacity of the sorbent i.e. the amount of active material will be reduced by a factor $(R_1/R_2)^3$. The result is given in the Tab. 10.2.

Table 10.2. Change of properties of the sorbent as a function of the ratio of the inner radius and the outer radius

R_1/R_2	0.3	0.4	0.5	0.6	0.7	0.8
Mass Transfer increase factor	1.1	1.25	1.51	2.03	3.14	6.22
Capacity Loss	3%	6%	13%	22%	34%	51%

It is desirable to improve the mass transfer performance of this sorbent particle by a factor 2. We can reduce the radius from 2 mm to 2 mm/$\sqrt{2}$ = 1.4 mm. The disadvantage is the increase in pressure drop, which is in most cases unacceptable. Or we can provide the sorbent particle with an inert core with a diameter of 0.6·2 mm = 1.2 mm and the same outside diameter of 2 mm. The disadvantage in this case is that the adsorption capacity is reduced by 22%. However using the same adsorption process unit, the process performance is of course increased considerably since the mass transfer increase factor is a good indication of the reduction in cycle time going from adsorption to regeneration.

10.4.3 Applications of Composite Granulated Sorbents to Gas Separations in Reduced- or Zero Emission Power Plants (RZEPP)

The present state-of-the-art power plants running on fossil fuels use air to burn the fuel. Since air consists of around 80% Nitrogen, the amount of gas running around in these power units could be reduced by a factor of more than four if most of the Nitrogen would be separated and removed. Running such plants on pure oxygen would generate an off gas mainly consisting of CO_2 ready to be captured. Due to the absence of nitrogen the burning of fuel is more efficient since less gas is handled. Another positive effect is that no NOx can be generated what also is an environmental pollution. The main reason, that this principle is not applied, is that the savings obtained at the moment are off-set by the electricity costs of producing the high purity (90%) oxygen. High purity oxygen can be produced by adsorption processes [2].

The use of wind or solar power for electricity generation is increasing. These relative small power plant units are distributed in larger numbers over larger areas in a country and have the disadvantage, that they are less reliable, since the wind or sunshine are not always available.

Running these small power plants based on wind or solar energy in combination with smaller power generators running on natural gas would be an interesting alternative if switching between the energies could happen fast enough. Here adsorption processes for oxygen production would be ideal to burn natural gas the clean way, if the operating costs to produce pure oxygen could be reduced. It has been shown above that new sorbents can be produced having a considerable increased diffusion rate. Since this diffusion rate is the limiting factor in the productivity of such an oxygen production unit, costs of oxygen production can be reduced significantly. Consequently electricity can be generated with this new technology and more reliably at considerably reduced costs, in combination with wind power and/or solar cell units. Even if the CO_2 produced is not captured the amount of CO_2 produced per kWh is considerably reduced. To prevent the combustion unit (f.i. a turbine) from overheating, part of the CO_2 produced is recycled and mixed with the high purity oxygen entering the combustion unit.

10.4.4 Applications to Catalysts

For catalysis the same principle plays a role. Diffusion rate of reactants and reaction products are important factors of the catalyst performance. With the rotor agglomeration unit nickel catalyst on a carbon carrier has been produced. Thanks to the layering technique the carbon carrier (cf. Fig. 10.4) can be made in such a way that the nickel metal particles are present only in the outer ring of the carbon carrier. However no Nickel is actually present on the outer surface of the granule. This has the advantage that no nickel will be lost due to abrasion in the product stream passing along the particle.

In the outer surface of the carrier openings can be seen (cf. Fig. 10.5), which allow a sufficient flow of reactants and products in and out of the granulated carrier. The Nickel is present in a very fine dispersion (cf. Fig. 10.6) on the carbon carrier. Metal catalyst on carbon carriers are of interest for Fischer-Tropsch process and hydrogenation processes which can take place during distillation. When the catalyst has lost part

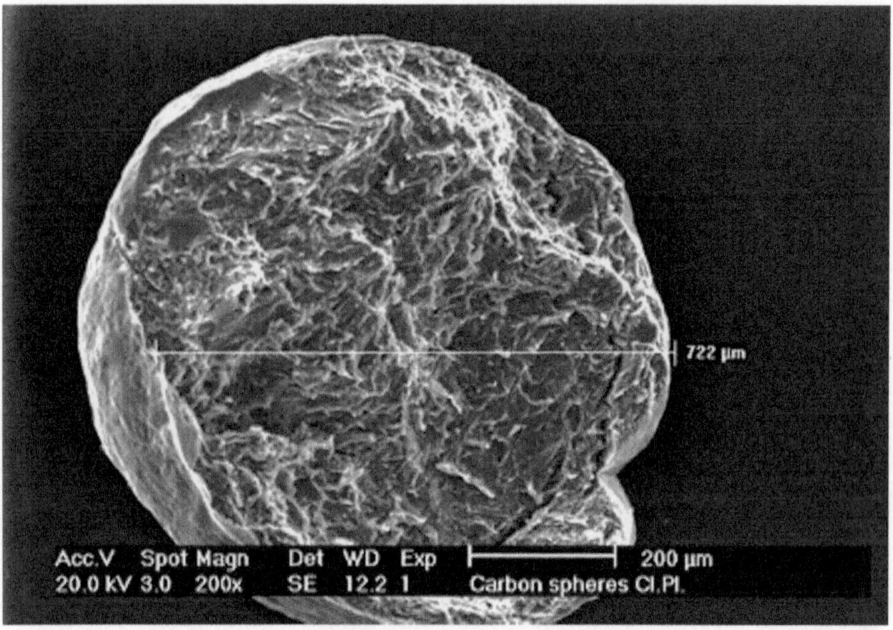

Fig. 10.4. Carbon carrier

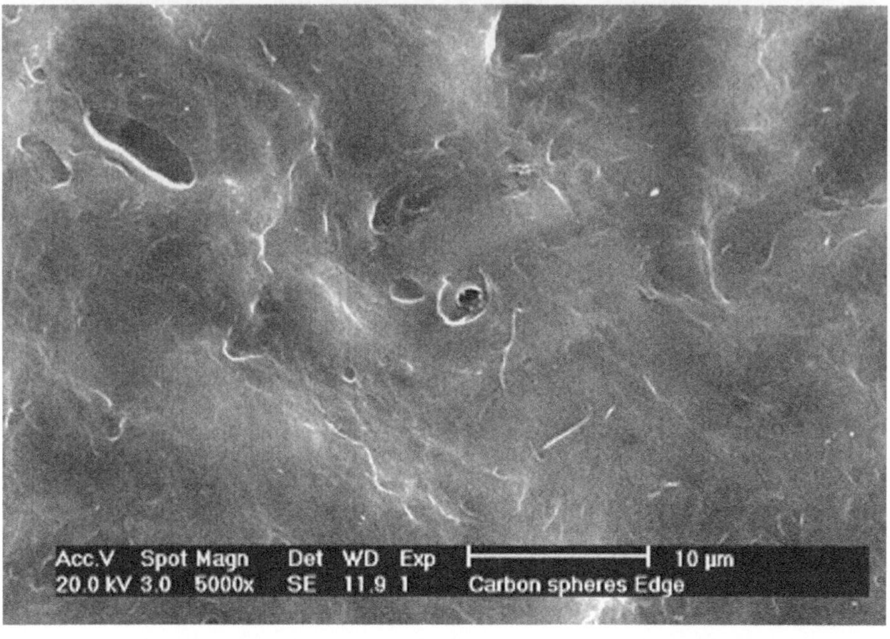

Fig. 10.5. Surface area of the carbon carrier with opening

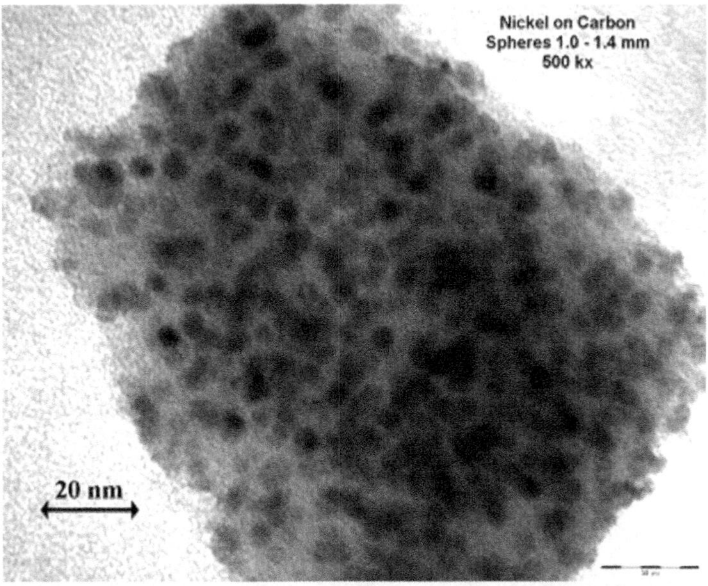

Fig. 10.6. Cut-off of the outer layer of the carbon carrier with finely dispersed Nickel

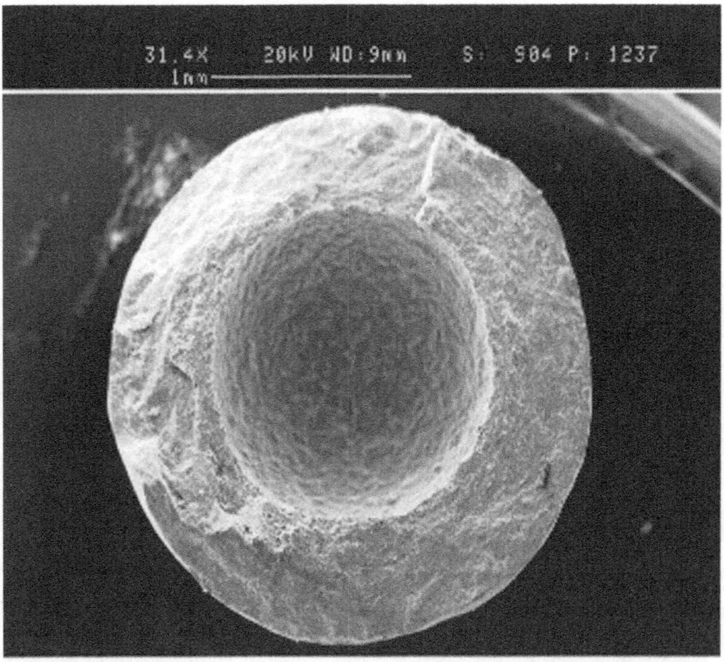

Fig. 10.7. Porous ceramic hollow sphere of 2 mm diameter

of its activity and needs to be regenerated, the carbon carrier can be burned and the expensive metal in the remaining ash can be reused without larger effort.

Looking at the series of figures 10.4 – 10.7, it becomes evident how the manufacturing process can help to formulate catalysts with properties that meet the customer specification.

The SEM picture 10.7 shows a 2 mm porous hollow sorbent sphere for practical applications. Using the described equipment, it is possible to manufacture such porous ceramic hollow spheres in large scale industrial production. Due to the manufacturing process it is also possible to donate the catalyst with metal.

10.5 Summary

In this chapter it has been shown how to design the dimensions of a sphere for optimal performance for gas separation processes using mathematical model equations. These design methods can also be applied for other applications such as catalysts, room perfuming or pharmaceutical dose systems. The latter two cases focus at desorption as the most important process phase. Glatt granulation equipment makes it possible to produce these adsorbent spheres according to design specifications.

References

[1] Glatt Systemtechnik Dresden, Kohlenstoff-Granulat, Verfahren zu dessen Herstellung und dessen Verwendung, DE 102006022866 (= WO 2007131795) (2006)

[2] Ruthven, D.M., Farooq, S., Knaebel, K.S.: Pressure Swing Adsorption. John Wiley & Sons, New York (1993)

11 Problems in the Identification of Application Areas of Hollow Spheres and Hollow Sphere Structures

Christian Augustin

hollomet GmbH, Grunaer Weg 26, D-01277 Dresden, Germany and
University of Konstanz, Faculty of Humanities, Department of History,
D-78457 Konstanz, Germany
Christian.Augustin@glatt.com

11.1 Time and Idea

Victor Hugo (1802-1865) is attributed the verifiably false quotation that nothing is as powerful as an idea, whose time has come. Be that as it may, this quotation is at first a causality statement, which combines the presence of a certain idea in connection with specific basic conditions (time) as almost inevitable for a certain successful development. At first thought this linkage, that if an idea finds its perfect timing it will be successful, seems to be evident. By "time" we not only mean the chronological period, but as an abstract category of environmental conditions that occur at this time at a specific place.

We tend to agree with this time-idea connection; particularly in the political history we see proof of this causality statement, for example with 1789 seeming to be the right time to convert the ideas developed in the Age of Enlightenment into political action. The idea of freedom, equality and fraternalism earlier found the right hotbed and developed a new era. Historians teach us that the French Revolution only came about at this particular time due to specific basic conditions (to mention only a few: Reform blockade, financial needs and price increases). With this combination it was possible to change the entire socio-economic environment. Even observers of the revolution seem to understand the combination of time and idea as core for the change. One author mentions that: "We crossed the area of three centuries in three days" [1]. Regarding this thesis (the linkage between time and place) there is hardly no other historical development possible at this time than that of the French Revolution.

The pseudo-Hugo quotation does however not stress to describe the link between time and ideas in the political sense, but rather something in general. This logic is cleverly used in many company brochures and in connection with product novelties, with the quotation regarded as a proof for the prospective positive development of the company or the new marked product. The self-image established means that the market will inquire the innovation, since in the future there will be a strong market demand. The time for the penetration of an innovation into the market is ripe.

Ideal examples of perfectly timed ideas for product development in the recent past are the Apple I-pod and the Tamagotschi, both consumer electronic examples with a strong market demand. There are however many other examples of perfect timing

A. Öchsner, C. Augustin (Eds.): Multifunctional Metallic Hollow Sphere Struct., EM, pp. 195–212.
springerlink.com

where – even good ideas need a lot of time to get onto the market and finally become successful products. Many product novelties however are not accepted and after short time disappear from the market, or require a considerable long time to become generally accepted. If one brings this into context with the pseudo Victor Hugo quotation there are two determinants, which decide on success and failure of a product innovation: Time and Idea. But what can be drawn from the many examples of product developments that succeed or fail on the market, for example most consumer products are not recognized by the market and disappeared later on. It looks like that the fitting of the two determinants is quite complex and hard to predict. Therefore one may get the impression that the future is evenly not an extrapolation of the past but an open development which produces factors that interfere with time and idea.

11.1.2 Is Product Innovation Just Gambling?

The determinants 'time' and 'idea' are however not indispensable developments, but can be effected by participants; especially in the industrial goods and semi-finished products range the empirical investigation shows that product- and process-development takes a long time, even if the product later on becomes very successful. It also seems to depend on the industry how long the conversion of new ideas are required.

Aluminum foams have for example been investigated by a larger Research Community since approximately the beginning of the eighties, with however the marked success just having been developed within the last years [2]. Consistent use of carbon fiber constructed units for the aircraft industry has also had a long production development history despite since long time the advantages of carbon fiber concepts for larger jets have been known for a long time with nowadays the new Boeing Dreamliner 787 [3] a stringent carbon fiber concept has been converted. This example illustrating the long period between developing and establishing a manufacturing technology to it finally becoming commercially availability – which caused problems, which no one expected.

Is the time and idea right for the Dreamliner or did the Dreamliner idea find the right time?

So, time and idea seems to initially fit, but at second glance, there are many problems to be solved until time and idea matches a commercially successful product. It is foreseeable nevertheless that the manufacturing of the Dreamliner will one day become reality without the initial problems, and the carbon fiber concept establishing itself as state of the art. This is what Boeing counts on and hopes to be first with this advantageous technology. Nevertheless, the future is not computable and a project that looks very promising today can be put out of business in the future due to circumstance, the ceteris paribus changes.

Time and idea depends thus even from each other, cause themselves and form themselves.

History teaches that it is very difficult to judge when the time is ripe for an idea and/or a technology, since social participants and environmental conditions can affect both on the side of the market and protagonists of an idea considerably the development. We cannot predict the future behavior of the market unerringly.

One tends to accept the product idea - particularly if it comes from oneself – to be genius and far beyond its time but successful at the market. Afterwards, due to psychological reasons, we tend to push the blame of a product not being accepted on the market on to being too advanced. "The time does not seem to be right for this innovative idea" - this argument may be of comfort for the inventor or the development department, but like we described however it is analytically wrong under the premises mentioned. The fact that the product does not match the time period does not necessarily mean that the product is too advanced but only, that the product is not accepted at this stage.

Only once a product is incorporated without later on undergoing any changes does it become a success, then we can talk about being an technology that has become "too early". However, then the question once again arises as to why the product did not succeed, leading the impression that the marketing or sales channels of the product may not have been adequate, the product itself was not adequate, etc.

Future is not fatalistic

The nexus of time and an idea, on the one hand success and/or failure is not an uncontrollable constant. As previously stated it could be worth taking time to theoretically reflect the more founded practical applicable manipulation possibilities of the mentioned determinants.

In order to understand and to affect the factors of influence on product- and process development better, hollomet choose a new way. In the context of a philosophical dissertation we tried to analyze successful and unsuccessful innovation processes within the range of the cellular materials, in accordance with the well known Lessing-Formula "history is not the memory to weight but to illuminate the mind". At first we looked at the historical genesis of the metallic Hollow Spheres, so we examined and evaluated the factors of time and idea over a one hundred-year period in the development. A shortened version of the historical genesis is presented in chapter 2 of this book. In a second step, on a more analytic-systematic level the historical development was abstracted. From this analysis in a second step concepts are to be developed, in order to develop products, mainly focusing on the effective control of development time and resources employment.

Three substantial conclusions could be drawn from the qualitative and quantitative patent analyses we used as a historiographical method to analyze the historical development of the Hollow Spheres over 100 years.

Replacement and destruction allow the creation of the new. New developments replace old ideas with different speed and penetration. In depth this concept was first described by economist Joseph Alois Schumpeter (1883-1950) as "creative destruction" [5]. Schumpeter is convinced that many economic developments based on the process of the creative destruction. The factors of production are rearranged by the destruction of old structures and combine with the new. The destruction is thus necessary (and not a system error), so that re-organization can take place and an efficient resource-allocation under new environmental conditions is possible. This principle should be understood as a foundation and enables technological developments in both, evolutionary and shock-wise forms of progress. The historical analysis also shows that evolutionary and shock-wise developments are structurally relatively homogenous. How developments are classified is a question of the perception category.

In the history of Hollow Spheres one can extract several schemata to classify the technological development.

Idea diffusion. Successful technological developments usually point a high interaction with market tendencies and a measurable interdependence to other technological developments. From our research we can conclude that the knowledge very often at the core of an invention is in use in another field of technology. Ideas and procedures are rarely entirety new. Usually one may find tracks of the idea in other patents, which may concern quite different fields of technology. These ideas are placed into a new context and changed where needed. From the quantitative patent analysis the conclusion can be drawn that the more pronounced the idea diffusion is, the „innovatively" the patents and the higher the potential benefit of the patents is. In quantitative patent analyses the idea diffusion can be measured by the number of citations and the relative distance to the International Patent Classes (IPC) of the citations to the IPC of the patent [4].

Gradual developments. Only in the rarest cases "The New" - this is at least the result of the patent analyses - appears directly as marketable product or technology. Mostly the nuclear idea is subject to a genesis. This principle is to be represented almost prototypically with the development of the Hollow Spheres. As example is to be referred here only to the patents of Jäckel, which incremental but however stringent describe the development of the Hollow Spheres, based on the nucleus idea of using powder metallurgy in combination with an advanced binder-system. Finally and in Jäckel's most recent patent (cf. Chapter 2) he lays the foundation stone to a modern procedure.

Fazit. The historical investigation based on the development of Hollow Spheres show that the applications of Hollow Spheres and Hollow Sphere technology are very strongly subject to an evolutionary technology and production development. New manufacturing processes displace older procedures. One may find that after a new procedure is established it can be found in several other patents, describing how the Hollow Spheres for a specific application may be manufactured. The more exactly one looks at the technological development on the basis of patents, the smaller the innovative aspects in the technological development appear. Quite often it seems as though an idea is taken into the scientific community and is integrated into existing routines. Some patents are promising, and some are not. Some of the promising ones became state of the art while the other – we can hardly give evidence why – promising ideas are never found their way into the market.

These three transformation categories could be identified as concepts in connection with ex post examined research we did in the field of the Hollow Spheres. The normative reverse, the acceptance that these three identified categories are promising to use them as heuristic must be individually empirically tested. Furthermore the three identified theses are not strong enough, to found general theoretical statements.

These references clarify the necessity for a general viewpoint of technical developments in the time-ideas-continuum, proving that an analysis of the innovative terminology used is helpful.

11.1.3 Invention and Innovation

Most research literature regarding product and process developments differentiates between the terms *Invention* and *Innovation* [6].

An invention can be defined as something newly, importantly and unexpectedly developed. An article or a procedure is usually regarded as an invention, with which the inventor succeeds in introducing a new, important and unexpected idea. A widely used definition speaks of the fact that: "Inventions are special, not obvious, material constructions or a procedure, which enables new and useful applications." [6] The invention is thus the nucleus of each new product or procedure. Based on this definition the question of what exactly does „new and useful" mean arises, how something obvious or not can be measured – as afterwards the inventions often look very simple and obvious.

Many great inventions are often first recognized as new/unexpected and important at a much later stage. The inventors of the laser had created something, which according to their own statement was new, but was this really unexpected for them? They had been in this field of invention for years and from their last step prior to invention of the laser there must have been a continuous process. They did not wake up one day and set up something completely unthinkable, but began with a definite problem and developed a solution step by step. During development however the future social and mercantile meaning of their invented technology was not recognized. The concept of new/important/unexpected is thus a predication in context with the respective environment, to measuring the actual situation of the idea evaluated at one time. What exactly was the "idea" behind the laser invention? It seems to be appropriate that the separation between invention and innovation is not empirically adequate, except one accepts platonic ideas as categories, which on the other hand then combines the term invention and innovation to a single concept. Another interesting question is if one can talk about unknown entities, or do we first develop a language about a new item and first then develop the idea behind the language we are using? [7] What may at first sight seem as ivory-tower philosophical problems may have a strong impact on the invention-term.

The second term, which is frequently mentioned in the framework of product and process development is the term of the *innovation*. This calls the penetration of an invention in the market. In the academic discourse these terms - if it to be at all separated - are treated by different disciplines. While the term invention is researched by psychologists, philosophers and more recently by neuroscientist and neuroscientifically affine thinking scientists, is the innovation thus - intersperse the penetration and marketing of an invention - accordingly primary the subject of management economics, which can avail themselves again of different disciplines such as brand management, psychology etc. [8] But if we use the historical development as a reference, one can clearly see, that innovation always involves invention and vice versa.

The historical analysis and the short expedition into some philosophical questions mentioned above however highlights that this linguistic and disciplinary separation, e.g. invention and innovation, raises some problems when one takes a closer look at the defacto processes of product- and process development. As above presented, it is not very common that a finished new product or procedure is born, which is then not further marketed nor developed. This can by be said for most cases, where the original idea requires a lot of further invention steps before any marketing and marked accepted solutions. So the "original" invention is in actual fact only one part of the productdevelopment. In the innovation process further inventions have to be added to the nuclear idea. What is labled as invention however is relative.

11.2 Hollomet Product Development

Due to these findings it made sense for the development heuristic, concentrating on Hollow Spheres, to reject the orthodox separation of invention and innovation as *the* division of a development process. To develop applications for Hollow Spheres we regard the production development as a structural problem solution process, which consists of the two components, invention and innovation. In this process we use the terms invention and innovation, however without wanting to give a precise separation of the respective extension and intension.

Step 1: Start from what you have

We proceed from the thesis that in most production development processes a concrete problem has to be solved. If we focus on a problem we already have a rough idea of what a solution could look like. At least we can - more or less exactly, that is determined by the knowledge we have of the problem – distinguish between an adequate solution and an inappropriate solution. If we start the problem-solution-development we then try to reach the adequate solution we have in mind by choosing means and ways from our repertoire (e.g. knowledge, abilities, available materials, economic possibilities).

Often we notice during this process that the solution we are seeking is not an appropriate one, which leads to the solving of our problem. Or the structure of the problem we have in mind is not the true problem. So even if we fail with our strategy at the first step, we gain a huge benefit: We attain knowledge over the nature of the problem. As a consequence we learn and modify the solution horizon and the strategies allegedly used, i.e. the initial question as well as the approach to our problem.

Therefore it can occur that the original idea is dismissed in the context of a problem solving process and something unexpected is created, something which was truly inconceivable from the time the problem arose. This is however no argument against the thesis that we proceed from a concrete problem definition. Participants always need a sense horizon, in order to become effective. This - Niklas Luhmann would probably speak of a double contingency [9] - sense supply is a premise of acting. Humans are not in the position to think completely free as they are bound and limited by language and tradition. Again: one can leave all this behind, one can define new concepts and terms but in each case one has to start from a point of reference. The best term in use to meet this holistic understanding is marketing, which combines the aspects of the invention- and innovation process and orientates closely at the problem to be solved, which can be formulated only against the background of the marked condition. As products are more than just idea or hardware they have to find their way into the economics process.

Due to this reason an implicit connection exists between the two ‘problem solving’ stages and the ‘problem’ we begin with. If no problem to be solved exists no solution development is required.

A certain knowledge point along with a clever idea is a sufficient starting point to achieve a solution to a problem. For this reason particularly good ideas usually look simple; because they perfectly fit into the time-and-place-continuum in which we were able to identify and talk about the problem. There can still be a better solution for the addressed problem, but the solution is sufficient at this time. Later on perhaps the solution is once again taken up, because it is no longer sufficient. If the knowledge horizon and/or the market conditions change, this can lead to the fact that the established solution is no longer sufficient and the solution becomes the basis of a problem to be solved. The state of research or the solution today is the potential problem, the potentially unsatisfactory of tomorrow, because change happens. The change can be driven e.g. by creative destruction, idea diffusion or gradual development as mentioned above.

It can recapitulatory be noted that we always start production developments from a concrete problem. The formulation of this problem can be completely different and range for example from "which applications are there for Hollow Spheres" to more specific questions like "Can we absorb high frequency tones more strongly with Hollow Spheres?" With the formulation of our initial starting point of the product or process development process, we already have in mind a specific concept how an adequate solution could look. Usually we do not know an adequate solution but we know that "singing loud in the shower" is not an adequate solution to the question of how "can high frequency tones be absorbed".

Already as with the precise explication and the speculation about possible solution types and conditions, an intensive and structured consideration takes place about the kind of the problem and its solution conditions. On the basis of the considerations explained above we developed an innovation and intention-schemata that explicates the intuitive Product-development-process and shall support the intuitive production development process.

In order to reach and formulate the solution, we become the inventor, with the participants striving to develop a creative solution. If the idea developed within this process is an adequate problem solution, the production development process is terminated.

However if the solution does not prove qualitatively sufficient, we learn something more about the nature of the problem with each missed solution. With the advancement of the problem solving pattern a broader knowledge spectrum is constantly achieved.

Step 2: try to learn

All parties concerned in the process generate more information and generate ever more knowledge at each failed step. The more knowledge is attained, the more precisely the defining problem can be fixed. By the way: If the knowledge is once reached and documented, then it can be imparted without imparting the process of the knowledge genesis.

Afterwards - the innovation process is easy to oversee and to understand - the transacted comprehensive product- or process development from problem collection, e.g. understanding, to problem solution appears as a single invention and not as a complex process. That is why retrospectively and shortening one sometimes speaks of a 'sneezing realm' or an 'ingenious solution'. This creative process of trial and error is usually made intuitively within a structured development process.

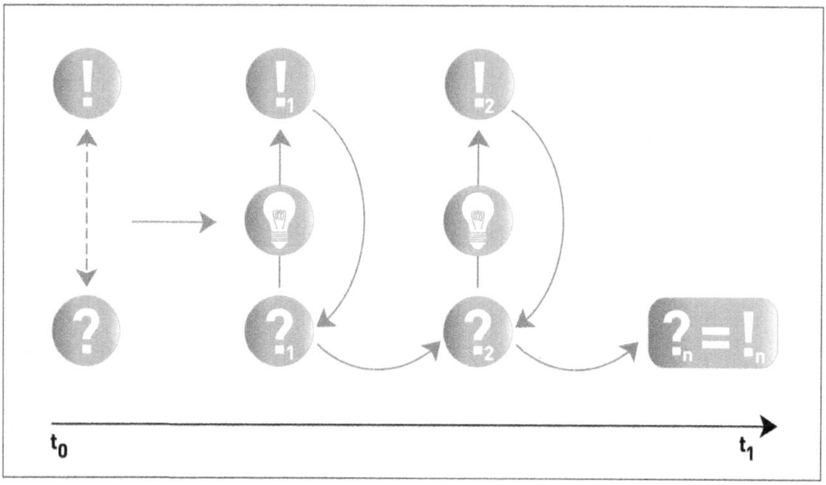

Fig. 11.1. Visualization of the Product and Process Development Process

An explication and communication of this pattern during the development-process in our expertise leads however to a more intense communication and a faster examination of implicit premises. Our findings were that implicit premises are often responsible for unnecessary redundant developments. The following sketch 11.1 illustrates the preceding remarks and summarizes.

In the course of the production development process (t_0 to t_1) we start at a concrete problem [?]. The horizon of the solution [!] is already known by exclusion, which does not appear to be a proper solution or by an unspecified idea what might be a proper solution. On the basis of these two fixed points the question arises, how [?] and [!] can be brought into a balance. This question is the core question of the production development. With creative power, i.e. research and development affords, a solution according to [?] = [!] is searched.

The exact mechanisms and heuristics that are suitable to get from [?] to [!] depend very much on the subject and the manpower. Based on what we call "creativity" a solution [!] is suggested. This solution [!] is in the sequal step being examined if it fulfills the requirements formulated in [?] = [!]. If the solution is regarded as problem-adequately the product development process comes to an end.

If this is not the case, then the knowledge that in the context of the past product development has been generated is used in order to specify both the problem level [$?_1$] as well as the solution level [$!_1$] again. However, now one may specify more precisely what [$?_1$] and [$!_1$] are all about. Thus, inevitably and with each step the knowledge of the participants and the fundamental abilities of the concomitantly concerning the selected problem rise to solve the problem in the future (at the time t_x).

In order to return to the initially stated considerations from "time" and "idea", the potential of bringing "idea" and "time" into a successful solution is improved, with each step, because in the long run problem solutions [!] and questions [?] are elements in the time; and an idea, a creative product development will soon or later lead towards a sufficient solution.

We used this product development process in numerous development projects, and together with customers. It was shown that sticking to this procedure above all possesses the following advantages:

- It makes a flexible and fast control of the project development process possible.
- It enables position-fixing which is an important starting point for a perceived identification of process points and executory project-points.
- Implicit premises, which can lead to incorrect or redundant results can be made more easily explicit.
- Contrary to Gantt Diagrams, which are beneficial, one can plan the actual next step using the hollomet procedure. Intermediate steps can be included dynamically in a pattern.
- The procedure makes a simple documentation possible of the project and the project progress.
- A visualisation is simply implemented, also in terms of software.

11.2.1 Invention and Creativity

A central problem with production- and process developments is the concept of ideas – I mentioned it as creative achievement above - which could be included to find a solution for the equation [?] = [!]. Creativity and creative research are very complicated and complex research fields, whose results can be used only in very few cases. At times the creative-researchers' suggestions may work, but sometimes not. Nevertheless it was also our goal to point out how "creativity" *can* be encouraged within this range of possibilities, of course without an exclusively normative claim.

It showed up in the development process that the application of 'Theory of History' or 'Philosophy of Science' theories and assumptions could be unusual, but fruitful; especially, if the participants defined targets, which were in such a manner complex that it was not possible to find a fast and simple solution.

The example describes, in which some basic assumptions of Paul Feyerabend, a science historian and philosopher come into place, which he in turn has drawn from his analytical philosophy and historical investigations [10,11]. His conclusion is that there are no strict criteria for science. Feyerabend then concludes that any kind of research strategies might be useful to attain a specific goal. The philosopher points out that the goals for the research are defined by society. His general philosophy claims that dogmatic theories may not be interpreted ex cathedra. The scientist should be creative to find a solution to his problems. To enhance the creativity and help the scientist to understand his subject better, he could use research strategies which have worked in the past to gain progress. In his research Paul Feyerabend identifies some 'rules of thumb' which at times coincided with progressive historical developments.

If the investigations are a starting point and bring together our concept we can say that the market and our tradition (language etc.) performs to define our goals. We are only able to think within our tradition.

Therefore, this means to recognize the following rules of thumb as a progressive science heuristic being used in the history:

Contra-induction: Confront established theories with contradicting facts. By using this strategy Feyerabend claims to build purposefully „untrue" facts into established theories or problem solution strategies. Thus one may – according to Feyerabend – learn a lot about the theory that is in use itself and about its weak points. So one haves the opportunity to improve.

Retort induction with established theories contradicting hypotheses. This research strategy also aims at identifying the strengths and the weaknesses of a theory. A theory is a complex thing with many basic assumptions and many theories around its core. Contradicting hypotheses could help to see the established theory in a new light.

Use of historical theories. This can help to get a systematically (evenly historically well-founded) alternative view on the theory; and thus help to experience more over the nature of the problem. Historic theories are not brainless according to Feyerabend. They source the phenomena in a different way. If we use historical theories to explain and reflect on our current theories we could also learn a lot about them and the kind of phenomena they are assorting.

Use of ad-hoc hypotheses. Using these ad-hoc hypotheses enables one to start thinking and sorting a new field of research. These are to found a provisional robust concept to address obvious problems.

Use of propaganda. Which seems to disqualify itself at first sight as a particularly unscientific procedure is very often a very effective strategy to develop new theories or solutions, as the use of propaganda is an attempt to defend a theory or a solution against any kind of objection. To succeed the propaganda used, one must always and frequently adapt the propaganda-theory. And thus support the development of a new theoretical basis. Finally this leads to the fact that the theory is constantly improving, as the nature of propaganda is primarily not to protect an idea, but however to raise influence using a theory. Indicating that in the words of product development the propaganda helps to form the theory, with propaganda always focusing on gaining power. And power can only be gained if it – that means the underlying theory - is successful. Only so the idea/theory has time to develop and become more than an ad-hoc-hypothesis. This concept Feyerabend sees both in the political history – e.g. in the history of the start of communism with Marx and Engels to Lenin and Trotzki - and in the history of science – one very common example is the development of the galileian astronomy – as a very strong tool.

These rules of thumb are not under any circumstances a guarantor for success, but a possibility, an option to be considered. The basis of all these strategies is the thought that there is no superior heuristic. For the applicative development process there the question of a superior heuristic theory is not of importance, as well the question of scientificness is of subordinate importance. The measure of quality here is problem-solving and market success.

We can use these strategies to reflect both our questions, our potential solutions as "theories" and investigate them under the different 'rules of thumb'.

11.3 Application Search Hollow Spheres

Since Hollow Spheres are semi-finished materials, with the production potential of a large product spectrum, a consistent search for applications is a central point in marketing and therefore also production development strategy. We know that the development of Hollow Spheres applications is a long track as the world was not waiting for our innovation. Hollow Spheres do not have a killer-application or a an outstanding character that is superior to others. Hollow Spheres are a material category, usually as good, if not better than existing solutions. Hollow Spheres are however multifunctional and can integrate many tasks in one solution.

It is necessary that existing development resources are significally designated in the early present state of the hollow-sphere development, since it cannot be assumed that there is an excessive market pull. This is not due to the quality of the product, but primarily the result of a communication deficit. The market does not "know" its needs and options with this material class because the Hollow Spheres are widely unknown. For most market players the product is unknown and thus not available as an option for the solution of their problems; it is an inconceivable option. The first step in the search for applications is therefore the identification of goal products, and the search for cooperation partners. The second step is communication highlighting the advantages of the new product and the generation of need.

Together with these partners the actual product innovation process is initiated. On the basis of the knowledge horizon (e.g. from the side of hollomet: material, production; from the side of our cooperation partner: requirements, price etc.) of the parties involved problems are defined and solutions are worked out; sometimes using the hollomet product development schemata.

Usually in a second step a prototype is manufactured. This contains all nuclear components of the late industry-scale produced product and serves a critical empirical examination. This critical empirical examination pursues the goal of proving the adequacy of the developed solution. If no agreement is to pointed out, then the hollomet development-schemata is very suitable to be a fast and efficient error tracing mechanism to accomplish and to plan outgoing improvements. If the use of the Hollow Spheres is successful in the prototype, then together with our partner further procedures are planned along with the potential customer. Since this depends on industry-sector or customer based standards (for instance size of series production, safety precautions, series tests) no general accepted strategy or standard operation procedure can be established by hollomet. With the construction and successful testing of a prototype, the industrial manufacturing, later on the development process is however not yet final. The production development is at best a permanently operating advancement of the existing technology.

To achieve this, the hollomet product development schemata can also be used. Forward as well as backwards-arranged (i.e. historical) analysis of the development steps can be supported, which can be of great use in improving an existing excellent product; particularly with larger development projects, documented and developed over several years and with a large number of developers involved, suggestions for further development steps can be received and redundancies avoided.

In the following I will describe actual development projects to present the multifunctional character of the Hollow Spheres. I will describe how this technology helps our customers to solve their problems in an easy and effective manner.

11.3.1 Projects in the Development at Hollomet

Together with our customers and on behalf of our own market research we develop several projects in which we use metallic Hollow Spheres to improve existing products. In the following points we describe some of these projects. In order to give a better overview over the projects I will not describe the used hollomet Product Development Process in detail but only the current stage of the projects. A more detailed insight into the procedure of using the hollomet Product Development Process can be found in [4].

11.3.1.1 Exhaust Gas System

In the customer order hollomet developed a combined muffler and catalyst system. Setting tasks were to unite the functions of muffler and catalyst in a singular unit. It was the aim to reduce production costs and more favorable thereby to meet new law requirements.

A number of potential cost cutters were identified by the customer at the construction unit, by implementing Hollow Spheres both as muffler and as a catalyst. On the basis of a specification sheet we developed a solution for the customer. Together with partners the hollomet accomplished an analytic optimization of chamber and tubing volumes by network modeling for the customer. Afterwards the simulation results were empirically examined. In addition, a three-chamber muffler kit was developed. Metallic Hollow Sphere structures with differing material parameters can be inserted into this kit and examined for their suitability. The following figure 11.2 shows a housing cut into the carrier component we used to examine the simulation results. Figure 11.3 shows several muffler and catalyst units which could be placed into the housing.

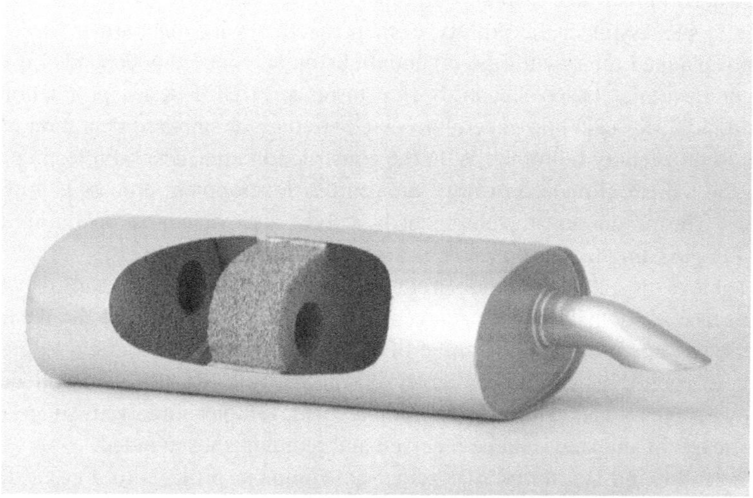

Fig. 11.2. Cut by a muffler component system. Source: Emcon Technologies LLC.

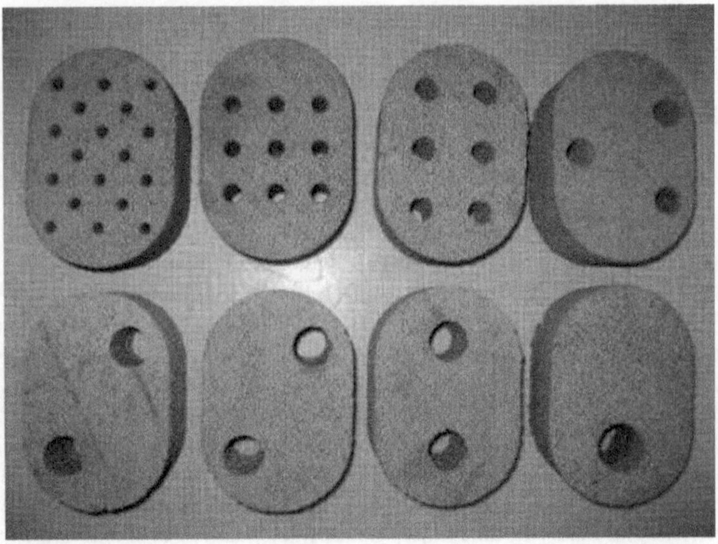

Fig. 11.3. Possible filling materials for in Fig. 11.2 shown muffler kit. Source: Akustikforschung Dresden GmbH.

Thus different muffler types can be implemented. By means of this component system pure absorption muffler or a combination of absorption and reflection mufflers can be sketched. The following sketches 11.4 show two possible combination options.

With the muffler component system not only sound pressure could be measured, but also pressure losses itself in the whole exhaust gas system. The customer did not allow a higher pressure loss than his existing solution has, otherwise negative effects were expected on the efficiency of the engine.

After identification of a suitable structure a muffler prototype was constructed, which could then be tested in a serial model. The muffler prototypes were subjected to further tests parallel to the serial model, such as the product submitted to a thermal shock permanent run with 900°C, which it passed without objections. Altogether a

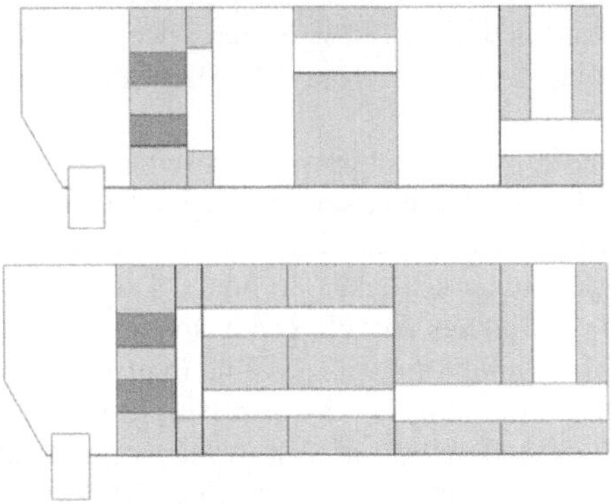

Fig. 11.4. Different configurations in the muffler component system schematically represented. Source: Akustikforschung Dresden GmbH.

reduction from 4 to 10 dB was achieved, dependent on respective muffler types and the respective rounds per minute of the engine used. These values were based both on the engine test stand as well as on the modified standard model. An interesting side effect was that in a configuration the efficiency of the engine could be increased around 10%: the muffler and catalyst unit provided a reduction of 5 dB sound pressure and a reduction of the pressure decrease in the entire exhaust system at the same time.

Together with a partner the Hollow Spheres were catalytically coated using a wash-coat process. The desired emission reduction could be proven likewise.

The principal item is thereby a sintered Hollow Sphere structure made of 1.4767 steel. For the construction of prototypes Hollow Sphere structures were intially sintered, secondly cutted by a water jet, and finally milled into their necessary geometric shape. It is intended to use a shaped part machine for industrial production, to bring the green spheres into the precise geometry and finally sinter them.

Positive cost effects occurred on the one hand from a better catalysis achievement per unit catalyst (improved selectivity), resulting in the shaping of the Hollow Spheres. On the other hand cost-effective savings arise as a result of the combination of the mufflers and catalytic unit functions. Further positive cost cuts are expected by a more favorable production process as well as by improved logistics. Reproaching of fewer components and housing is necessary. Besides some components can be used modular in different products, saving the purchase department further expenditure. Further advantages is the fact that the catalyst/muffler attachment unit could be approved, its optimized weight reduction and environmental compatibility. No fibrous materials should loosen themselves in the life cycle of the muffler and non should be carried with the exhaust gas into the environment. Besides this, the product is easy to recycle. Altogether the required tasks could be solved using metallic Hollow Spheres.

11.3.1.2 Forming Tool for the Deformation of Plastic

Forming tools for the deformation of plastics are made the conventional way of forms with ceramic(s) or aluminum foam as a filler. The customer's requirement was to achieve a high firmness of the forming tool and massive weight saving for a complex two meters long free form. With metallic Hollow Spheres we could fulfill the prerequisites.

First the hull geometry of the free forming tool was provided by using rapid prototyping. In a second step metallic Hollow Spheres were filled into this free forming and the rapid prototyping form was removed. After smaller mechanical reworks the form could be used.

The test showed that the weight not only decreased from 50 to 60%, but the production process could also be significantly simplified. To deform plastic plates the forming tools are usually heated in a furnace, which takes some time to guarantee a wholly heated form. In this specific case it takes 36 hours. Due to the advantageous thermal conduction of the Hollow Spheres as well as the smaller total masses, which had to be heated, the heating of the whole form was possible in just 5 hours.

Altogether the project was judged very positively by the customer. hollomet succeeded in reducing the manufacture price as well as three times more heating and cooling cycles per time unit. With the use of hollomet Hollow Spheres it is possible to reduce costs and improve handling. The following picture 11.5 shows such a complex forming tool.

Fig. 11.5. Complex forming tool. Source: Ebalta GmbH.

11.3.1.3 Laser Mirror Carriage

Altogether the fulfillment of rapid prototype structures seems to be an interesting application field for metallic Hollow Sphere structures. Also in this example, a lasermirror carriage was manufactured by means of rapidly prototyping. Such laser carriages are high-complex construction units exposed to high kinetic energy.

First a form was constructed directly including providing liaison vehicles such as mounting plates and nuts. In the second step a Hollow Sphere epoxy resin mixture was filled into the form, which hardens to a high-duty and very light structure. Similar results regarding firmness and weight ratio can otherwise only be attained with carbon. This is a substantially more difficult manufacturing process and thus more

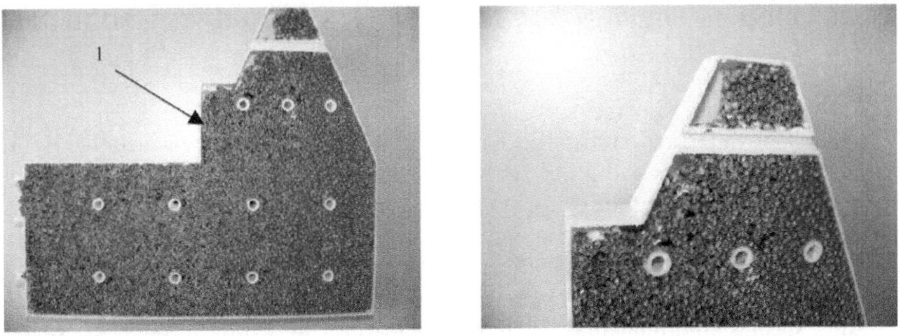

Fig. 11.6 Cut by means of rapidly prototyping manufactured and with Hollow Sphere brazed laser carriage. Source: SITEC GmbH, Dresden / Innozellmet.

cost-intensive. Under the small weight of the Hollow Sphere structure recesses can be avoided. Former they had to be removed for weight reasons. Removing them once again improves the complexity of the construction unit and also the simulation of the stability of the laser carriage substantially. Picture 11.6 shows a cut through the Laser Mirror Carriage. One can see clearly the filled rapid prototyping form in which metallic Hollow Spheres have been filled. Parts that are not subject to high strain can be left without hollow spheres (e.g. marked with 1, shown more in detail at the right picture).

Picture 11.7 shows the complex geometrical structure of the laser carriage with all fastening parts.

Fig. 11.7. Laser mirror carriage with fastening parts. Source: SITEC GmbH, Dresden / Innozellmet.

11.3.1.4 Hydrogen Storage

A further innovative research project is the storage of hydrogen in metallic Hollow Spheres. This project is still at its first stages, but first attempts have been promising. The nuclear idea is therefore that metallic Hollow Spheres are used as hydrogen high pressure storage containers without valves. The Hollow Spheres can be manufactured with pressure strength of up to 1000 bar. Under pressure and heat the hydrogen can be brought into the spheres and be again taken out under warmth.

Using special environmental conditions hydrogen can diffuse through the shell of the metallic Hollow Spheres. The H_2-Storage capacity can reach weight percentages of between 4% - 6%, which is a very promising result.

11.3.1.5 Cast Inlays

Besides this, hollomet as well as partners work on the production of light and multifunctional casting construction units.

Constructed Hollow Spheres shall be injected into so-called inlays in iron casts. These inlay Hollow Spheres remain in the lead cast to provide weight savings from 20% to 30% opposite to a solid construction unit. The goal is to reach the same firmness and rigidity values as being provided by a massive construction unit and to receive at least equal absorption characteristics. Promising results are present to some combinations of cast and Hollow Sphere materials, in particular the structure-borne sound could effectively be reduced. The structure-borne sound absorption is the ability of the material to convert mechanical oscillation energy by the internal friction in a body into warmth. Unwanted noise and oscillations are weakened by the transformation of warmth. The following photograph 11.8 shows a cut cast part. The Hollow Spheres in the casting can be clearly recognized.

Fig 11.8. Cut through a cast with Hollow Spheres

A potential advantage is that the casting conditions of construction units can sometimes be handled better. Also the "core" can remain in the cast because Hollow Spheres may also be used as a "core".

However large developments are still necessary. In particular Hollow Sphere manufacturing processes, geometry parameters of the sphere such as sphere density, sphere porosity, sphere diameter, structural density as well as possible coating of the sphere have to be adopted due to the specific use of this technique.

11.4 Conclusion

The development projects represented before were developed in co-operation with hollomet customers and partners. In accordance with the represented innovative pattern much was learned about the material and the used technologies. In context of these innovation projects various further ideas were developed for the use of the Hollow Spheres. It was shown that innovation is primarily a question of the company culture and - like Gunter Dueck [12] would say- "not a department". To integrate this firm culture it is necessary to define how one may proceed and communicate that culture. The hollomet product development process is however one possibility to start the ever ongoing product development process.

References

[1] Thamer, H.-U.: Die Französische Revolution. Freiheit, Gleichheit, Brüderlichkeit. Gerstenberg, Hildesheim, 9 (2007)
[2] Ashby, M., Evans, T., Fleck, N.A., Gibson, L.J., Hutchinson, J.W., Wadley, H.N.G.: Metal Foams: A Design Guide. Butterworth-Heinemann, Boston (2000)
[3] Wagner, M., Norris, G.: Boeing 787 Dreamliner. Motorbooks Intl., Osceola WI (2008)
[4] Augustin, C.: Prozess und Produktinnovationen aus philosophisch-historischer Sicht am Beispiel Nicolaus Copernicus und metallischer Hohlkugeln. University of Konstanz, Konstanz (2009)
[5] Schumpeter, A., Schäfer, A.: Die Kraft der schöpferischen Zerstörung. Campus, Frankfurt/Main (2008)
[6] Innovation and Invention (accessed May 2008), `http://wikipedia.de`
[7] Mittelstraß, J.: Das Undenkbare denken. Über den Umgang mit dem Undenkbaren und Unvorstellbaren in der Wissenschaft. UVK, Konstanz (1998)
[8] Mittelstraß, J., von Graevenitz, G.: Kreativität ohne Fesseln. Konstanzer Universitätsreden Band 1. UVK, Konstanz (2008)
[9] Baraldi, C., Corsi, G., Esposito, E.: GLU-Glossar zu Niklas Luhmanns Theorie sozialer Systeme. Suhrkamp, Frankfurt (2008)
[10] Augustin, C.: Aber ein Paul hilft doch dem anderen. Passagen, Wien (2009)
[11] Feyerabend, P.: Wider den Methodenzwang. Suhrkamp, Frankfurt (2003)
[12] Dueck, G.: Wild Duck. Empirische Philosophie der Mensch-Computer-Vernetzung. Springer, Berlin (2002)

12

Application of Hollow Sphere Structures and Composites in Processing Machines

Martin Schöler[1], Marc Mauermann[2], and Jens-Peter Majschak[3]

[1] Technische Universität Dresden, Chair of Processing Machines/Processing Technology, 01062 Dresden, Germany
martin.schoeler@tu-dresden.de

[2] Fraunhofer Application Centre for Processing Machines and Packaging Technology, Heidelberger Strasse 20, 01189 Dresden, Germany
marc.mauermann@avv.fhg.de

[3] Technische Universität Dresden, Chair of Processing Machines/Processing Technology, 01062 Dresden, Germany
vmvat@tu-dresden.de

Abstract. The term processing machines refers to a large variety of machines and equipment for the mass production of consumer goods. Processing machines are often specially tailored according to the particular processing task they have to fulfill, such as pasteurising milk or packing goods. Furthermore the vast majority is characterised by large production outputs, demanding high processing velocities and fast moving parts. A special application is that of the food and drug processing area. Due to high quality standards there are special requirements for material surfaces to be considered if surfaces are in contact with the manufactured product. This chapter deals with the special demands concerning dedicated engineering materials that occur from state of the art processing machines. Specific requirements ranging from high damping abilities for support components to high stiffness and low weight for fast moving tools are presented by means of a general classification of processing machines into four functional areas. Accordingly, the proposed hollow sphere structures and similar material classes are presented and applied on each of these four areas.

Keywords: hollow sphere structures, processing machines, metallic hollow spheres, hygienic design, food processing.

12.1 Introduction

Assigning specific material to a construction-related task and, thus, to a specific machine part still poses a problem, with design, engineering and material development facing this problem with respect to new materials in processing machines.

Unlike sectors such as the automotive industry the development and the design of processing machines is markedly different in terms of choice and use of materials. While long development processes are invested into a product in the automotive industry, tailor-made development is common in the field of processing machinery due to small quantities. Thus, complex tests and optimisations of individual components are more exception than rule. Clearly, this poses a problem to the use of new materials.

A. Öchsner, C. Augustin (Eds.): Multifunctional Metallic Hollow Sphere Struct., EM, pp. 213–222.
springerlink.com

Design engineers select a group of materials primarily on the basis of knowledge gained by experience. Novel groups of materials will only be put to use if the realisation of specific parts poses problems that cannot be tackled with conventional methods. Accordingly, the use of novel materials with an evidently superior material performance in non-critical parts is often neglected. However, these applications would provide an enormous potential due to their higher quantities. The following chapter will therefore introduce a new methodology that simplifies assigning novel materials such as metallic hollow sphere composites to specific groups of materials, in which the substitute material's scope of properties will be ideally used. Vice versa, the categorising of material groups may also serve to ascertain optimised aims in material development.

12.2 Requirements for Materials in Processing Machines

As previously mentioned, the requirements concerning construction materials in processing machines are different from other sectors. A simplified method of assigning different material classes to specific applications without extra in-depth knowledge and legal requirements for food processing will subsequently be introduced. Finally, an individual feature profile will be described for each defined material class. The following chapter deals with specifically assigning metallic hollow spheres (MHS) composite to individual material classes.

12.2.1 Functional Classes in Processing Machines

The classification of the constructive assembly of processing machines concerning material use draws on the system model of a processing machine [9] and the system of material selection [13], as described by a number of different authors.

According to Reuter the process of material selection is based upon the compilation of material parameters that are vital for the respective task [13]. This compilation provides the respective combination of parameters relevant for each individual machine part. Thus, these steps have to be carried for each new part. If parts, however, are combined in appropriate groups, this classification becomes predictable.

According to Goldhahn [9] a processing machine can be divided into the functional areas: *matter*, *casing*, *energy*, and *signal*. The functional area *matter* includes all components directly or indirectly in contact with processing goods (foodstuff, pharmaceutical products, etc.). The functional area *energy* refers to all the parts that directly serve the transmission or conversion of mechanical energy between the positions of drive and effect. The functional area *casing* describes the parts that neither have contact with the product nor serve an energy conversion function. The functional area *signal*, finally, refers to all elements serving the processing/ forwarding of information and are not explicitly included in the following consideration. By now choosing the features that satisfy the above-mentioned functional areas according to Reuter's compilation [13], a first detailed level of the selection method would be obtained.

Another criterion for the description of parts movement will be introduced to ensure a sufficient characterisation of the elements of a processing machine, namely the division into stationary parts and parts in motion. Parts in motion are parts that change their position frequently or periodically under normal operational use and whose movement is accelerating. This means that the part's resulting inertia forces are significant in relation to its weight. The requirements concerning the stationary parts remain unchanged. Parts in motion, however, require a lot concerning the lightweight construction and, thus, their mass-specific material characteristics.

This compilation of parts criteria for the functional areas of a processing machine and the movement of parts provides a definition of six material classes. As the criterion for parts in motion cannot be fulfilled by *casing* elements (e.g. doors, coverings), there is only one single material class for this functional area. The same applies to the functional area *energy*. Also, according to definition, all stationary parts (that neither transmit nor convert energy) would automatically be assigned to the class *casing*. Thus, the number of material classes in question is reduced to four (cf. Table 12.1). Clear feature profiles can now be assigned to these four classes. Furthermore, these will be analysed with respect to their general application and use of MHS.

Table 12.1. Material classification in processing machines subject to part motion

	Functional class matter	*Functional class casing*	*Functional class energy*
Stationary parts	material class I	material class III	-
Parts in motion	material class II	-	material class IV

12.2.2 Hygienic Requirements for Surfaces with Food Contact

High priority is given to the production of high-quality and hygienic products in the foodstuff industry since it is required by law and a key customer demand. Internationally, this is based on the Codex Alimentarius of the WHO [3] and reflected in standards and national laws [4,5,6,7]. These guidelines provide clear requirements for the design of food processing machines concerning the surface texture of parts, which will be analysed subsequently.

From a hygienic point of view a processing machine can be divided into non-product contact surfaces and product contact surfaces according to [7], the latter including splash contact surfaces where product may accumulate but cannot drop back in the main product. With respect to its risk potential the processing good may have different requirements concerning material and surface. General criteria include:

- Material surfaces and surface coatings must be durable, cleanable and, if necessary, able to be disinfected
- Material surfaces/surface coatings must be solid against mechanical influence
- Material surfaces/surface coatings must not absorb any undesired matters

According to the additional requirements stated in [7] materials and surface coatings in the foodstuff industry must be corrosion-resistant, atoxic and non-absorptive (only if it cannot be prevented for technical reasons). They must not transmit any undesired smell, artificial colours and/or flavour additives onto the food. Neither should they support any contamination nor negative impact on the food. Furthermore, surfaces must be smooth; they must not show any cracks in which product remains may be left. Unintended product removal from the foodstuff area and its return is not admissible.

A surface roughness of $R_a \leq 0.8$ µm for product contact surfaces is recommended. Unprotected surfaces in the non-product area must be made of corrosion-resistant or coated material. In addition it must be cleanable, easy to disinfect if necessary and not able to negatively influence the food. All this, in turn, results in great demands concerning the material in question. Composite materials, in particular, must be adapted onto the respective conditions due to their mainly heterogeneous surface texture. Construction and production-specific measures will facilitate the adjustment.

12.2.3 Requirements for Materials of Class I

Material class I includes all parts that can be assigned to the functional area *matter* and are stationary. Of vital importance is the surface texture. Clearly, a smooth and even surface is desired to guarantee sufficient cleanability of the elements. Equally, it is crucial that the elements are resistant to aggressive media e.g. alkaline solution used for cleaning. The weight in this case can be neglected as there is no accelerated movement. The main features can, therefore, be summarised as follows:

- Achievable surface roughness ($R_a \leq 0.8$) for product contact surfaces
- Good cleanability
- Good chemical resistance (alkaline / acid solution (ph2-13); Chlorine Dioxide)

Material selection for stationary parts in the functional area *matter* is mainly affected by the requirements for easy cleanability and chemical resistance. Thus, alloyed stainless steel (e.g. X 5 CrNi18-10/AISI 304 or for higher standards X 2 CrNiMo 17-12-2/AISI 316L) is used. For gaskets it is also possible to use plastics, for example Polyetheretherketone (PEEK) or Ethylenepropylenedienemonomer (EPDM). An overview of material recommended for use in food contact can be found with the European Hygienic Engineering and Design Group (EHEDG) [8] and the United States Food and Drug Administration [2]. Examples for parts of class I comprise guide rails, retainer jaws, flight conveyors, funnels, and vessels.

12.2.4 Requirements for Materials of Class II

Material class II includes all elements of the functional area *matter* whose movement, in contrast to material class I, is accelerating. Therefore, the focus in this material class is on lightweight construction and on hygienic design requirements. Accordingly, the specific requirements for material use are:

- High specific stiffness, if necessary good buckling performance
- Achievable surface roughness ($R_a \leq 0.8$) for product contact surfaces
- Good cleanability
- Good chemical resistance (alkaline/acid solution (ph2-13); Chlorine Dioxide)

Due to the specific, restrictive needs for hygienic design material, alloyed stainless steel can be also used as well although they do not have a better stiffness performance than those mentioned in material class I. The use of composite material such as carbon fibre reinforced plastic (CFRP) or glass-reinforced plastic (GRP) would generally be more suitable due to its high specific stiffness.

Due to their epoxy resin surface these composite materials, however, do not show the required features concerning cleanability and chemical resistance. Thus, they should not be used for hygienic design application. An exception is coated material. This procedure would enable the requirements for the part's surface and could be combined with the favourable features of the core material. Applications would be positioning devices, levers, transport or dispenser devices, and folding elements.

12.2.5 Requirements for Materials of Class III

Material class III comprises all elements of functional area *casing*. This class' parts serve as the storage and support of other functional classes and, thus, bear the stress and strain. Furthermore they serve the mechanical protection of the process from outside the machine. A high strength performance of the parts is crucial. Depending on the part's function the self-damping behaviour of the material is also relevant to avoid any impact from possible vibrations outside the machine onto the machine devices. It is also relevant to avoid any dynamic stress from a machine device onto the entire machine. Another aspect of this material class is the production-related possibility of functional integration. I.e. the machine frame is a geometrically complex part with a number of different functions that can be simplified by respective selection of material and processing technology. Due to the various possible applications it is rather difficult to find sufficiently requirements for this class. Given, however, pure frame parts as very frequent element of this class, the requirements are as follows:

- High part strength
- Good damping behaviour
- Production-related possibility of functional integration

The same, strict hygienic requirements of class I and II do not apply to this class. Still, the unprotected surfaces must be made of corrosion-resistant or coated material to allow for easy cleaning and disinfection without negatively affecting the product. This calls for manufacturing processes that allow for a combination of different materials. Classic material in this respect is grey cast iron (e.g. EN-JL 1050/ AISI A48-45B), as well as construction steel (S 355 JR) or stainless steel (X 5 CrNi18-10/ AISI 304) for welded frames. Also common are reaction resin-bound minerals or mineral casting [10]. These materials are preferred due to their specific self-damping behaviour and are especially used for casted machine frames.

12.2.6 Requirements for Materials of Class IV

Material class IV refers to all elements assigned to the functional area *energy*. The transmission or conversion of mechanical energy requires both a high degree of strength and stiffness to avoid losses during the energy transmission due to forming of the parts. The requirements for the surface texture are solely conditioned by the occurring mechanical stress. Hygienic aspects and chemical resistance are to be neglected. Thus, the materials have to fulfil the following requirements:

- High specific strength
- High specific stiffness, if necessary high buckling performance

High-strength steel and stainless steel alloys are mainly used as they have a comparatively high specific strength in contrast to other material classes. This applies in particular to cases with rotating moving parts (shafts, cog wheels). Regarding translational movements composite materials such as CFRP or GRP may also be occasionally used.

12.3 Application of MHS in Processing Machines

The major benefit of MHS with a reactive epoxy resin binder is their scope of processing possibilities from the design engineer's perspective. They can be classified concerning their different manufacturing methods as also described by Neumann [12]. The general suitability of these three processes will briefly be analysed with respect to the use of the parts.

The *solo construction* is the most frequently used manufacturing method. Here a mixture of liquid epoxy resin and different hollow spheres as well as aggregates is cast into a lost or a permanent mould. The benefit ascribed to this construction method is its relatively simple cold casting method. Unfavourable, at times may be the surface texture as both the matrix material and the aggregates appear on the surface. This is not beneficial for a hygienic design of parts in the product area. Furthermore, this manufacturing method needs to find a way to compensate the poorer results of the material under tensile load.

The *composite construction* is a method in which functional elements not made of MHS (guide rails, facing, threaded bolt, media cables) are combined by an auxiliary construction (cf. Fig. 12.1). This either serves only the exact positioning of parts to one another or it serves to tolerate specific forms of stress and strain. The open surfaces of this so-called 'skeleton' are afterwards closed by cast parts and the entire form is filled with MHS. The result are parts of high precision (no shrinkage due to combination of functional elements via 'skeleton') and of combined material features (good damping behaviour, high stiffness).

Fig. 12.1. Scheme of a Composite Structure comprising (1) sheet metal surface, (2) MHS, (3) skeleton and (4) mounting plate

The *grouting construction* represents a particularly promising manufacturing method for processing engineering. To produce this, an almost closed case construction (e.g. made of stainless steel) is filled with MHS. This combination exploits the potentials provided by both the use of MHS and the properties of stainless steel, as inert external surfaces are especially required in the foodstuff industry. Furthermore, the frame of the part may be laid out in such a way that stress is distributed onto different elements.

12.3.1 Application of MHS in Material Class I

In areas critical from a hygienic point of view the need for suitable surface textures is particularly important. The heterogeneous structure of pure composite material makes it impossible to produce a homogeneous surface suitable for cleaning. This is the reason why the third manufacturing method, the *grouting construction*, is recommended for the application of MHS in class I. If, regarding this, a case structure is selected from an authorised material (e.g. X 5 CrNi18-10/ AISI 304), then this can be carried out comparatively easily and filled with MHS.

A further option is the use of hollow sphere structures with metal matrix, given the matrix material meets the mentioned requirements. MHS would be technologically suitable for building elements still positioned within the product area. These do not have to be completely encased. Instead, it is sufficient to outfit the surfaces within the product area inertly. The remaining surfaces, however, could be manufactured differently in a more simplified way.

The substitution of stainless steel parts by *grouting constructions* using stainless steel sheets and MHS is also very interesting from an economic point of view. As material prices for stainless steel have recently soared, MHS gets more and more attractive as an alternative since.

12.3.2 Application of MHS in Material Class II

Restrictions are similar to those of class I in the case of fast moving machine parts in areas that are critical from a hygienic point of view. A complete encasement will mostly become necessary. Furthermore, the dynamic stress on the machine part and, thus, the requirements for lightweight material play a very crucial role. Consequently, the machine parts manufactured with MHS in *composite constructions* or *grouting constructions* must also have a high stiffness performance and be lightweight. Presently, however, these two requirements are technologically challenging concerning the use of MHS but represent a vital starting point for future composite developments. Some applications are an exception: These are applications in which the stress occurring on the machine part mainly results in buckling, not in bending. In a case like this MHS can be given priority due to its little density and weight. Another field of application in this material class would be transport units in processing machines used for handling of complex formed processing goods such as cups or flacons. There is only minimal mechanical stress on these elements. They must be relatively easy to produce in complex geometries, and finally, they should be comparatively lightweight to reduce energy loss especially with intermittently working machine units.

12.3.3 Application in Material Class III

Presently, the use of MHS as construction material in the functional area *casing* is the most promising in the field of processing machinery. Based on experience with industrial use of mineral casting the success of MHS seems quite likely, as it combines the manufacture-related benefits and the high self-damping performance of mineral casting with low specific weight due to the use of hollow metallic spheres. This statement is supported by machine parts that have already been realised in the field of tool and die production [1]. Components for product pick-up or entire machine beds were mainly manufactured. In analogy, the use for processing machines is feasible. Furthermore, a number of companies already use mineral casting for machine frames in packaging systems. This would also clearly highlight the advantage of functional integration. By processing MHS as cold cast material, it is also possible to cast other elements (plastic cables for liquid/ gasiform media, electrical lines) into the machine part. Accordingly, there are no restrictions in material class III from a manufacture-related point of view. Machine parts could be realised as *solo construction, composite construction* as well as *grouting construction*, given the boundary conditions are met such as corrosion resistance or cleanability described in chapter 12.1.2.5. Especially for machine parts with large surfaces whose weight could be improved and optimised besides the benefits already mentioned. As a matter of fact elements of material class III are stationary. Nevertheless, the total weight of the machine remains crucial for its transport to the customer and costs resulting from this transport. A further advantage of MHS lies in its excellent self-damping behaviour, which is common for almost all cellular metals [11]. This performance is crucial in this material class, chiefly to realise dynamic decoupling of individual functional elements through the frame. The most promising potential of MHS, however, lies in its easy processing. A financial reduction, at last, is undeniable for the MHS technology, especially with regard to complex machine parts that have so far required a variety of different processing steps.

12.3.4 Application of MHS in Material Class IV

In material class IV the focus is primarily on a high specific strength and stiffness performance. Therefore, similar to class II, the main objective for further development is the increase of the specific stiffness of MHS. Generally speaking, the benefit of MHS concerning manufacturing methods realised by *composite construction* is the integration of functional elements (cogging, sliding surfaces, etc.).

Additionally, the cold cast method used for the manufacture of reaction resin-bound MHS provides the possibility of integrating functional elements directly into the machine part. Provided the above mentioned criteria such as high specific strength and stiffness performance are fulfilled, this could open an even larger scope of possible applications. It would, for instance, be possible to use strain gauges or pressure sensors to log relevant process parameters or machine part safety in real-time. This function has become increasingly popular with security-related sectors such as the automotive industry. This is, however, difficult to be realised in case of metallic material in the conventional cast method.

12.4 Conclusion

Paving the "industrial" way for novel construction material is a rather challenging process. This is not easy especially in an industry, shaped by tailor-made manufacture for customers such as the processing machinery engineering. Besides the suitability of the material itself, this requires support with decision-making concerning material to provide further incentives for the design engineer. These incentives are primarily tools for the calculation and dimensioning of material. Also required are simple and clear information as to which machine parts are best suited for the novel material with its specific scope of properties. A possible approach has been introduced by the classification of the system processing machine into specific material classes. Assigning general scopes of properties to the classes resulting from the classification provides a helpful tool for the materials development. Assigning construction methods and application examples of reaction resin-bound MHS to material classes emphasises the promising possibilities of the material itself and the functionality of the system. Generally it has to be stated that MHS have a particularly interesting scope of properties for the use in material classes II und III from which a number of other applications could be derived. Especially the cold cast method used for producing MHS parts marks a clear advantage in the processing machinery industry, where small manufacturing quantities arrogate easy to handle prototype-like manufacturing technologies. This is supported by the established use of comparable material such as mineral casting, also clearly emphasising the need to promote these benefits to the users in the industry.

Acknowledgments. The project based on this chapter was funded by the Federal Ministry of Education and Research of Germany, grant identification number 03WKO01F. The authors assume full responsibility for the contents of this article.

References

[1] Baumeister, E., Kläger, S.: Advanced New Lightweight Materials: Hollow-Sphere Composites (HSCs) for Mechanical Engineering Applications. Adv. Eng. Mat. 5, 673–677 (2003)
[2] Code of Federal Regulations, Title 21 (21 CFR): Food and Drugs. United States Food and Drugs Administration. Rockville, USA, Parts 170–199 (2006)
[3] Codex Alimentarius Commission, Recommended international code of practice – general principles of food hygiene. CAC/RCP 1-1969, Rev. 4. World Health Organization, Geneve (2003)
[4] EN ISO 12100-1:2003: Safety of machinery—Basic concepts, general principles for design. Part 1: Basic terminology, methodology.
[5] EN ISO 12100-2:2003: Safety of machinery—Basic concepts, general principles for design. Part 2: Technical principles.
[6] EN ISO 14159:2002: (E) Safety of machinery—Hygiene requirements for the design of machinery
[7] EN 1672-2:2005: Food processing machinery—Basic concepts—Part 2: Hygienic Requirements

[8] European Hygienic Engineering and Design Group, Hygienic equipment design criteria. Guideline Doc. 8, 2nd edn. EHEDG, Brussels (2004)
[9] Goldhahn, H.: Grundlagen für Maschinensysteme der Stoffverarbeitung. In: Beitz, W. (ed.) Dubbel, 20th edn. Springer, Berlin (2003)
[10] Jackisch, U.-V.: Mineralguss für den Maschinenbau. Verlag Moderne Industrie, Landsberg (2002)
[11] Neugebauer, R., Hipke, T., Hohlfeld, J., Thümmler, R.: Metal Foam as a Combination of Leightweight Engineering and Damping. In: Altstädt, V., Münstedt, H. (eds.) Cellular Metals and Polymers. Trans Tech Publications, Stafa-Zürich (2005)
[12] Neumann, M.: Werkstoffgerechte Gestaltung und wirtschaftliche Fertigung dauerhaltbarer Maschinenbauteile aus Mineralguss. Ph.D. thesis, Rheinisch- Westfälische Technische Hochschule Aachen, Aachen (1985)
[13] Reuter, M.: Methodik der Werkstoffauswahl. Hanser, Munich (2007)

13

Screening of Available Hollow Structures in Regard to Noise Reduction in Automobile Applications

Burkhard Heine

University of Applied Sciences Aalen, Beethovenstrasse 1
73430 Aalen, Germany
burkhard.heine@htw-aalen.de

Abstract. The pressure to improve the efficiency of lightweight automobile concepts is permanent. Therefore, relevant automotive structural materials are progressively being optimized, with some materials seeming to have nearly reached their properties limits. Thus, slight improvements to their potential light weight can only be realized through considerable efforts.

In the recent past, materials with controllable porosity were proposed to open new light weight potential by reducing density. This chapter gives an overview on discussed metallic hollow structures and their evaluation in regard to automobile applications, using the criteria mechanical properties and processability. Due to acoustic properties becoming an evermore issue in cars of the future, this chapter concentrates on their evaluation.

Let us first look at possible ways in which to improve the acoustic properties of a component by selecting materials that are more effective than the most commonly used ones in automobile production, high strength steel.

13.1 Selection of Material

A sound wave that reaches the surface of a component generally reacts in two different ways with the surface (Fig. 13.1). One part of the wave is reflected. The remaining part initiates vibrations of the surface layer of atoms as well as the layers of atoms, which are representing the volume of the component. The initiation of vibrations requires kinetic energy, taken from the transmitting part of the sound wave. As a consequence, the frequency as well as the amplitude of the transmitted part of the sound wave is changing.

The processes initiated by transmitting part of a sound wave can be described by two phenomena, the sound velocity and the sound damping. Sound velocity is the measurement of time that passes from the moment a noise reaches the surface of a component to the moment we become aware of a noise. In most typical automobile cases the noise is permanent, thus the sound velocity is of lower importance.

Sound damping, a measurement to modify frequency and amplitude of the transmitted part of a sound wave that reaches our ear is the most important factor regarding automobile applications. However let us first take a look at the influence of applied material on the sound velocity.

A. Öchsner, C. Augustin (Eds.): Multifunctional Metallic Hollow Sphere Struct., EM, pp. 223–239.
springerlink.com

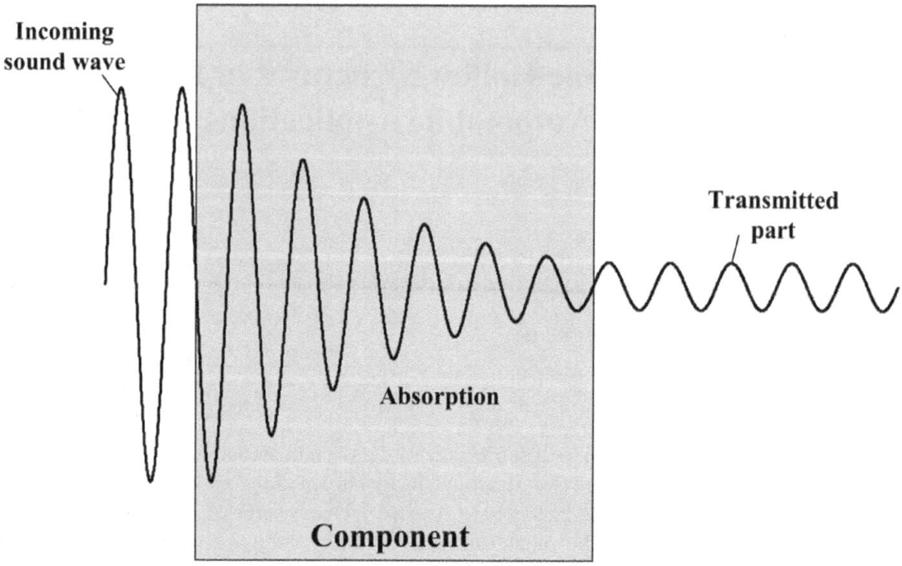

Fig. 13.1. A surface appearing sound wave reacts in two different ways with the surface

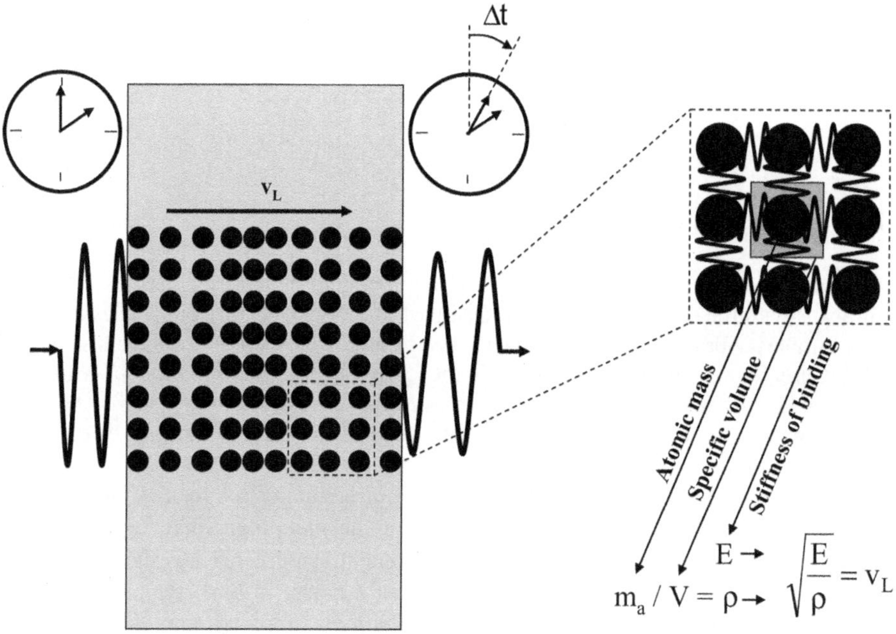

Fig. 13.2. Dependence between sound velocity and binding stiffness, atomic mass as well as specific volume of an atom

13.2 Sound Velocity

A sound wave can be treated as a mechanical wave. To quantify sound velocity in a solid solution we need to analyze the spreading of mechanical waves. Neglecting the dependence of the sound velocity v_L of longitudinal waves in solid solutions the sound velocity can be expressed by the Young's modulus E and the density ρ of the material according

$$v_L = \sqrt{\frac{E}{\rho}} \,. \tag{13.1}$$

Background is the dependence of the sound velocity on the binding stiffness E and the atomic mass m_a as well as the specific volume V of the single atom which leads to the density (Fig. 13.2).

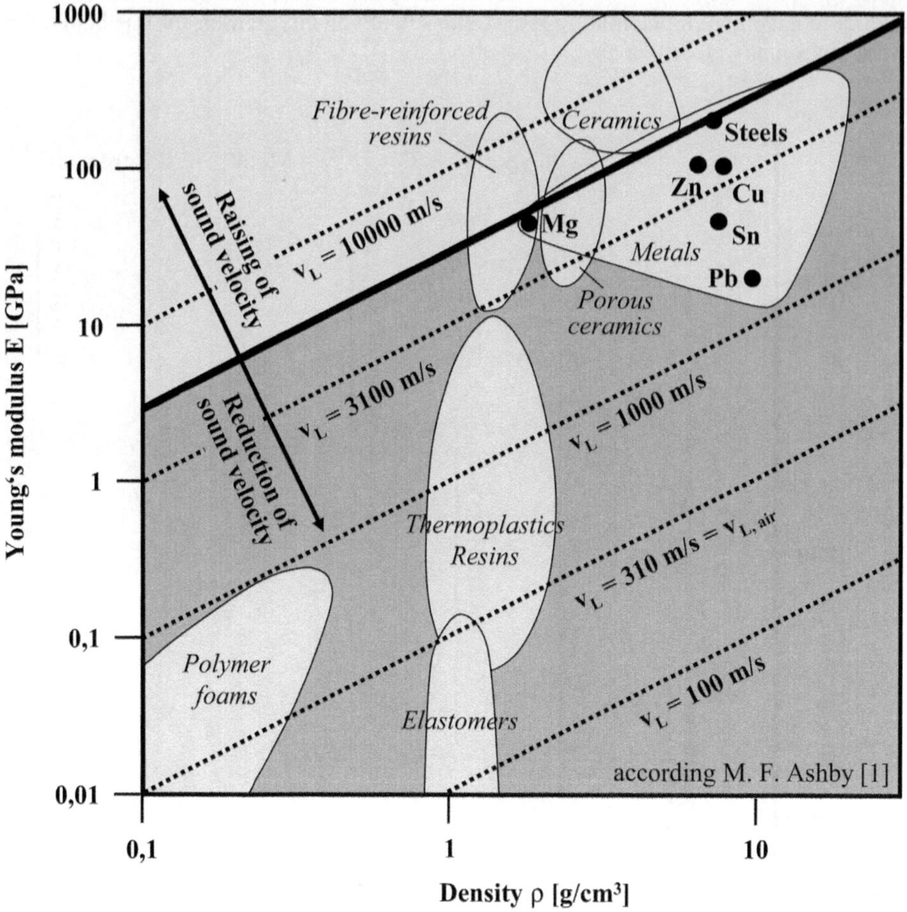

Fig. 13.3. Double logarithmic plot of Young's modulus E of materials versus their density ρ

To give a component a definite sound velocity by selecting the suitable material it would be helpful to be able to compare the materials in regard to their sound velocity. We can narrow the aim by modifying Eq. (13.1) according to the proposal of Ashby [1] to

$$\log E = \log \rho + 2 \cdot \log v_L \tag{13.2}$$

Using a double logarithmic plot of Young's modulus E of materials versus the density ρ of these materials (Fig. 13.3), Eq. (13.2) represents a line with a slope of 1. All materials positioned on the same line have the same sound velocity. A master line is shown in Fig. 13.3. Moving the master line upwards the materials now positioned on the line have a raised sound velocity and vice versa.

As previously mentioned, the most commonly used material for the construction of car body parts is high strength steel. Due to this the materials ranked in Fig. 13.4 were orientated on this with regard to their sound velocity. Taking our main focus away from monolithic materials and broaden our horizons to constructed materials such as heterogeneous materials with "air" as second phase like porous materials, we furthermore have a high potential to reduce the sound velocity, reason for this being, the very low sound velocity in air.

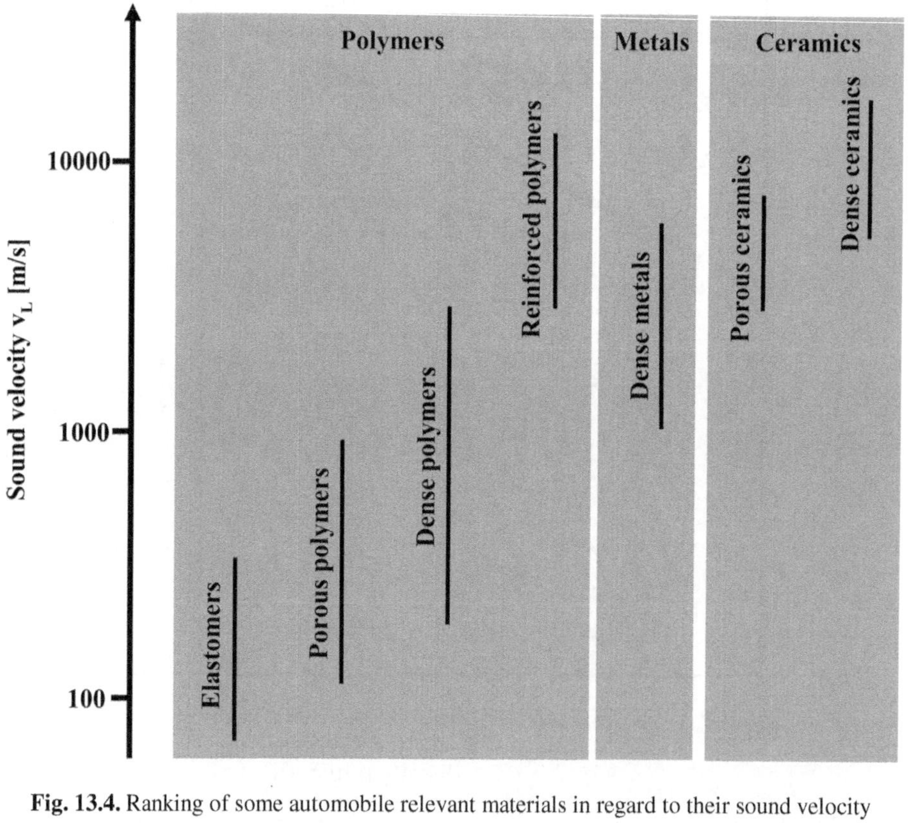

Fig. 13.4. Ranking of some automobile relevant materials in regard to their sound velocity

13.3 Sound Damping

While a component is oscillating at a frequency which promotes internal friction processes in the material, anelastic behaviour is presented: Each elongation consists of a linear-elastic part as spontaneous reaction to the exterior load and additionally of an anelastic part which rises during the time the component is stressed by an exterior load. The resulting stress-strain-curves are similar to those shown in Fig. 13.5. To initiate oscillations with amplitude, which do not change during time, it is necessary to compensate the volume dependent energy ΔW_v which dissipates as heat.

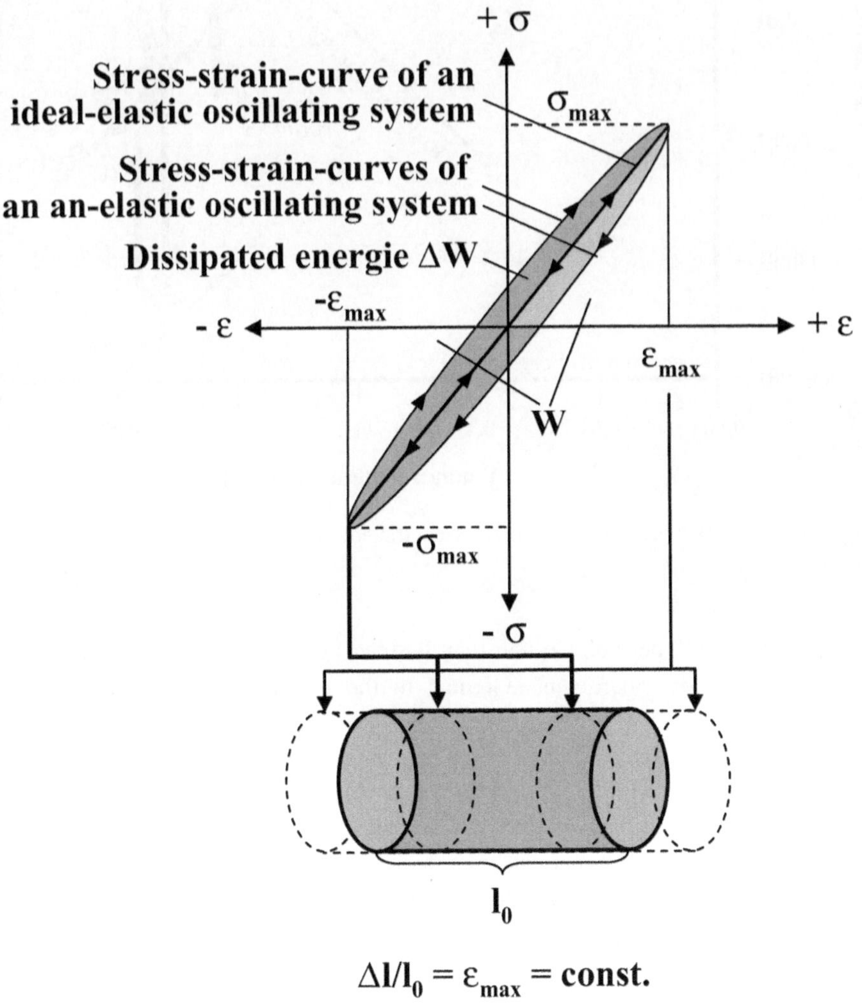

Fig. 13.5. Energy loss in an anelastic oscillating system

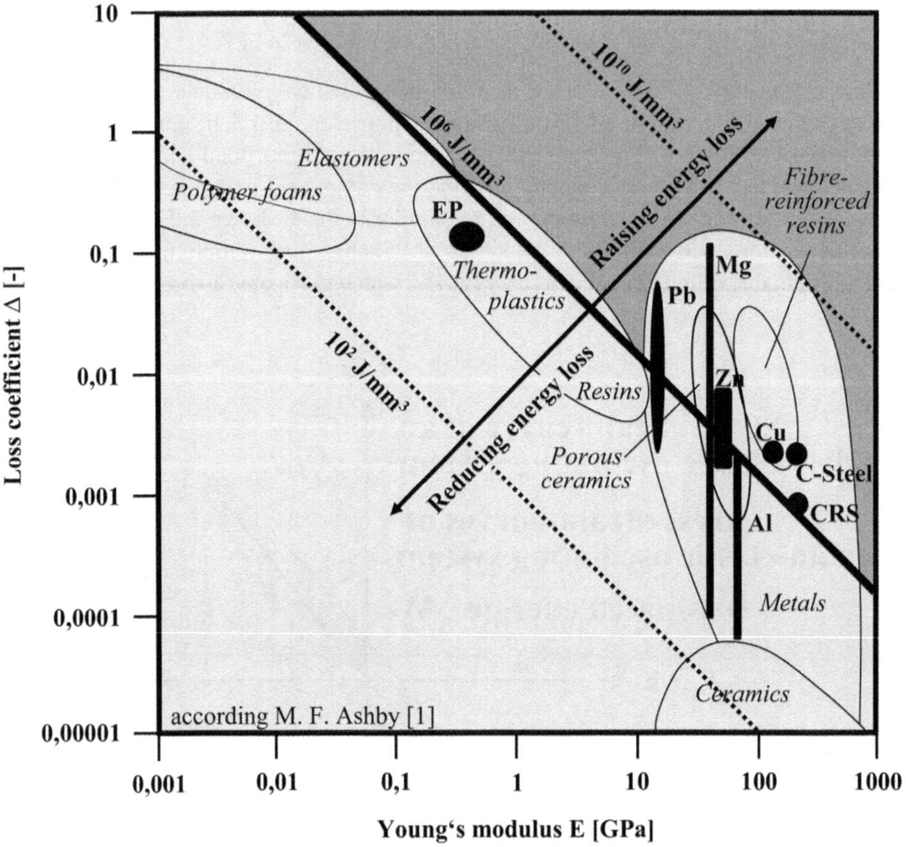

Fig. 13.6. Double logarithmic plot of loss coefficient Δ of materials versus their Young's modulus E

The energy ΔW_v per volume unit that dissipates during one oscillation can be described by the loss coefficient Δ (equal to the log decrement D divided by π) according to

$$\Delta = \frac{\Delta W}{\pi \cdot W} \quad . \tag{13.3}$$

The elastic deformation energy, imported per volume unit during one oscillation period can be calculated by using Hooke's law:

$$W = 2 \cdot \left(\frac{1}{2} \cdot \sigma_{max} \cdot \varepsilon_{max} \right) = E \cdot \varepsilon_{max}{}^2 , \tag{13.4}$$

and Eq. (13.3) can be written as

$$\Delta = \frac{1}{E} \cdot \frac{\Delta W}{\pi \cdot \varepsilon_{max}^{2}} \,. \tag{13.5}$$

To give a component a definite sound damping by selecting the suitable material it would be helpful to be able to compare the materials with regard to their sound damping. To narrow this aim we look upon the propagation of a sound wave in a solid solution as the propagation of a mechanical wave. Modifying Eq. (13.5) again according Ashby [1] to

$$\log \Delta = -\log E + \log \frac{\Delta W}{\pi \cdot \varepsilon_{max}^{2}} \,. \tag{13.6}$$

Using a double logarithmic plot of loss coefficient Δ of materials versus the Young's modulus E of the materials equation (13.6) represents a line with a slope of -1 (Fig. 13.6). All materials positioned on the same line have the same damping efficiency. A master line is shown in Fig. 13.6. Moving the master line upwards the materials now positioned on the line have raised efficiency of damping and vice versa.

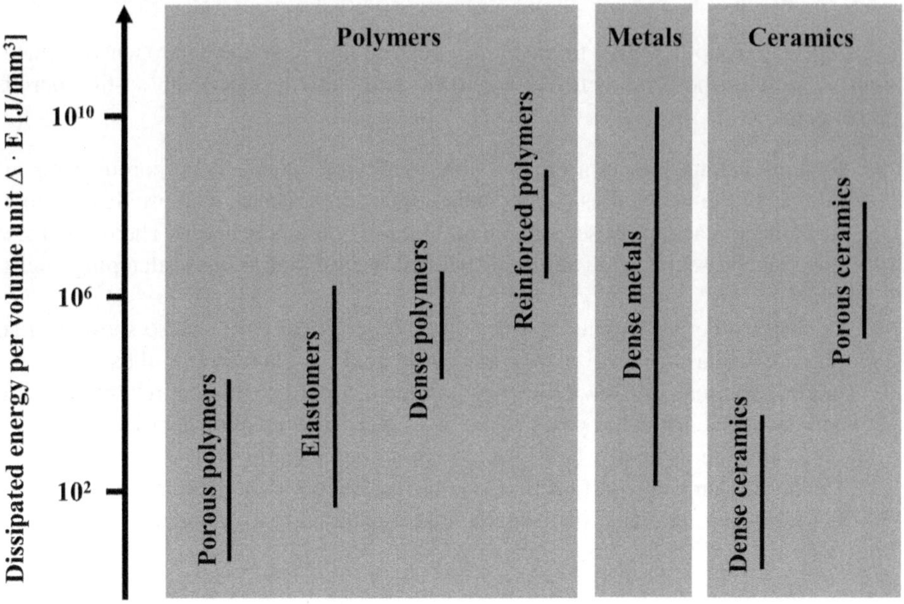

Fig. 13.7. Ranking of some automobile relevant materials in regard to their loss coefficient Δ

Figure 13.7 demonstrates a ranking of the loss coefficient of automobile relevant materials. For the monolithic materials we find the following correlation of microstructural properties and degree of damping:

- Due to high stability of the chemical bindings in the case of monolithic ceramics their internal friction and, as a consequence their damping potential in comparison to other materials can be neglected.
- In the case of metallic materials, the movement of the dislocations, which cause microplasticity and is the main reason for internal friction. The lower the amount of these movements in pure metals and bcc alloys (e.g. aluminium and austenitic corrosion resistant steels [CRS]) the higher degree solid solution and precipitation hardened high alloyed metals (e.g. bronze and high strength steels) can be found. The second reason for internal friction in the case of metallic materials can be strain-induced cooperative movements of complete groups of atoms (e.g. forming of martensite in carbon containing steels and TR(ansformation)I(nduced)P(lasticity)-steels as well as twin-formation in Mg-alloys and in TW(in)I(nduced)P(lasticity)-steels).
- Above the glass temperature the van-der-Waals-bindings of monolithic thermoplastics are solved and the sliding of the macromolecules along each other can occur intensely. Below the glass temperature the van-der-Waals-bindings are fixed, considerably reducing the slip-along of the macromolecules. Corresponding to the quantity of solved bindings is the value of the loss coefficient. The higher the quantity of solved bindings the higher the loss coefficient is.

A special group in regard to the ability of damping, form the heterogeneous materials, especially the heterogeneous materials with "air" as second phase like porous materials:

- Regarding some cast iron groups, fibre reinforced polymers and porous ceramics internal phase-boundaries are in contact. As a consequence, with oscillating vibrations the contacting phases slip lateral to these phase boundaries. Therefore, these materials present a relatively high internal friction and a raised damping coefficient.
- In contrary porous structures with elliptic or spherical cavities also show internal phase-boundaries, however they are not in contact. Therefore, in those cases the internal friction is of low efficiency and the damping coefficient relatively small. Moreover, the air in the pores has an extremely low damping coefficient in comparison to the material which would replace the air in the case of dense material. Hence the damping coefficient of porous materials with non-contacting interfaces is lower than the value of the same dense material that means material without pores.

Materials that in comparison to corrosion resistant steels (CRS) offer a more effective damping are fibre-reinforced resins, some non reinforced polymers (in car-body construction used as adhesives), copper alloys, lead alloys and tin alloys (the last two only applied as base elements of brazes), magnesium alloys and zinc alloys (both with additional potential in regard to light weight structures).

13.4 Structures with Internal Cavities

The strong sound velocity lowering in the case of "air" as second phase was an additional motivation for the development of hollow structures. Besides this we must also keep our focus on internal friction processes, which we need for a sufficiently high sound damping coefficient. Figure 13.8 shows possible hollow structures schematically. Some of them are already on the market, whereas others are still prototypes of research laboratories.

In general, we can roughly divide the possible hollow structures into two groups. The "natural" hollow structures can be produced with flat surfaces but are mostly neither gas nor liquid-tight, and raise problems concerning the joining of components. The natural hollow structures only fulfil their demand for tightness after covering their surface with sheets. These skinned natural hollow structures belong to the second group, the "constructed" hollow structures. Beside these skinned natural hollow structures, we know a lot of skinned constructed hollow structures, which in the core consist of constructed hollow structures. In general, the constructed hollow structures are gas and liquid-tight and exhibit a flat surface due to their covering sheets.

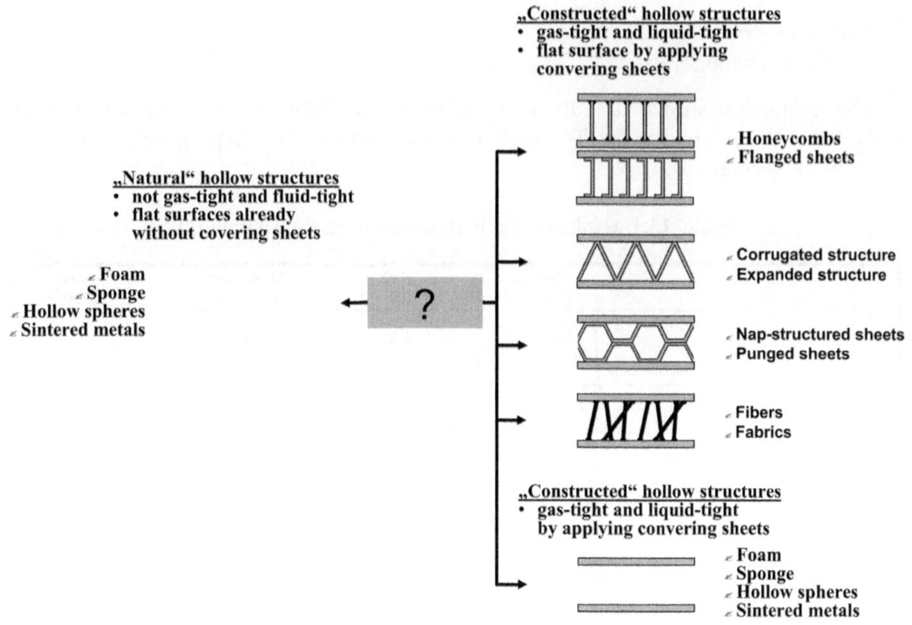

Fig. 13.8. Possible hollow structures schematically

Two general points are of great importance for each engineer dealing with hollow structures:

1. In comparison to massive structures, hollow structures have a reduced tolerance to deformation. Almost after a small amount of deformation the structure is damaged to such an extent that it can not fulfil its tasks anymore. As a consequence, bending a flat structure, applying low bending radii and deep-drawing such a structure is nearly impossible.
2. Most hollow structures must either be joined with each other or with a massive structure. In order to receive an acceptable stress situation and to reduce fabrication costs fusion-welding is desired. When the integration of bushes becomes necessary the fabrication costs rise and stress is concentrated on the area of the joining points.

Now we want to try to evaluate the different available hollow structures with regard to car-body applications using the criteria add a hard stop:

Physical properties

- Sound velocity reduction and sound damping
- Calculability of the acoustic properties of real components
- Gas and liquid-tightness

Mechanical properties

- 3-dimensional isotropic mechanical behaviour
- Lateral isotropic mechanical behaviour

Technical properties

- Fusion weldability

The evaluation shown in Tab. 13.1 allows marks between "- -" (very low; very bad) over "-" (low; bad) over "0" (without influence) and "+" (high; good) up to "++" (very high; very good).

Table. 13.1. Evaluation of hollow structures shown in Fig. 13.8

	Physical properties			**Mechanical properties**		**Technical properties**
	Sound velocity reduction sound damping	Calculability of the acoustic properties of real components	Gas-tight and liquid-tight	3-dimensional isotropic mechanical behaviour	Lateral isotropic mechanical behaviour	Fusion weldability
Foams	+	-	- -	+	+	-
Sponges	+	-	- -	+	+	-
Hollow spheres	++	-	- -	+	+	-
Open sintered metals	0	-	- -	+	+	-
Skinned foams	+	-	++	-	+	+
Skinned sponges	+	-	++	-	+	+
Skinned hollow spheres	++	-	++	-	+	+
Skinned open sintered metals	0	-	++	-	+	+
Skinned honeycombs	++	+	++	-	+	+
Skinned flanged sheets	++	+	++	-	-	+
Skinned corrugated structure	++	+	++	-	-	+
Skinned expanded structure	++	+	++	-	+	+
Skinned nap-structured sheets	++	+	++	-	+	+
Skinned punged sheets	++	+	++	-	+	+
Skins with fibers as spacers	++	-	++	-	+	+
Skins with fabrics as spacers	++	-	++	-	+	+

13.5 Foams

Foams are produced in several ways. One of these examples being, a solid component is added to a melt shortly before being cast into a mold. Between liquidus-temperature and solidus-temperature the added component releases an expanding gas. The gas bubbles will be included in the freezing melt. Another way of production operates with a mixture of the solid component and a metal powder. After compression a sintering process follows. Slightly below solidus-temperature or between liquidus-temperature and solidus-temperature the added component again releases an expanding gas and bubbles are formed.

Foams are known on the basis of a lot of base materials. In regard to car body production aluminium and steels, in regard to engine and powertrain production also nickel is of main interest.

There is great variety in the shapes and sizes of foam pores (Fig. 13.9). Beside this their geometry and distribution varies from component to component, causing the modelling of the acoustic properties nearly to be impossible. The 3-dimensional and the lateral isotropic mechanical behaviour are given, also a sound velocity reduction. Due to the continuous metallic skeleton the sound damping is only moderate.

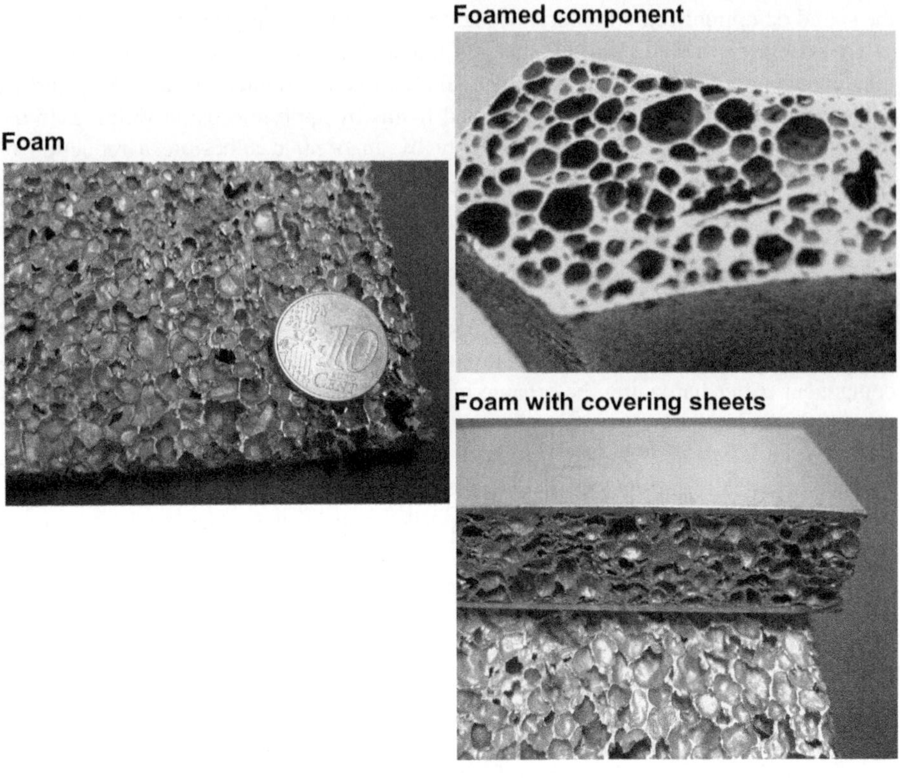

Fig. 13.9. Foam

By skinning a metallic foam, gas and liquid-tightness can be reached, but unfortunately by loosing the 3-dimensional isotropic mechanical behaviour. A further advantage is the possibility to join skinned foams by applying fusion welding. In the case of assembling the skin with the core by an organic adhesion a reduction of sound velocity and a raising of sound damping can be achieved.

13.6 Sponges

Sponges are also produced in several ways. One method is by squeeze-casting a melt in a mold, in which a preform of refractory spheres is positioned. The melt fills the space between the spheres. After freezing and cooling to room-temperature the refractory spheres are removed by etching. Another technique is by casting a melt according to the lost-foam-procedure. The melt penetrates into an open-cellular PUR-foam along its grain-boundaries and freezes. Simultaneously the PUR burns out.

In these ways sponges of aluminium, steel and nickel can be produced.

Regarding shapes and sizes of sponge pores, we can repeat the remarks we already mentioned by discussing the properties of foams. So the modelling of the acoustic properties in this case is nearly impossible. The 3-dimensional and the lateral isotropic mechanical behaviour are given, also a sound velocity reduction. Once again the sound damping is only moderate because of the continuous metallic skeleton.

By skinning a metallic sponge gas and liquid-tightness can be reached, but unfortunately again by loosing the 3-dimensional isotropic mechanical behaviour. A further advantage is the possibility to join skinned foams by applying fusion welding. In the case of assembling the skin with the core by an organic adhesion, a reduction of sound velocity and a raised sound damping can be achieved.

13.7 Hollow Spheres

One way to produce hollow spheres operates with pre-expanded EP-spheres. The spheres are coated with a semi-liquid mixture of metal powder and an organic binding component. After hardening the organic binding component, the EP of the EP-spheres as well as the binder are burned out in a pyrolysis step. Most of the gaseous reaction products can leave the hollow sphere because it is not yet dense (some of the reaction products, especially carbon, diffuse into the metallic shells during the following sintering step). The following sintering step compacts the shells and closes them. The joining of the hollow spheres can be carried out by sintering, brazing and by using an adhesive.

This is an example of how hollow spheres of carbon steel and corrosion resistant steel could be produced.

Theoretically with spheres of the same size a close-packed and reproducible geometry could be created. To do so, each of the spheres has to be exactly positioned according to a hexagonal close-packed crystal structure. Using spheres with a diameter of several millimetres a lot of time to produce a component with realistic dimensions would be required. To shorten the production time a disordered package, where the neighbourhood varies from place to place is generally used. (Fig. 13.10).

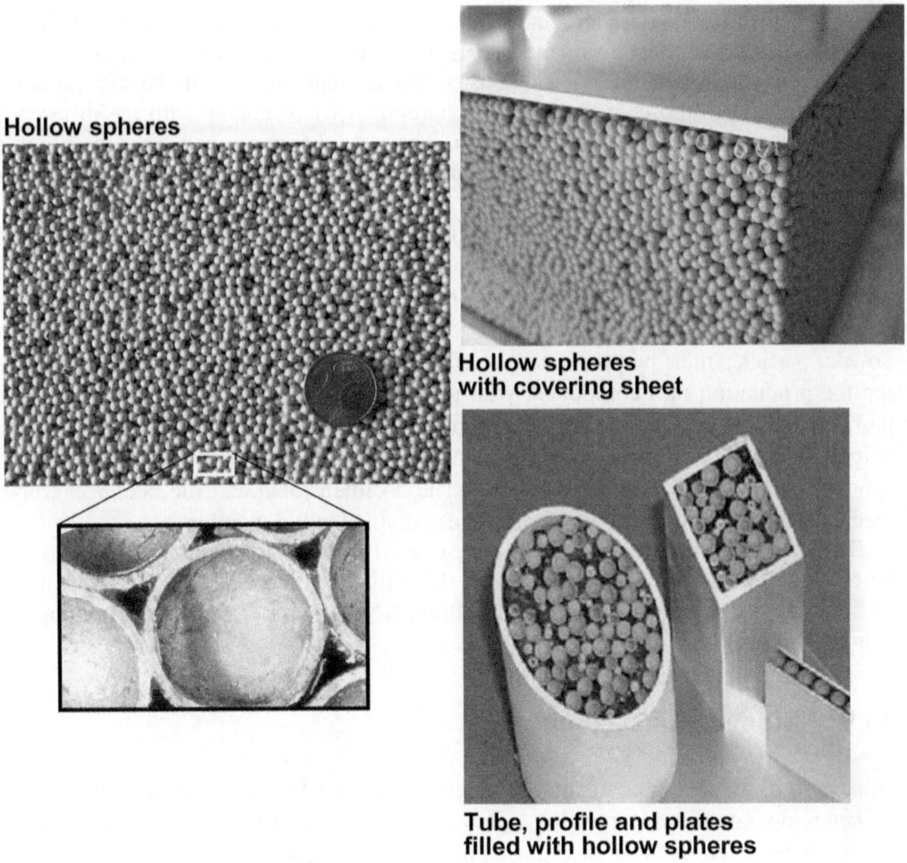

Fig. 13.10. Hollow spheres

Therefore the modelling of acoustic properties is nearly impossible. The 3-dimensional and the lateral isotropic mechanical behaviour are given, also a sound velocity reduction. In the case of joining the hollow spheres by an organic adhesion a further reduction of sound velocity and a raised sound damping can be achieved.

By skinning a structure of metallic hollow spheres gas and liquid-tightness can be reached, but unfortunately looses the 3-dimensional isotropic mechanical behaviour through this process. A further advantage is the possibility of joining skinned hollow spheres by applying fusion welding. In the case of assembling the skin with the core by organic adhesion, a further reduction of sound velocity and a raised sound damping can be achieved.

13.8 Open Sintered Metals

The ways in which to produce open sintered metals have been well known and practiced for decades. Metal powder in a first step is compacted and in a second step sintered at temperatures near the solidus-temperature. At these temperatures surface-diffusion processes are dominant and reduce the air-filled volume of the powder package. To achieve an open sintered metal the sinter process must be stopped well before reaching the theoretical density. The arrangements of the powder particles are comparable to those of the hollow spheres in 13.7 of course with the difference that the spheres are not hollow but solid. The next difference concerns the powder particle diameter: In most cases the diameters of the powder particles are much lower than those of the hollow spheres.

In this way for example open sintered structures of aluminium, titanium, carbon steel and corrosion resistant steel as well as nickel can be produced.

As mentioned previously, hollow spheres with powder particles of the same size a close-packed and reproducible geometry could also be created. To do so, each of the powder particles must be positioned exactly. In comparison to hollow spheres hindering the production of a component with realistic dimensions is the very low powder particle diameter. In most cases we therefore use a disordered package, where the neighbourhood varies from place to place. As a consequence the modelling of the acoustic properties is nearly impossible. The 3-dimensional and the lateral isotropic mechanical behaviour occures, as well as a sound velocity reduction.

By skinning a structure made of open sintered metals gas and liquid-tightness can be reached, but unfortunately by loosing the 3-dimensional isotropic mechanical behaviour. A further advantage is the possibility to join skinned foams by applying fusion welding.

13.9 Skinned Honeycombs

Honeycombs are available in several materials. In regard to sufficient strength even at higher temperatures only metallic honeycombs are a possible solution. The honeycombs have to be skinned in order to form a gas / liquid-tight as well as a fusion

Fig. 13.11. Honeycombs

weldable surface. Aluminium was applied with reference to Fig. 13.11. Besides aluminuium, honeycombs are also produced of titanium, steel and nickel.

The high stiffness of the honeycombs lateral to their walls and the low stiffness perpendicular to their walls results even in the case of a skinned honeycomb in a not given 3-dimensional isotropic mechanical behaviour. In contrast to this a lateral isotropic mechanical behaviour is realized. The given sound velocity reduction and sound damping raising can be improved by joining the honeycombs and the skinning sheets by an organic adhesion. The geometric periodicity of the honeycombs enables modelling of their acoustic properties.

13.10 Flanged Sheets, Skinned Corrugated Structures, Expanded Structures, Nap-Structured Sheets and Punged Sheets

The remarks mentioned regarding skinned honeycombs can also be carried out by describing the properties of flanged sheets, skinned corrugated structures (Fig. 13.12), expanded structures, nap-structured sheets (Fig. 13.13; without skins) and punged sheets. As core materials and materials for skins, all combinations as well as all joining techniques imaginable are possible.

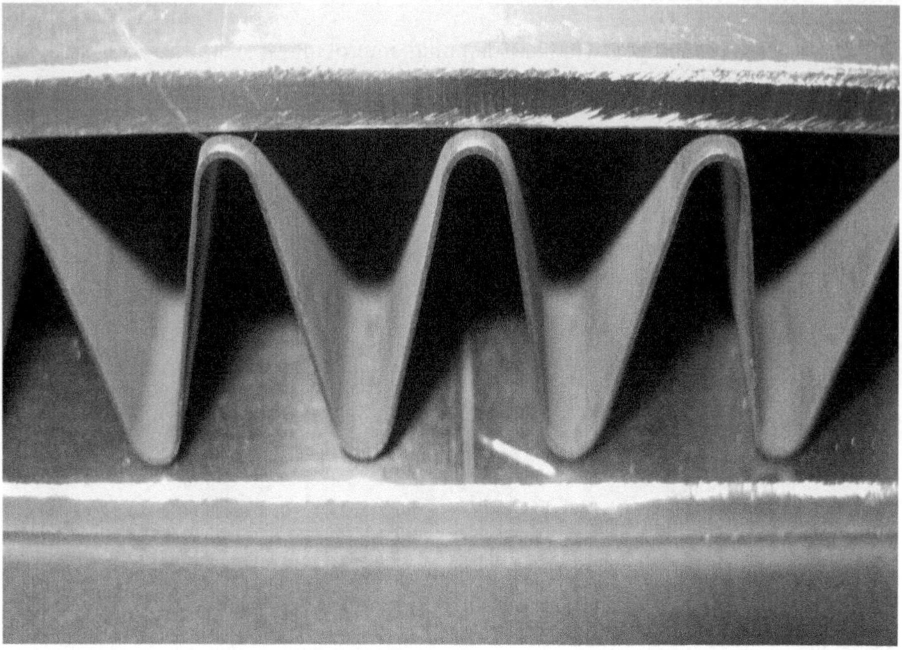

Corrugated structure with covering sheets

Fig. 13.12. Corrugated structure

Fig. 13.13. Nap-structured sheets

13.11 Skins with Fibres and Fabrics as Spacers

Sound velocity reducing and sound damping improving spacers between skinning sheets fibres and fabrics can be integrated (Fig. 13.14). For lower temperatures of application a thermoplastic adhesive or a resin, both filled with short glass- or aramid-fibres can be positioned between the skinning sheets. In this case the fibres are orientated parallel to the covering sheets. For higher temperatures the materials of skins

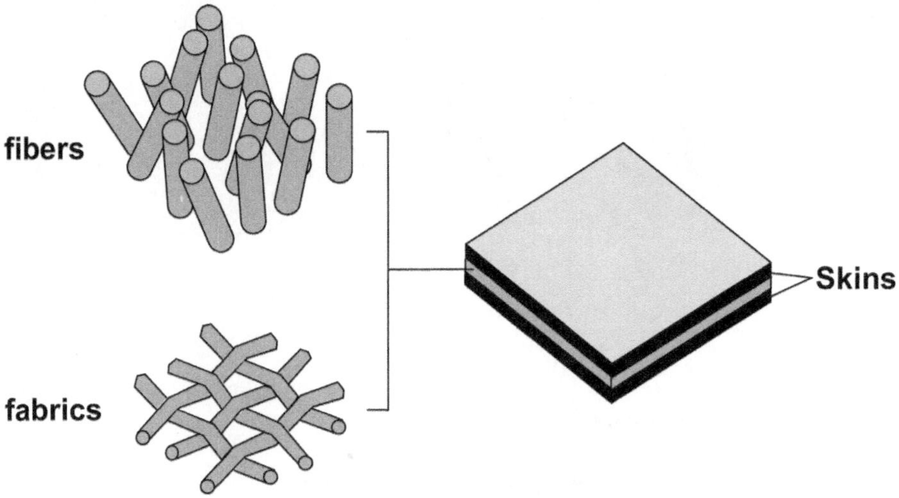

Fig. 13.14. Fibres and fabrics

and fibres and fabrics must be changed. Possible materials for the skin are aluminium, titanium, steel or nickel. Metallic fibres can be produced by melt-extraction and with a look at necessary strength may consist of titanium and steel. Their orientation is mostly vertical to the sheet plane. Long fibres of aluminium, titanium, steel or nickel are used for the production of fabrics. Fibres and fabrics can be joined to the skins by using adhesives and brazes.

For fibres and fabrics the mechanical properties of a skinned structure are not 3-dimensional isotropic. Due to the random distribution of the fibres in sheet plane we find lateral isotropic mechanical properties. In the case of joining the fibres and fabrics to the skinning sheets by an organic adhesion results in a further reduction of sound velocity and a raise of sound damping can be achieved. Unfortunately the lateral distribution of the fibres is not homogeneous and most of the fibres can not be positioned in an exact vertical position. Therefore modelling acoustic properties is close to impossible.

Summary

As a consequence of the content of Tab. 13.1 we can state that

- skinned hollow spheres (when it is possible to position the spheres periodically)
- skinned honeycomb
- skinned nap-structured sheets
- skinned punged sheets

have well-balanced properties and so the highest potential for the future. When we integrate the material selection in regard to reducing the sound velocity and raising the sound damping the application of fibre-reinforced resins (limited to lower temperatures), magnesium alloys and zinc alloys as base materials and adhesives (limited to lower temperatures) and brazes on the basis of lead, tin, zinc and copper as joining elements must be preferred.

Literature

[1] Ashby, M.F.: On the Engineering Properties of Materials. Acta Metal. 37, 1273–1293 (1989)

14

Commented Literature for Hollow Spheres and Hollow Sphere Structures

Christian Augustin[1] and Andreas Öchsner[2]

[1] Christian Augustin
hollomet GmbH, Grunaer Weg 26, D-01277 Dresden, Germany and
University of Konstanz, Faculty of Humanities, Department of History,
D-78457 Konstanz, Germany
Christian.Augustin@glatt.com

[2] Andreas Öchsner
Technical University of Malaysia, Faculty of Mechanical Engineering, 81310 UTM Skudai, Johor, Malaysia and
The University of Newcastle, Callaghan, New South Wales 2308, Australia
Andreas.Oechsner@gmail.com

This section collects major references for hollow spheres (HS) and hollow sphere structures (HSS). In addition to journal references and patents related to HS and HSS, a bit more general reference books and conference proceedings related to cellular materials in general are given which allow a first introduction to the topic. The references in each section are given in chronological sequence in order to indicate the development in this area.

14.1 Monographs and Textbooks Related to Cellular Materials

1. Hilyard NC (1982) Mechanics of cellular plastics. Macmillan Publishing, New York
2. Gibson LJ, Ashby MF (1997) Cellular solids: structures & properties. Cambridge University Press, Cambridge
3. Ashby MF, Evans A, Fleck NA, Gibson LJ, Hutchinson JW, Wadley HNG (2000) Metal foams: a design guide. Butterworth-Heinemann, Boston
4. Degischer H-J, Kriszt B (2002) Handbook of cellular metals: production, processing, applications. Wiley-VCH, Weinheim
5. Scheffler M, Colombo P (2005) Cellular ceramics. structure, manufacturing, properties and applications. Wiley-VCH, Weinheim
6. Mills N (2007) Polymer foams handbook: engineering and biomechanics applications and design guide. Butterworth-Heinemann, Oxford
7. Hipke T, Lange L, Poss R (2007) Taschenbuch für Aluminiumschäume. Aluminium-Verlag, Düsseldorf
8. Öchsner A, Murch GE, de Lemos JS (2008) Cellular and porous materials: thermal properties simulation and prediction. Wiley-VCH, Weinheim
9. Körner C (2008) Integral foam molding of light metals: technology, foam physics and foam simulation. Springer, Berlin

A. Öchsner, C. Augustin (Eds.): Multifunctional Metallic Hollow Sphere Struc., EM, pp. 241–252.
springerlink.com

14.2 Conference Proceedings Books Related to Cellular Materials

1. Sieradzki K, Green DJ, Gibson LJ (eds) (1991) Mechanical properties of porous and cellular materials - Materials Research Society symposium proceedings volume 207. Materials Research Society (MRS), Pittsburgh
2. Schwartz DS, Shih DS, Evans AG, Wadley HNG (eds) (1998) Porous and cellular materials for structural applications - Materials Research Society symposium proceedings vol 521. Materials Research Society (MRS), Warrendale
3. Banhart J, Eifert H (eds) (1998) Metal foams (Fraunhofer USA Metal Foam Symposium 1997). MIT-Verlag, Bremen
4. Banhart J, Ashby MF, Fleck NA (eds) (1999) Metal foams and porous metal structures (MetFoam 1999). MIT-Verlag, Bremen
5. Clyne TW, Simancik F (eds) (2000) Metal matrix composites and metallic foams. Wiley-VCH, Weinheim
6. Banhart J, Ashby MF, Fleck NA (eds) (2001) Cellular metals and metal foaming technology (MetFoam 2001), MIT-Verlag, Bremen
7. Ghosh AK, Sanders Jr. TH, Claar TD (eds) (2002) Processing and properties of lightweight cellular metals and structures. The Minerals, Metals & Materials Society (TMS), Warrendale
8. Banhart J, Fleck NA, Mortensen A (eds) (2003) Cellular metals: manufacure, properties and applications (MetFoam 2003). MIT-Verlag, Bremen
9. Singer RF, Körner C, Altstädt V, Münstedt H (eds) (2005) Cellular metals and polymers. Trans Tech Publications, Uetikon-Zuerich
10. Stephani G, Kieback B (eds) (2005) Cellular metals for structural and functional applications (CELLMET 2005). Fraunhofer IRB Verlag, Stuttgart
11. Nakajima H, Kanetake N (eds) (2006) Porous metals and metal foaming technology (MetFoam 2005). The Japan Institute of Metals, Sendai
12. Lefebvre LP, Banhart J, Dunand DC (2008) Porous metals and metallic foams (MetFoam 2007). DEStech Publications, Lancaster
13. Zhao H, Fleck NA (eds) (2009) IUTAM Symposium on mechanical properties of cellular materials. Springer, Dordrecht

14.3 Patents Related to Hollow Spheres and Hollow Sphere Structures

1. Spahr O (1906) Method or process of making hollow metal balls. US 861403
2. Schatz JW (1910) Process of making hollow metallic balls. US 955698
3. Seamless Hollow Ball Company, Gruver JH (1917) Sheet metal hollow ball and method of making the same. US Patent 1278914, cf. (1920) Verfahren zur Herstellung von Hohlkugeln. AT 92463
4. Krone A (1921) Maschine zum Verreiben und gleichzeitigen Mischen von Stoffen in einer Hohlkugel, welche gleichzeitig um zwei senkrecht zueinander stehende Achsen gedreht wird. DE 371505
5. Schwindt JM (1922) Herstellung von Würfeln aus einer Hohlkugel. DE 379264
6. Seamless Hollow Ball Company (1924) Verfahren und Einrichtung zur Herstellung von metallenen Hohlkugeln. CH 112433

7. Seamless Hollow Ball Company (1924) Verfahren und Vorrichtung zur Herstellung von Hohlkugeln aus Metall. AT 103454
8. Champion Spark Plug Company (1932) Verfahren zur Herstellung von Formlingen, insbesondere aus keramischem Stoff. German Reichspatent 604691
9. Fried. Krupp Grusonwerk (1935) Dünnwandige Hohlkugel als Füllkörper. DE 642672
10. Universal Oil Products Company (1935) Verfahren zur Herstellung von hochaktiven festen, kalzinierten Katalysatoren. DE 893048
11. Toyo Tokushu Immono Kabushiki Kaisha (1938) Verfahren und Vorrichtung zum Herstellen metallener Hohlkugeln. German Reichspatent 694903
12. Carborundum Company, Geiger CF (1946) Method for making refractory bodies and products thereof. US 2553759
13. Stark Ceramics, Ramsay WS (1947) Method of producing porous, lightweight, burned argillaceous material. US 2543987
14. Baumat AG (1948) Verfahren zur Herstellung von als Zuschlagstoffe für die Bauindustrie verwendbaren Hohlkugeln aus keramischen Stoffen. CH 266 315
15. Deutsche Gold- und Silber-Scheideanstalt, Wagner W (1950) Verfahren zur Herstellung von porösen Körnungen aus hochschmelzenden Oxyden. DE 828521
16. Heinrich JB (1950) Rotierende Hohlkugeln mit in derselben angeordneter, ebenfalls in Rotation zu versetzender Bodenscheibe zur Aufnahme von Zuschauern. DE 905346
17. Merremann E (1950) Hohlkugel als Friseuraushängeschild aus dünnem Bleich mit leicht lösbarer Verbindung der Halbkugeln. DE 1617181
18. Standard Oil Development Company (1954) Verfahren zur Herstellung von hochaktiven Reformierungskatalysatoren. DE 963980
19. Franke E, Franke K (1958) Als Hohlkugel ausgebildeter Schwimmkörper für Fischnetze, Reusen u. dgl. DE 1784004
20. Stolle L (1958) Vorrichtung zur Herstellung von Hohlkugeln. DE 1092868
21. Schmidt M (1960) Schweißvorrichtung zum Verbinden von zwei Hohlkugelhälften, z.B. zu Christbaumkugeln od. dgl. DE 1168058
22. Badische Anilin- & Soda-Fabrik Aktiengesellschaft, Bär K, Sperber H, Göhre O, Leibner G (1963) Verfahren zur Herstellung von Katalysatorträgern in Form von Hohlkugeln. DE 1278411
23. Rosenthal Aktiengesellschaft, Bock P (1963) Verfahren zur Herstellung von kleinen keramischen Hohlkugeln. DE 1227377
24. Badische Anilin- & Soda-Fabrik Aktiengesellschaft, Bär K, Sperber H, Göhre O, Leibner G (1964) Verfahren zur Herstellung von Katalysatorträgern in Form von Hohlkugeln. DE 1442567
25. Nukem Nuklear Chemie und Metallurgie GmbH (1964) Device for a production of a hollow sphere. GB 1083993
26. Kroyer KKK (1965) Verfahren zum Heben eines gesunkenen Schiffes. DE 1247893
27. Société Le Carbone-Lorraine (1966) Graphit-Hohlkugel zur Verwendung in der Kerntechnik. DE 1960237

28. Weibel & Dall'O, Weibel B (1967) Verfahren zur Herstellung von Hohlkugeln aus keramischem Material und Vorrichtung zur Durchführung des Verfahrens. CH 470949
29. Jenaer Glaswerk Schott & Gen., Dutz H, Meyer H (1969) Verfahren zur Herstellung von Hohlkugeln aus Glas. DE 1961628
30. Novak P (1970) Hollow sphere and structural of element for constructing same. US 3691704
31. Koreha Kagaku Kogyo KK, Toyo Boseki KK, Amagi Y, Shiiki Z, Nakao M, Urawa S, Bunrin H (1972) Zusammengesetzter Schaum aus anorganischen Hohlkugeln in einer Metallmatrix und Verfahren und Vorrichtung zur Herstellung derartiger Schäume. DE 2218455
32. Hoechst AG, Stenzel J, Hinz A (1973) Verfahren zur Herstellung eines aus stabilen Hohlkugeln bestehenden Materials. DE 2353631
33. Pallana GmbH, Schmidt M (1973) Verfahren und Vorrichtung zur Herstellung von Plastik-Hohlkugeln, insbesondere Christbaum- oder Dekorationskugeln. CH 554739
34. Schwerin H (1974) Neue Konstruktionselemente in Form von Hohlkugel-Zweieck-Körpern und deren Teilstücke. DE 2433357
35. SKF Industrial Trading and Development Co.B.V., Martinie HM (1974) Hohlkugel für Kugellager. DE 2407479
36. Mylaeus A (1976) Hohlkugeln sowie Verfahren zu ihrer Herstellung und Vorrichtung zur Durchführung des Verfahrens. DE 2631120
37. Akademie der Wissenschaften der DDR, Köcher P, Frischbutter E, Voges N, Günther U (1977) Verfahren zur Herstellung hochtonerdehaltiger Hohlkugeln. GDR 134638 (= DD 134638)
38. Jaeckel M (1982) Metallische, im wesentlichen kugelförmige Leichtkörperteilchen, sowie Verwendung und Verfahren zu ihrer Herstellung. DE 3210770
39. Kombinat VEB Elektronische Bauelemente, Kaspar E (1982) Verfahren zur Herstellung von Ventilmetallkörpern für Elektrolytkondensatoren. DE 3227726
40. Metco Inc., Longo FN, Bader NF, Dorfman MR (1983) Hollow Sphere ceramic particles for abradable coatings. EP 86938
41. Commissariat à l'Énergie Atomique, Bonceur M, Prigent Y (1984) Method of Manufacturing highly porous bodies from inorganic material which may be used for preparing hygroscopic, inflammable or radioactive articles. EP 119913
42. VEB Ingenieurbüro für Rationalisierung Kosmetik, Kühn R, Umlauf K (1984) Hohlkugel für Roll-on-Erzeugnisse. GDR 231541 (= DD 231541)
43. Kawasaki Steel Corp., Shosuke T, Kazuhiro H (1986) Production of hollow pitch sphere. JP 62195080
44. Minnesota Mining and Manufacturing Co., Douden DK (1986) Keramische Hohlkugeln und Verfahren zu deren Herstellung. DE 3628813
45. Université Toulouse III, Blottiere Y, Bonino J-P, Rousset A, Rossignol C, Gossart INH (1986) Verfahren zur Herstellung von geschlossenen und nahtlosen Hohlköpern, hergestellte Hohlkörper und Vorrichtung zur Herstellung von Hohlkugeln. AT 0212712
46. Kirin Brewery Co Ltd, Ryoichi K (1987) Hollow sphere of calcium. JP 63198970

47. Norddeutsche Affinerie AG, Jaeckel M, Smigilski H (1987) Verfahren zum Herstellen von metallischen oder keramischen Hohlkugeln. DE 3724156
48. Norddeutsche Affinerie AG (1988) Verfahren zur Herstellung von Hohlkugeln oder deren Verbunden mit Wandungen erhöhter Festigkeit. DE 3640586
49. Hitachi Chem Co Ltd, Koji K, Kuniteru M, Yoshimasa S (1989) Production of hollow fine metal sphere. JP 02305970
50. Rohr Industries Inc, Norris B, Gojny FJ (1989) Hollow metal sphere filled stabilized skin structures and method of making. US 4925740
51. Shinagawa Refract Co Ltd, Hajime A, Hiroshi T, Hideo M, Shunsuke S (1989) Production of refractory hollow sphere. JP 02277544
52. Ateca RDM, Blottiere Y, Chapuis P, Valaud M (1991) Kunststoff-Hohlkugeln, Verfahren zu ihrer Herstellung und ihrer Verwendung. DE 69102775 (= FR 9000484)
53. Mitsubishi Materials Corp, Yasuyuki O (1991) Production of ceramic hollow sphere. JP 04265150
54. Annawerk Keramische Betriebe GmbH, Fickel A, Konz R (1992) Feuerfeste keramische Leichtwerkstoffe und Bauteile daraus. DE 4208155
55. Audi AG, Haldenwanger H-G, Klose P (1993) Kraftfahrzeug mit Deformationselementen. DE 4340349
56. Sanwa Seimitsu Kogo KK, Tadashi M (1994) Hollow polyhedral sphere and its production. JP 07241204
57. Carl Zeiss, Kaufmann P, Bingel U (1995) Metallmatrix-Hohlkugel-Kompositwerkstoff. DE 19537137 (= EP 0767246)
58. Comtrade Ltd., Mevissen K (1995) Verfahren zur Herstellung von wärmeisolierenden Formkörpern, nach dem Verfahren hergestellte wärmeisolierte Formkörper sowie ihre Verwertung. DE 19538667
59. Hoechst AG, Kühn H, Brockmeyer A, Escher C, Voßberg D, Süßbrich S (1996) Anorganische Hohlkugeln, Verfahren zu ihrer Herstellung und ihre Verwendung. DE 19603196
60. Glatt Systemtechnik Dresden GmbH, KWS Kleinwanzlebener Saatzucht AG, Bretschneider F, Bruno P, Brückner J (1997) Einrichtung und Verfahren zur Herstellung eines schüttfähigen Produktes mit einer Hüllschicht. DE 19750042 C2
61. Hunkemöller P (1998) Leichtbauelement mit Hohlkugeln. DE 19817959
62. Unterländer J (1998) Fisch-Lockfutter-Hohlkugel. DE 19846784
63. Daimler-Chrysler AG, Fußnegger W, Thoms V (1999) Konstruktionsbauteil und Verfahren zu dessen Herstellung. DE 19937375
64. Fraunhofer-Gesellschaft zur Förderung der angewandten Forschung e.V., Nabertherm GmbH & Co. Industrieofenbau KG, Boehm A, Korhammer C, Scholl R (1999) Thermal Insulation to be inserted between two insulating structures. WO 99/54655
65. Fraunhofer-Gesellschaft zur Förderung der angewandten Forschung e.V., Glatt Systemtechnik Dresden GmbH, Bretschneider F, Stephan H, Brückner J, Andersen O, Kieback B, Stephani G, Schneider L, Waag U (1999) Verfahren zur Herstellung metallischer, oxydischer oder keramischer Hohlkugeln. DE 19929760

66. Laboure JC (1999) Procede pour realisier des spheres creuses (boules de petanque) homogenes en durete sur toute la surface par assemblage des deux demi-coquilles constituent la boule au moyen d'un faisceau laser. FR 2795010
67. Fraunhofer-Gesellschaft zur Förderung der angewandten Forschung e.V., Hochschule für Technik und Wirtschaft Dresden, Glatt Systemtechnik Dresden GmbH, Stephani G, Andersen O, Zikoridse G, Bach E, Bretschneider F (2000) Vorrichtung und Verfahren zur Behandlung von Verbrennungskraftmaschinenabgasen. DE 10058580 (= EP 133603 = WO 2002/040837)
68. Glatt Systemtechnik Dresden GmbH, Fraunhofer-Gesellschaft zur Förderung der angewandten Forschung e.V., Bretschneider F, Stephan H, Brückner J, Schneider L, Stephani G, Waag U (2000) Verfahren zur Herstellung eines Leichtbauelementes und Leichtbauelement. DE 10003175 (= EP 1251985 = WO 2001/054846 et al.)
69. Glatt Systemtechnik Dresden GmbH, Fraunhofer-Gesellschaft zur Förderung der angewandten Forschung e.V., Bretschneider F, Stephan H, Brückner J, Schneider L, Stephani G, Waag U (2000) Metallische miniaturisierte hohle Formkörper und Verfahren zur Herstellung derartiger Formkörper. DE 10018501 (= EP 1272300 = WO 2001/078923)
70. Glatt Systemtechnik Dresden GmbH, Fraunhofer-Gesellschaft zur Förderung der angewandten Forschung e.V., Bretschneider F, Stephan H, Brückner J, Schneider L, Stephani G, Waag U, Hunkemöller P (2000) Hohlkugel und Verfahren zur Herstellung von Leichtbauteilen mit Hohlkugeln. DE 10046174 (= EP 1251985 = WO 2001/054846)
71. Glatt Systemtechnik Dresden GmbH, Fraunhofer-Gesellschaft zur Förderung der angewandten Forschung e.V., Stephani G, Andersen O, Schneider L, Waag U, Bretschneider F, Brückner J, Stephan H (2000) Verfahren zur Herstellung von Leichtbauteilen. DE 10048412 (= EP 1251985 = WO 2001/054846 Method for producing leightweight structural components)
72. Glatt Systemtechnik Dresden GmbH, Fraunhofer-Gesellschaft zur Förderung der angewandten Forschung e.V., Bretschneider F, Stephan H, Brückner J, Schneider L, Stephani G, Waag U, Hunkemöller P (2000) Hohlkugel und Verfahren zur Herstellung von Leichtbauteilen mit Hohlkugeln. DE 10046174 (= EP 1251985 = WO 2001/054846)
73. Institut für Fügetechnik und Werkstoffprüfung GmbH, Müller H, Neuhäuser M, Bliedtner J (2000) Verfahren zur Herstellung von Hohlkugeln. DE 10039320
74. Glatt Systemtechnik Dresden GmbH, Fraunhofer-Gesellschaft zur Förderung der angewandten Forschung e.V., Scholl R, Böhm A, Stephani G, Bretschneider F (2001) Thermisch isolierende Hohlkugeln und Verfahren zur Herstellung. DE 10160640
75. Oikos Co. Ltd., Bae JY, Kim H, Lee CH, Lee GH (2001) Fabrication method of ceramic hollow sphere by emulsion/extraction and ceramic hollow sphere manufactured by the method. KR 1020030050414
76. Prefecture K, Bizen H, Uezono T, Ito H, Sodeyama K, Hamaishi K (2001) Fine metal-combined hollow sphere and its manufacturing method. JP 2002241805

77. Glatt Systemtechnik Dresden GmbH, Fraunhofer-Gesellschaft zur Förderung der angewandten Forschung e.V., Kieback B, Waag U, Löthmann P, Kostmann C, Schneider L, Reinfried M, Bretschneider F (2002) Verfahren zur Herstellung gesinterter Formkörper und so hergestellter Formkörper. DE 10256221 (= EP 1917116 = WO 2007/01984 6 Calcinated or sintered hollow body comprising a spherically curved surface)
78. Kim WB, Kim KB, Kim YS, Lee GJ, Chun YJ, Jung CU (2002) Synthetic fiber containing powders with the shape of hollow sphere. CA 2485226
79. Kim YB, Jung ES, Lim MS (2002) Hollow sphere and tube manufactured using iron (III) oxide and manufacturing method thereof. KR 1020030072836
80. Miwa Kako:kk, Nishida N (2002) Plastic-made hollow sphere and its manufacturing method. JP 2004230876
81. Fraunhofer-Gesellschaft zur Förderung der angewandten Forschung e.V., Hübelt J, Kostmann C, Waag U, Stephani G, Lotze G (2003), Schallabsorber. DE 10347226 (= EP 16735 21 = WO 2005/035950 = AT 405730)
82. Glatt Systemtechnik Dresden GmbH, Plansee AG, Schaumaplast Sachsen GmbH, Fraunhofer-Gesellschaft zur Förderung der angewandten Forschung e.V., Schneidereit H, Kretzschmar R, Reinfried M, Kostman C, Waag U, Stephani G, Färber J, Zobel K, Schulze F (2003) Vorprodukt und Verfahren zur Herstellung von Grünkörpern für gesinterte Leichtbauteile. DE 10355298
83. Aerospace Exploration Agency, Kitazono K, Sato H, Kuribayashi K, Kikuchi Y, Kakehi K (2004) Hollow metal sphere made from magnesium alloy. JP 2006124755
84. Daimler-Chrysler AG, Bjekovic R, Fussnegger W, Gerick A, Lempenauer K, Wortberg D, Langenbucher T, Anderson K (2004) Hollow sphere-filled deformation components or crash structures having a gradient. WO 2005/028261
85. Fraunhofer-Gesellschaft zur Förderung der angewandten Forschung e.V., Glatt Systemtechnik Dresden GmbH, Hungerbach W, Schneidereit H, Kretzschmar R, Schneider L, Waag U, Stephani G (2004) Schallabsorbierende Struktur. DE 102004003507 (= EP 1557819)
86. Glatt Systemtechnik Dresden GmbH, Cornelius HD (2004) Verfahren zur Herstellung eines Leichtbauelementes aus Hohlkugeln mit dichter Packung. DE 10259153
87. Greiwe R (2004) Verfahren zur Herstellung eines Leichtbauelements aus Hohlkugeln. DE 102004063489
88. LG Electronics Inc., Hong SE, Kim GD, Kim YB (2004) Adiabatic structure of refrigerator cabinet using micro hollow sphere. KR 1020060069744
89. Glatt Systemtechnik Dresden GmbH, Hungerbach W, Kretzschmar R, Bingel G (2005) Verfahren zur Herstellung von Hohlkörpern mit sphärisch gekrümmter Oberfläche, mit dem Verfahren hergestellte Hohlkörper und deren Verwendung. DE 102005040599
90. National University Corporation Kagawa University, Mihara Y, Yoshimura K, Murakami M (2005) Hollow metallic sphere structure body, hollow metallic sphere formed body, and method of producing hollow metallic sphere structure body. WO 2005/084854
91. Papierfabrik August Koehler AG, Horn M, Boschert M (2005) Wärmeempfindliches Aufzeichnungsmaterial, DE 102005050418

92. Sunhanmnt Co. Ltd., Chung H, Lee KC, Lee SH (2005) Hollow glass sphere with the coated meso porous titanium dioxide thin film and manufacturing method thereof having the pollution prevention, deodorization and antibacterial Functions, KR 1020070031151
93. Fraunhofer-Gesellschaft zur Förderung der angewandten Forschung e.V., Lawrence Livermore National Laboratory, Wild C, Wörner E, Biener J, Hamza A (2006) Hollow sphere with a coating and method and device for the production thereof. WO 2007/045479

14.4 Journal Papers Related to Hollow Spheres and Hollow Sphere Structures

14.4.1 Manufacturing

1. Andersen O, Waag U, Schneider L, Stephani G, Kieback B (2000) Novel metallic hollow sphere structures. Adv Eng Mater 2:192-195
2. Körner C, Singer RF (2000) Processing of metal foams - challenges and opportunities. Adv Eng Mater 2:159-165
3. Waag U, Schneider L, Löthman P, Stephani G (2000) Metallic hollow spheres - materials for the future. Met Powder Rep 55:29-33
4. Banhart J (2001) Manufacture, characterisation and application of cellular metals and metal foams. Prog Mater Sci 46:559-632
5. Palmer RA, Gao K, Doan TM, Green L, Cavallaro G (2007) Pressure infiltrated syntactic foams - process development and mechanical properties. Mater Sci Eng A-Struct 464:85-92

14.4.2 Structural Characterisation

1. Li ZW, Wei ZJ, Wang HW, Zhu ZJ (2007) Microstructure and compressive properties of INCO-617 alloy hollow sphere foam. Rare Metals 26:129-133
2. Veyhl C, Winkler R, Merkel M, Öchsner A (2008) Structural characterisation of diffusion-bonded hollow sphere structures. Defect Diffus Forum 280-281:105-112

14.4.3 Mechanical Properties

1. Bonino JP, Rousset A, Rosignol C, Blottière Y (1990) Composite material with hollow sphereical modules made of microcrystallized or amorphus nickel alloys. Matériaux et Techniques 78:25-28
2. Ozgur M, Mullen RL, Welsch G (1996) Analysis of closed cell metal composites. Acta Mater 44:2115-2126
3. Löthman PA, Andersen O, Gründer J, Landgraf G (2000) Mechanische Eigenschaften metallischer Hohlkugelformkörper und Charakterisierung ihrer Zellwandeigenschaften. Materialwiss Werkst 31:529-530
4. Barrett KE, Gotts AC (2002) Finite element analysis of a compressible dynamic viscoelastic sphere using FFT. Comput Struct 80:1615-1625

5. Lim T-J, Smith B, McDowell DL (2002) Behavior of a random hollow sphere metal foam. Acta Mater 50:2867-2879
6. Gasser S, Paun F, Cayzeele A, Bréchet Y (2003) Uniaxial tensile elastic properties of a regular stacking of brazed hollow spheres. Scripta Mater 48:1617-1623
7. Sanders WS, Gibson LJ (2003) Mechanics of BCC and FCC hollow-sphere foams. Mater Sci Eng A-Struct 352:150-161
8. Sanders WS, Gibson LJ (2003) Mechanics of hollow sphere foams. Mater Sci Eng A-Struct 347:70-85
9. Baumeister E, Klaeger S, Kaldos A (2004) Lightweight, hollow-sphere-composite (HSC) materials for mechanical engineering applications. J Mater Process Tech 155-156:1839-1846
10. Gasser S, Paun F, Bréchet Y (2004) Finite elements computation for the elastic properties of a regular stacking of hollow spheres. Mater Sci Eng A-Struct 379:240-244
11. Gasser S, Paun F, Riffard L, Bréchet Y (2004) Microplastic yield condition for a periodic stacking of hollow spheres. Scripta Mater 50:401-405
12. Abdennadher S, Zhao H (2005) On the rate sensitivity of metallic cellular solids. Int J Vehicle Des 37:156-174
13. Baumeister E, Klaeger S, Kaldos A (2005) Characterization and application of hollow-sphere-composite lightweight materials. P I Mech Eng L-J Mat 219:207-216
14. Fiedler T, Öchsner A, Grácio J, Kuhn G (2005) Modelling the mechanical behaviour of adhesively bonded and sintered hollow-sphere structure. Mech Compos Mater 41:559-570
15. Motz C, Friedl O, Pippan R (2005) Fatigue crack propagation in cellular metals. Int J Fatigue 27:1571-1581
16. Öchsner A, Grácio J (2005) On the macroscopic thermal properties of syntactic metal foams. Multidiscip Model Mat Str 1:171-181
17. Yamada Y, Wen CE, Asahina T, Kato K, Sonoda T, Mabuchi M (2005) Compressive properties and energy absorption of hollow sphere aluminum. Mater Sci Forum 475-479:333-336
18. Zhao H, Elnasri I, Abdennadher S (2005) An experimental study on the behaviour under impact loading of metallic cellular materials. Int J Mech Sci 47:757-774
19. Fiedler T, Öchsner A, Grácio J (2006) The uniaxial strain test – a simple method for the characterization of porous materials. Struct Eng Mech 22:17-32
20. Fiedler T, Sturm B, Öchsner A, Gracio J, Kuhn G (2006) Modelling the mechanical behaviour of adhesively bonded and sintered hollow-sphere structures. Mech Compos Mater 42:559-570
21. Karagiozova D, Yu TX, Gao ZY (2006) Modelling of MHS cellular solid in large strains. Int J Mech Sci 48:1273-1286
22. Klaus H, Huber O, Kuhn G (2006) Materialverhalten zellularer Kugelverbundwerkstoffe im linear-viskoelastischen Bereich. Tech Mech 26:71-84
23. Koo J-M, Jung S-B (2006) Mechanical properties of copper hollow sphere manufactured by sintering process. Mater Sci Forum 510-511:730-733

24. Ruan HH, Gao ZY, Yu TX (2006) Crushing of thin-walled spheres and sphere arrays. Int J Mech Sci 48:117-133
25. Adrien J, Maire E, Gimenez N, Sauvant-Moynot V (2007) Experimental study of the compression behaviour of syntactic foams by in situ X-ray tomography. Acta Mater 55:1667-1679
26. Elnasri I, Pattofatto S, Zhao H, Tsitsiris H, Hild F, Girard Y (2007) Shock enhancement of cellular structures under impact loading: Part I experiments. J Mech Phys Solids 55:2652-2671
27. Fiedler T, Öchsner A, Grácio J (2007) Uniaxial elasto-plastic behaviour of adhesively bonded hollow sphere structures (hss): numerical simulations and experiments. Mater Sci Forum 539-543:1874-1879
28. Fiedler T, Bingel G, Öchsner A, Kuhn G (2007) Numerische Analyse des Leichtbaupotenzials von metallischen Hohlkugelstrukturen (Numerical analysis of the lightweight potential of metallic hollow sphere structures). Konstruktion 1/2:67-71
29. Gao ZY, Yu TX, Karagiozova D (2007) Finite element simulations on the mechanical properties of MHS materials. Acta Mech Sinica 23:65-75
30. Karagiozova D, Yu TX, Gao ZY (2007) Stress-strain relationship for metal hollow sphere materials as a function of their relative density. J Appl Mech-T ASME 74:898-907
31. Kari S, Berger H, Rodriguez-Ramos R, Gabbert U (2007) Computational evaluation of effective material properties of composites reinforced by randomly distributed spherical particles. Compos Struct 77:223-231
32. Vesenjak M, Öchsner A, Hribersek M, Ren R (2007) Behaviour of Cellular Structures with Fluid Fillers under Impact Conditions. Int J Multiphys 1:101-122
33. Vesenjak M, Öchsner A, Ren Z (2007) Thermal post-impact behavior of closed-cell cellular structures with fillers – Part I. Mater Sci Forum 553:196-201
34. Vesenjak M, Žunič Z, Öchsner A, Ren Z (2007) Thermal post-impact behavior of closed-cell cellular structures with fillers – Part II. Mater Sci Forum 553:202-207
35. Yu TX, Karagiozova D, Gao ZY (2007) The dependence of the energy-absorption capacity of metal hollow sphere materials on their relative density. Key Eng Mat 340-341:389-396
36. Caty O, Maire E, Bouchet R (2008) Fatigue of metal hollow spheres structures. Adv Eng Mater 10:185-191
37. Fiedler T, Öchsner A (2008) Experimental analysis of the flexural properties of sandwich panels with cellular core materials. Materialwiss Werkst 39:121-124
38. Fiedler T, Öchsner A (2008) On the anisotropy of adhesively bonded metallic hollow sphere structures. Scripta Mater 58:695-698
39. Friedl O, Motz C, Peterlik H, Puchegger S, Reger N, Pippan R (2008) Experimental investigation of mechanical properties of metallic hollow sphere structures. Metall Mater Trans B 39:135-146
40. Ren Z, Vesenjak M, Öchsner A (2008) Behaviour of cellular structures under impact loading - a computational study. Mater Sci Forum 566:53-60
41. Vesenjak M, Fielder T, Ren Z, Öchsner A (2008) Behaviour of syntactic and partial hollow sphere structures under dynamic loading. Adv Eng Mater 10:185-191

42. Veyhl C, Winkler R, Merkel M, Öchsner A (2008) Mechanical testing of diffusion bonded metallic hollow sphere structure (mhss). Defect Diffus Forum 280-281:85-96
43. Viot P, Shankar K, Bernard D (2008) Effect of strain rate and density on dynamic behaviour of syntactic foam. Compos Struct 86:314-327
44. Winkler R, Merkel M, Öchsner A, Günter W (2008) On the vibration analysis of adhesively bonded hollow sphere structures. Materialwiss Werkst 39:139-142
45. Neville BP, Rabiei A (2008) Composite metal foams processed through powder metallurgy. Mater Design 29:388-396

14.4.4 Thermal Properties

1. Fiedler T, Pesetskaya E, Öchsner A, Grácio J (2006) Calculation of the thermal conductivity of porous materials. Mater Sci Forum 514-516:754-758
2. Fiedler T, Öchsner A (2007) On the thermal conductivity of adhesively bonded and sintered hollow sphere structures (hss). Mater Sci Forum 553:39-44
3. Fiedler T, Öchsner A (2007) Influence of the morphology of joining on the heat transfer properties of periodic metal hollow-sphere-structures. Mater Sci Forum 553:45-50
4. Fiedler T, Öchsner A (2007) Finite element analysis of the temperature dependent conductivity of metallic hollow sphere structures. Int J Multiphys 1:283-289
5. Fiedler T, Öchsner A, Belova IV, Murch GE (2008) Recent Advances in the Prediction of the Thermal Properties of Syntactic Metallic Hollow Sphere Structures. Adv Eng Mater 10:361-365
6. Fiedler T, Öchsner A, Belova IV, Murch GE (2008) Calculations of the effective thermal conductivity in a model of syntactic metallic hollow sphere structures using a lattice monte carlo method. Defect Diffus Forum 273-276:216-221
7. Fiedler T, Solórzano E, Öchsner A (2008) Numerical and experimental analysis of the thermal conductivity of metallic hollow sphere structures. Mater Lett 62:1204-1207

14.5 Doctor Theses

1. Sanders WS (2002) Mechanical behavior of closed-cell and hollow-sphere metallic foams. Massachusetts Institute of Technology, USA
2. Gasser S (2003) Etude des propriétés acoustiques et mécaniques d'un matériau métallique poreux modèle à base de sphères creuses de nickel. Institut National Polytechnique de Grenoble (INPG), France
3. Baumeister E (2004) Hohlkugelkomposit - Charakterisierung thermischer und mechanischer Eigenschaften eines neuen Leichtbauwerkstoffes. Otto-von-Guericke-Universität Magdeburg, Germany
4. Friedl O (2007) Mechanische Eigenschaften von metallischen Hohlkugelstrukturen. Montanuniversität Leoben, Austria
5. Mamoud F (2007) Modélisation de l'élasticité et du fluage des empilements gravitaires de sphères creuses métalliques brasées. Institut National Polytechnique de Grenoble (INPG), France

6. Fiedler T (2008) Numerical and experimental investigation of hollow sphere structures in sandwich panels. University of Aveiro, Portugal
7. Caty O (2008) Fatigue des empilements de sphères creuses metalliques. L'Institut National des Sciences Appliquées (INSA) de Lyon, France.
8. Fallet A (2008) Structure et propriétés mécaniques d'empilements aléatoires de sphères creuses: caractérisation et modélisation. Institut National Polytechnique de Grenoble (INPG), France

14.6 Other Communications Related to Hollow Spheres and Hollow Sphere Structure

1. Kohler S, Fitzgerald TJ, Singer RF (1995) Herstellung und Eigenschaften von zellularem Magnesium. VDI Berichte Nr. 1235:305-319

Index

Author Index

GPSR Compliance

The European Union's (EU) General Product Safety Regulation (GPSR) is a set of rules that requires consumer products to be safe and our obligations to ensure this.

If you have any concerns about our products, you can contact us on ProductSafety@springernature.com

In case Publisher is established outside the EU, the EU authorized representative is:

Springer Nature Customer Service Center GmbH
Europaplatz 3
69115 Heidelberg, Germany

Batch number: 10761599

Printed by Printforce, the Netherlands